ORIGIN AFRICA

ORIGIN AFRICA

AFRICA

A NATURAL HISTORY

JONATHAN KINGDON

Princeton University Press

Princeton and Oxford

DEDICATION

In homage to loving mothers, to scientists, naturalists, artists
and musicians in a one-day-to-be-united Africa.

Mammal Madonna.

First published in Great Britain by William Collins in
2023
An imprint of HarperCollins*Publishers*

First published in the United States and Canada by
Princeton University Press in 2023

Princeton University Press
41 William Street,
Princeton, New Jersey, 08540
press.princeton.edu

10 9 8 7 6 5 4 3 2 1

Copyright © 2023 by Jonathan Kingdon

Jonathan Kingdon asserts the moral right to be
identified as the author of this work in accordance
with the Copyright, Designs and Patents Act 1988

British Library Cataloging-in-Publication Data is
available

ISBN: 978-0-691-22853-2
Ebook ISBN: 978-0-691-24903-2
Library of Congress Control Number: 2022945328

Princeton University Press is committed to the
protection of copyright and the intellectual property
our authors entrust to us. Copyright promotes the
progress and integrity of knowledge. Thank you for
supporting free speech and the global exchange of
ideas by purchasing an authorized edition of this
book. If you wish to reproduce or distribute any part
of it in any form, please obtain permission.

Requests for permission to reproduce material from
this work should be sent to permissions@press.
princeton.edu

Cover image: *Primate Patterns* by Jonathan Kingdon

Set in Latino URW

Typeset and designed by D & N Publishing,
Baydon, Wiltshire

Printed and bound in Bosnia & Herzegovina by
GPS Group

CONTENTS

All illustrations, unless attributed otherwise, are by Jonathan Kingdon.

ACKNOWLEDGEMENTS

This book, written during the COVID-19 pandemic, owes debts that long precede that tragic yet galvanising plague. A primary source of inspiration has been our ancient (probably pre-Neolithic) concept of *ubuntu*, which knows and embraces our origins as creatures of the African continent. As in the foraging communities from which we descend, every member of any group must act in the interest of the youngest child and the weakest adult. That is *ubuntu*. Here my tutor, in earliest childhood, was Saidi Batale Abisa. A fellowship at STIAS, 'A Creative Space for the Mind', introduced me to Edward Kirumira, Abdullah Daar and a score and more of international scholars in that admirable Institute for Advanced Studies (near Cape Town, South Africa). The Institute helped reinforce the flow of ideas that animate this book. Surgeon, idealist, friend and fellow-native of Tanzania, Abdullah Daar, sees too many governments and long-established institutions as significant sources of humanity's problems and urges us to become active agents of change, each on our own initiative. Edward and Abdullah shared with me the foundational influence of East Africa's Makerere University, where I taught for 13 years. Gregory Maloba, Mike Adams, Julian Huxley, Reg Moreau, Desmond Vesey-Fitzgerald, Tag El Sir Ahmed, Leonard Beadle, Tim Clutton-Brock, Stephen Tomkins, Bill Banage, Ibrahim Salahi, Peter Miller and Alan Walker were all formative influences over those years, as were Makerere's sister universities in Nairobi and Dar es Salaam.

I have also had the good fortune to have almost a lifetime's association with the National Museums of Tanzania, Uganda and Kenya. The Oxford University Zoology (now Biology) and Biological Anthropology departments and Ruskin School of Art offered me shelter after Idi Amin Dada drove the great exodus from Uganda. Many friends, colleagues and institutions have stimulated my work. Peter Williams of the Wellcome Foundation and Chris Barr and Patricia Shanley of Woods and Wayside have provided vital financial support over crucial periods. At Oxford University, David MacDonald, Dawn Burnham, Mark Stanley-Price, Fritz Vollrath, Leslie Scott, Bill (W. D.) Hamilton, Richard Dawkins, Niko Tinbergen, Morris Gosling, Ryk Ward and John Fanshawe have all been precious colleagues and supporters. Caroline Weaver, director and producer of two BBC television films that explored my work, is gratefully remembered. In Germany, Dieter Kock at the Senkenberg Natural History Museum helped host an ambitious exhibition of drawings and prints. François Bourlière at the Muséum national d'histoire naturelle, Paris, supported the publication, in French, of my pocket guide to African mammals. Jordi Sabater–Pi at Barcelona University, Gloria Svampa and Luigi Boitani at the Natural History Museum in Rome, and Francesco Rovero and Osvaldo Negra and a large part of all the staff at MUSE in Trento, Italy, have all been stimulating friends as well as commissioning the single most comprehensive display of my work. In Australia, Graham Harrington at CSIRO and Tim Flannery at the Australian Museum in Sydney have been generous hosts as well as passionate advocates of the need, throughout contemporary cultures, for more ecologically informed and humane attitudes. In America, John Harris curated major exhibitions of my drawings in the County Museum in Los Angeles and around the grand Rotunda of the Washington D. C. Natural History Museum. Steve Wainwright, Knut Schmidt-Nielsen, Dan Livingston and the Cartmill family illuminated a year at Duke University, as did Tadahisa Kuroda at Skidmore College in Saratoga Springs. In Japan, Shiro Kondo introduced me to Kyoto University, and also to the rudiments of Shinto! For more comprehensive lists of colleagues and the scientific universes I have inhabited, please refer to acknowledgements in the books listed under Further Reading. My most immediate debts are to my family, my *jamaa*.

Two selfless and talented parents, Dorothy and Teddy, raised me in Tanganyika. Elena and our sons and daughters, translator Pepita, curator and scholar Zachary, film producer Zuleika and artist poet Afra have all been a constant source of pride and inspiration, while Rungwe and Claude have built and cultivated creative spaces and communities in Africa (Rwenzori sculpture foundation) and Europe (Pangolin foundry and galleries). In these spaces, communities continuously define and redefine themselves, and expert crafters and artists join forces to assert cultural values that are all too quickly lost where industry and commerce (or military and political power) define and dictate culture. Laura has been my steadfast companion, ever-ready helpmeet and wife throughout this book's genesis and production. Our sister, Ann, hosted us for four fertile months in Mexico's Yucatan.

In about 1989 a young publisher at Collins, Myles Archibald, was my enthusiastic editor. More than thirty years later it has been my great good fortune to renew our one-time working association and friendship with the crafting of *Origin Africa*. Equally serendipitous, Namrita and David Price-Goodfellow have undertaken the formidable task of designing and producing the layout and maps for this quite sizable and visually challenging work. To all involved in the evolution and completion of this work, a heartfelt thank you.

PREFACE

Creatures of a Continent

The pages that follow are informed by a new, exploratory language, loaded with messages from our ancestors. We are still learning how to translate those messages, most of them, like us, originating in Africa.

The molecules of genes, themselves coded messages, offer us new grammars, a new syntax that is embedded in all living beings. It is a language that was decoded while I was in my teens by men I have met.

The modern science of biology came into being during my grandfather's lifetime with the publication of a book entitled *On the Origin of Species*. I and all my readers belong to just one of the species to which Charles Darwin referred. For the human species, Origin means Africa.

These languages and their messages are new. Yet the lessons we must learn require a complete reframing of knowledge itself. Knowingly or not, we have entered an age of learning in which everyone's lifelong commitment to ever-expanding knowledge has become the defining quest of contemporary culture. Furthermore, we must act on this always imperfect but always growing knowledge, not only for ourselves, but for our children's children. Defiant, willful ignorance threatened civilisation twice in the last century, while new disciplines of thought predict that all organisms, including us, will soon suffer unimaginable consequences from today's feckless industries and still worse politics.

Our subject, who looks back at us from the mirror, is also new, newly evolved, yet her ancestry takes us back to the beginnings of life on this planet. Our much shorter African history is calibrated against the ruined architecture of fossils in crumbling landscapes of rock and dust, yet its central value is life itself, the durability, diversity and abundance of ever-evolving, ever-changing life. Science, like life, must be alive and free to be true. Parasites and decomposers are components of ecosystems. It does not follow that human values should mimic and be built upon cadavers, minerals, tombstones or the social histories they hallow. The pages that follow reject those values of death and dollars that headstrong commerce, industry and political institutions seek to impose upon us.

This clash of values is nothing new and was symbolised in the story (as told by my grandmother, Bibi Minnie) of farmer Cain slaying his free-ranging, nomad brother Abel, a theme that will return throughout these pages.

OPPOSITE: Novice Among Elders.

Africa is the most misrepresented continent on Earth, with the most de-humanised and abused of histories. Much of what follows is motivated by trying to contradict such perversions and to engage with human beings as the fragile mammals we are. The pages of this book are couched in a language that engages with 'Life in Time'.

I seek a grasp on 'deep time' to shape this new language. Every paragraph is governed by time, while the fractal nature of genes creates multi-dimensional structures not native to traditional, verbal and mostly linear, modes of thinking and writing. Linked to ancient traditions and philosophies, rich in poetry and fable, languages such as Arabic, Greek, Hindi, Latin and Mandarin have not accommodated easily to the vocabularies of contemporary science.

The essays that follow represent sketches of Africa drawn from just one of the rapidly evolving dialects of science – evolutionary biology. As the pace of discovery in our science reaches overwhelming proportions, it can be a struggle to keep up. However, the veracity sought by science is a reassuring foil against the malevolence of current political, nationalistic and 'development' mythologies.

I write and sketch with a sense of urgency because Africa is no longer isolated from the external world. Our coasts may no longer be raided for slaves, but industrialised powers now pillage and poison the land, seas and air around us. These same powers now impose modes of mechanised, monocultural agriculture on our soils that scarcely differ from those worked for several centuries by labour stolen from Africa – slave-dependent plantation agriculture.

In concert with the degradation of our people, we watch the annihilation of our fauna, our flora and the natural communities that they form – assaults driven by global anarchy. Likewise, domineering industrial nations have stolen the dignity and reduced the worth of human beings, while turning the world's climate against us in what feels more and more like an implacable rampage.

Scientists have been allowed to maintain lofty detachment, up to now. We have enjoyed emancipation from many of the coercions that force others into new forms of servitude.

Behind our studied scientific detachment, there stand phalanxes of all-too-fallible individuals, with far-from-objective, emotional, instinctive, biological and social needs. For these reasons, most of my fellow scientists consciously choose to keep their scientific and private lives separate. In the following pages, that separation will be put to the test. If science has the significance we claim, then we cannot escape our obligation to make it inform both our private and political lives.

This book was conceived, written and published during a global pandemic which, together with the climate crisis, reprimands humanity for the mess that industry and politics have made of our small, blue planet. The climate crisis and Covid 19 demand some profound changes in our behaviour and the way we try to understand our place in nature. A preoccupation with disease, climate and the natural history of animals and plants is timely, but these themes also embrace the full evolution of humans.

ABOVE AND BELOW: *Strutting their stuff, living and laughing – a scatter of prehistoric cartoons.*

Africa has properties that ensure most of human evolution could have occurred nowhere else. A greater diversity of life-forms has forced more and more species to squeeze into narrower and narrower niches. Human complexity evolved in response to this, the most complex of continents, at first with narrow niches. Eventually, stealing niches became our niche.

On offer here is an intensely personal portrait of a continent bolstered by my own, excited animal senses. This sense-igniting mother-continent is not just my personal motherland, but the birthplace of all humanity.

I offer ways of looking and thinking about Africa that fellow scientists and some inner teacher have bid me try translate, try to share. Fellow people were ornamenting their skins, shelters and tools long before the earliest remaining traces (in Africa at about 30,000 years ago). The dramas of their lives were painted on the walls of rock shelters and caves.

I know of rock paintings that are essentially cartoons, one an amalgam of baboon and human, another a few limbs and a head radiating from a circular hole in the rock, depicting its subject as one big orifice. In both, perhaps humour triggered the impulse to create. In others the outlines of people, other animals and their activities were accurately observed and rendered. Many depict ceremonies, rituals, hair-dressing and costumes.

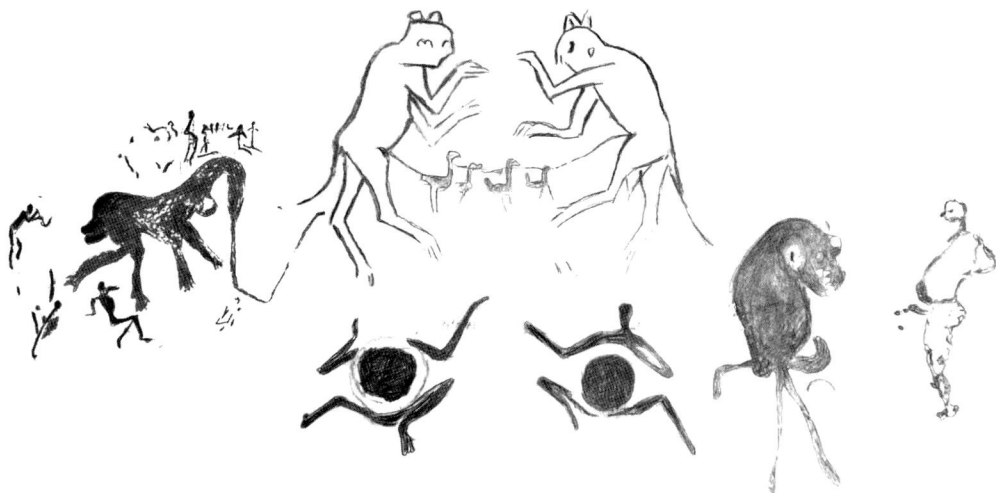

Even a record of their own extermination was left in the fastnesses of the Drakensberg, where Khoisan painters left us the silhouettes of mounted strangers firing their rifles at the fleeing artists as if they were 'game' (*see* above).

After a lifetime of teaching in African and other universities, I am not alone in being convinced that Africa, its inhabitants, its ecology, its history, natural and otherwise, its fossils, its scientists, its artists and its musicians have so much more to teach and share with the world than any body of knowledge, other than the universal gift of science, can ever offer Africa.

I am convinced that this didactic role depends upon survival of the still-living fabric of all those natural communities that once nurtured us on our evolutionary journey.

Why engage with all the time-consuming complexities and disciplines involved in exploring a continent's past and its multifarious ecological communities? Should we not be focused on 'development' and the integration of our continent into broader economies? These economies, incidentally, are represented as 'global' but are actually devised and operated by the interests of authorities that are foreign to our shores, and to our intellects and instincts.

Have we tried to understand, or even to glimpse, how our ancestors weathered all the hazards that separated *their* survival from extinction?

Among the privileges as well as obligations of being African are all the opportunities we have for observing, contemplating and learning to recognise the very processes that shaped and made us the African animals we are.

Today's emancipated student can now link her intrinsic curiosity about nature with the central riddles of existence. Along the way, she will gain some insight into how an elephant, an eland or a baobab came to be – free bycatches within a self-centred education bedded in history, but with and for a future. Education as a lifelong process, not a penance dedicated to indoctrination by self-appointed and so-called 'elders and betters', bereft of science.

Today's student may know or hear how her predecessors, even her parents, were once marshalled like shoals of fish, to memorise and repeat verses and passages from pre-fabricated, pre-scripted books, rites and inductions. However poetic or picturesque they may have been, those eras are over.

Even while knowing, even deploring their limitations, we have no right to mock the beliefs and wisdoms of the past. They had fewer resources and existed within boundaries not of their own making. Along the way they conjured magnificent fables of gods and heroes – Horus, Seth, Kintu, Shango, Likube, Ngai, Jehovah – all of them local, all combining splendid symbolic legends with fealty to tribe or sect. Some remained local, some spread more widely, some died out, others endured. Born into a world of intercontinental, inter-planetary rockets, satellites and landings, we are among the first generations of novices, students, mental voyagers, all in search of *universal* human histories, guided by new exploratory values.

Without any obvious intention, evolutionary thinking has found its continental home within today's institutions, our parks, reserves, universities, research institutes and the African Union. Proving to be a lot more original, creative and collaborative than most political players, our scientists, their colleagues and troops of committed field assistants are founding genetic institutes and excavating treasure troves of data (including fossil hominins), many of them close to our capital cities. Our artists, many of supreme talent, are everywhere.

Engulfed as we are in wave upon wave of scientific discovery, this sketchbook-cum-notebook-cum-scrapbook tries to share something of our generation's exhilaration. It is a passion tempered by near-despair at humanity's disinterest, even rejection, of science over our shared origins in Africa.

For glimpses of the innovative thinking that belongs to a 20th- and 21st-century upbringing I need only examine my own childhood. From an unusually early age, I was advised to look out over landscapes of mountain, valley and lake and imagine their histories and their inhabitants in deep time. My sketchbooks and print folios are replete with craggy panoramas, stacks of boulders and suggestive assemblages of ancient rocks.

ABOVE: *Lolui rocks.*

LEFT: *Waterfall zigzag, Mobuku valley (etchings).*

My schoolboy imagination was challenged to think in terms of millions of years – 1 million years being a very different measure from 10 or 100 million years, each marked out by fossil or geological milestones, each with time-shaped stories to tell. Such vistas of time are essential for the fabric of a continent to be seen to rise and fall under an eternity of dawns, dusks and droughts.

These were lands gnawed away by rodent rivers of a million years ago, and swept by cyclones that have disinterred the bones of animals that once wandered, hopped, burrowed, gallumphed and flew through landscapes once their own – now, for a passing moment, 'ours'.

Only knowledge and the people who put some store by knowledge are new in this ancient land. This is partly because people, the only source and medium for shareable knowledge, are themselves 'new', but mainly because the unstoppable quest for knowledge creates new disciplines of thought along the way. Every question begets more questions, as well as new techniques to try and answer them.

Details of our ever-more-systematic quest for self-knowledge in Africa reveal that, long before we arrived, long lines of ancestors walked these lands. We are the very latest, prodigal heirs to a motherland of matchless beauty and fecundity, but we behave like feckless mercenaries, the Ruga-ruga of an invading warlord's army.

I belong to the first generation of researchers and teachers charged with revealing and creating the lingos, the concepts, the vistas whereby we can share the thrills of discovery – discovery as individuals, as a species and as novice members of rich natural communities that long predate our oh-so-recent arrival. At the same time we must acknowledge and warn how little we know; how little we shall ever know.

OPPOSITE BELOW: *Portrait sculptures of Mikairi Wamala (left) and Ham Mukasa (right) by Gregory Maloba.*

RIGHT: *Two drawings of Gregory Maloba.*

BELOW: *African elephants (*Loxodonta sp.*), drawn beside Kazinga Channel.*

'You're going to find that difficult to put across in classes used to the certainties of mission schools.' Gregory Maloba, veteran teacher and celebrated sculptor, was briefing me, the novice teacher at Makerere University in eastern Africa, as we watched and I sketched some elephants. 'Now I see you shrink that *ndovu* into a few slashes of charcoal on paper, and much as I admire your skills I don't want you to domesticate that terrifying giant. I need real, untamed elephants to shrink me, not them, down to size.'

At that moment, *ndovu* suddenly entered depths deeper than it was tall. From a swirl of water the tip of its trunk emerged like a periscope. Gregory, turning his pebble-glass lenses on me, said, 'I think it's telling us that most of what *ndovus* are remains hidden. I guess much the same can be said of humans and of our knowledge about *them* or *us*. Like that *ndovu*, we are submerged, swimming in lakes of ignorance. What do *you* think?'

Gregory has joined our ancestors and I will follow soon enough, but he was right about the submerged genealogies of elephants and humans, as he was about humility in the face of nature and the shortcomings of knowledge itself.

This book is written in compliance and fellowship with dearest Gregory's intuitions. It samples something of what I've learned since that 1960 afternoon on the banks of the Kazinga Channel, Uganda, but it is also a belated effort to answer his question.

Like a forager's midden or passerine's nest, minds and thinking can be way-ward, tangled things, resistant to being unpicked and categorised. Nonetheless, I am enough of a scientist to try to discipline my thoughts and align them in the pages that follow along a broadly historical sequence, into essays that hint at a saga of continental proportions.

Africa is the least domesticated of continents, but the belief that it can and should be tamed and harnessed has a long history, deriving mainly, but not entirely, from outside our shores. In Rome, one-time capital of a part-African empire, maps have been drawn up in which every hectare of Africa, every hec-tare of every continent is designated as potential farmland. There is no room for elephants in the library on Via dei Terme di Caracalla.

A month or two after our safari in western Uganda, Gregory shared our con-templation of another, very different, giant, as Ikimuga's body lay in state in Makerere Medical School. 'Just look at his nobility, even in frozen surrender,' observed Gregory.

Determined to meet Ikimuga's family, I sought out Ruben Rwagasira, who had helped carry his body down the mountain-side. The only possible go-between, only possible Rafiq, Ruben led me out of Kisoro before dawn, our pathway rising steeply through misty fields of beans, peas and bananas. The sun came up just as we reached the forest, where swathes of herbs and mosses covered landslides, or punctuated groves of lichen-hung *kosso* trees and tracts of bamboo.

Ruben led me to where the family had dossed down for the night. Kneeling to press the back of his hand against a compress of crushed herbs and horse-like

Ikimuga lies in state, missing fingers and testicles.

manure, he beckoned me to follow his example. The air still held last night's chill, yet there was no denying that my wrist registered a hint of warmth at the base of that night-soiled nest.

Plagued by diarrhoea and in a rush of excuses, I scurried behind a small bush to relieve my bowels, unaware of a stock-still audience beyond. As I tugged down my trousers there was a roar of outrage, and I glimpsed a silverback gorilla crashing off through the undergrowth, accompanied by other black forms fleeing through the briars and wild celery.

Last night's septic supper had spoiled my first glimpse of living wild gorillas. That protesting bellower was the alpha male in Ikimuga's troop.

In the course of drawing Ikimuga's corpse, I learned how an unusually large Leopard had seized him by the throat while he slept. As a youngster, all the fingers of Ikimuga's left hand had been cut away by a steel wire snare, hence his local name 'cripple', which helped explain the apparent ease with which he had been killed. As if to register his contempt for such a feeble victim, the Leopard only deigned to eat Ikimuga's testicles.

Over the years I have watched and interacted with many primates, mostly on their terms and on their own home ground. Gorillas, in particular, have tolerated my vulgar stares as I tried to plumb the subtleties of dark, shy eyes and interpret expressive lips in quick sketches or broad sheets of studies. These primates express numerous ephemeral details of psychology and activity, and offer us countless glimpses of shared heritage.

Gorilla faces.

Denial of our relatedness only becomes comprehensible when we remember that loutish warriors cantering over near-lifeless desert sands, together with pirates riding waves and seas, founded the bombastic cultural traditions that have impoverished, even perverted, too many pre-scientific perceptions.

Some years after my safari with Gregory I joined a party of friends and visitors in another of Africa's celebrated parks where a visiting New Yorker, looking up from her field guide, asked, 'What are the deeper realities you claim wild animals in Africa express?'

'Did you notice that all the tracks our Land Rover followed were made by elephants?' I replied. 'Long before humans evolved, elephant tracks linked strategically placed "Ele-stations", arenas of compacted soil which we would call "communication hubs". Elephant needs preceded our movements, our tracks today – our soft-toed feet have always had their trails to follow, otherwise, like satyrs, we would have hooves. We now know that humans evolved and learned to accommodate within ready-made infrastructures partly created by the elephants (and other animals) that preceded us.

'Then take the very question you ask. You are the only sort of animal able to ask and I am the only animal to try to answer. We belong to the same new, upstart species and it is a plain, verifiable fact that the human ancestral lineage mainly evolved in Africa.

'That makes our continent and its fauna and flora especially relevant to both of us and to our ability to ask and find answers to questions. Most of the many stages our lineage went through can only be retrieved from African soils, or inferred from living natural communities. Right here, beneath your camp chair, East African ground apes, your ancestors, once foraged over a scatter of small but nutritious morsels interlaced with a witch's brew of poisonous mushrooms, deadly seeds and noxious animals.'

Warming to a challenge I had only just begun to explore for myself, I went on. 'That diet and the need to discriminate between a tasty or a toxic supper selected for observant eyes, intelligent brains and fingers that became a lot more adept than those of apes as we know them today. Co-ordinating hands with brains became a driving force in our evolution. Freeing up hands was what led to our becoming upright, bipedal and an inventor and user of tools. Our skeletons, our genes, our soils and our artefacts all bear the imprint of that history.'

The Yankee gal, who liked puns, muttered 'Hmmm! I like the idea of Trunk Roads but there's not much room, even for elephants, on 42nd Street, where we say "elephant in the room" to signify a big presence that everyone goes to great lengths to ignore. Back there, most of what you've been saying would likely be met with deafening silence and a change of subject.'

'Yes, here too. I say the elephant in the room is Evolution – yet it is manifest in every living thing around us. The concept of natural selection involves ways of thinking about, visualising and coming to terms with our surroundings which, right here, right now, includes the elephants we watched today. They and we humans share evolutionary origins in the material fabric of Africa. We and they *are* that fabric. It's quite some story, most of it still to be understood, let alone told, but a lot of people want to look the other way, block their ears and change the subject.'

The woman from New York got up. 'Phew! Anyone for a beer?'

I re-tell these bush-camp dialogues because they hint at some of the questions I shall try to address, about a shallowly misrepresented continent and its most grievously abused and least-known animals, which includes people – our people.

At a more personal level I seek to put some perspective around my preoccupation with finding language (not just in words) to communicate both discovery and pleasure in the colours, shapes, patterns and vivacity of a nature that has embraced me from infancy.

Exploring Africa's borderlands, even its no-man's-lands, has become one of the purposes of this book. Gregory and tropical Africa offer a different vision in which humans have to acknowledge their small scale, their fellowship with more mammals than can be counted, and their inborn lack of colour within a rainbow of exuberant nature.

Back in the 1960s, with Gregory and others, we begged, borrowed or stole the vitality of more colourful beings in sculpted or pictorial praise-songs, in costumes, concerts and dances. We could but mimic the flamboyance of monkeys, birds, beetles and fish. Yet, as actors and artists know, within any act of mimicry

Three postage stamps designed by my mother, Dorothy.

we may hope to share some understanding of principle and process, amid the relics of long-lost histories.

Some pleasure can be taken in that, especially if our senses can be deliberately, artfully engaged to speak, sing, dance, hand-paint, hand-shape – even hand-write our ideas: ideas given shape, and some, like the postage stamps my mother once designed, signifying origin in Africa.

In what came to be called 'The Independence Decade', Gregory, our colleagues and our students lived through a period of unprecedented exploration, of ourselves, of our time, of our continent. Meanwhile, the outlines of Africa's unique role in the evolution of humans and other animals continued and continues to unfold, all too slowly.

To be the guardians and explorers of such a heritage is a duty – these pages find their meaning in that duty.

Independence *monument by Gregory Maloba.*

WHAT IS AFRICA?

Asking questions. In which some scientific parameters of our continent take their measure beside the limits of human language and comprehension. Three mind-sets and more. Love letters to a mother. Pioneering ancestral histories and their settings.

'What is Africa?' Geologists might begin with it being mainly a more than 100 km-thick raft of Precambrian rock floating between three seas – the Mediterranean, the Atlantic and the Indian Ocean. Historians might first trace its name to some pre-Roman word for Tunisia's immediate hinterland. For buccaneers, it was no more than shores to be raided or traded for ivory, minerals, slaves, labour, land or lumber.

In 72 BCE Pliny wrote 'always something new out of Africa'. In 1352, Ibn Batuta described Africans as having a greater abhorrence of injustice than any other people. In 1550, Leo Africanus found men of learning greatly honoured in Timbuktu. Three centuries of systematic, eventually industrial trade in slaves followed, then conquest and colonisation by Europeans. In 1959, in response to independence for newly established nation-states, Kwame Nkrumah called for new social customs and new attitudes to life that would invigorate education and creativity.

In that year, I tried to explore a few of Africa's many dimensions (including hints of some then current stereotypes) in five painted panoramic panels. Among other things, my panels drew attention to the need for a broad-based education in a now global, decolonising, but still poverty-stricken world, in which human populations were growing faster than ever before.

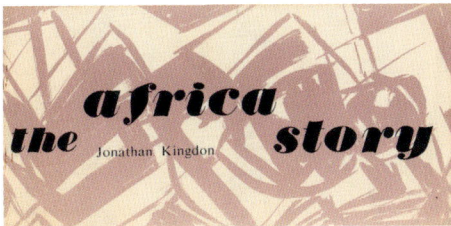

ABOVE: The Africa Story, *1959 (pamphlet cover, top left; panels 1, 3 and 5).*
OPPOSITE: Mama Afrika.

For me Africa is quite simply my birthland, my lifeland, so before seeking any sorts of insights into the tangled thickets of history, science and a still barbarous cockpit of politics, here is some personal background.

I benefited from unusually early exposure to the fundamentally different pre-occupations people bring to the most banal of experiences, and the simplest of questions.

My teacher mother, Dorothy, called my four-year-old self 'the question machine'.

'Why don't we have tails? Why are zebras stripy? Why are there boys and girls?', and, after a fierce storm, 'How come leaves are so strong?'

Her 'answer machine' was rudimentary, inherited from many previous generations – it went, 'Well, God made it so.'

'So what does God look like?'

'WELLL, [how I loved the solemnity Dorothy brought to that punctuating, delay-making word!] I suppose he's got a face that's all kind and smiley.'

Encouraged to speculate, I replied, 'WELLL – *I* s'pose he's got a face like a badger.'

'Why are there people and what are they for?' was more difficult, but Dorothy replied, word-for-word as her own mother, Bibi Minnie, had to her, at prayer-time, less than 30 years before. 'WELL, God made us as much like himself as he could, and told us to try to be as good as he is.'

It was September 1939, I was four years old and the adults talked non-stop about that week's declaration of war and what a wicked monster Mr Hitler was. The fact that Hitler had vowed to win back Germany's African colonies (including the one where we lived) gave an anxious edge to the conversation. It was an issue on which the lugubrious invalid Mr Chamberlain, Britain's then Prime Minister, had very nearly capitulated.

Listening in, I waited until prayer-time to quiz Dorothy. 'Did God make ALL kinds of people?'

'Yes, of course!'

'Even Germans?'

'Of course!'

'Did God make Mr Hitler?'

'Umm, er, well, er, yes.'

'WELLL, he must be very sorry now!'

I still like to examine propositions and follow through on suppositions, but it was a surprise to find such an early, word-by-word record of my interrogations in Dorothy's bequest to me of her notes, papers and letters. It was also surprising to learn that four-year-olds could have such a fluent grasp of sceptic logic, and could long for consistency.

Being inherently curious, and loving those few teachers that engaged with my curiosity, I became a teacher myself. You need to explore and search before you can answer questions. Besides, it is the exploring and searching that are the most fun.

In my beginnings, all social life found expression in Swahili and/or English. Down at the PWD workshops, Ali the pious foreman punctuated every one of his pronouncements with an emphatic 'w'Allahi', while another of his colleagues, an ex-soldier, laced language with 'fok', 'fok-in' and 'fok-it'. Being a born mimic, I immediately adopted the speech rhythms and expressions of these admirably expressive individuals.

Back at the house, these expletives did not go down well and my parents had to explain that 'you really should not use words you don't know the meaning of'. It had not occurred to me that such embellishments had much substance beyond being easily copied ornaments of personality, but the quest for meaning in words opened new lines of enquiry.

The things I most enjoyed doing were often labelled games or *mchezo* so it seemed logical to ask, 'Why are elephants and gorillas called "Games?"' At the time, Teddy, my father, probably had the words to explain this vestige of medieval European notions but, even in a nightmare, that would have been way beyond my imagination, let alone comprehension; I have forgotten his answer. So, if some Eurasian concepts and words are inexplicable to an African child, African concepts and the lilt of African languages, song, dance and social insights are probably just as closed to some non-Africans.

It was as a teacher that I was at the receiving end of 'what is Africa?', as blunt a question as any that my poor mother was burdened with.

Lydia James made the best pecan pie on sale at the weekly farmers' market in Chapel Hill, North Carolina. I had become her regular customer when she startled me with this same question. She was uncertain whether Africa was a country, a city or an island but she did know that some of her ancestors came from there and that several generations of them were once enslaved by others of her ancestors who weren't from there. I suppose that her question could have been answered in terms of a place, an idea, or even a concept, perhaps the auras surrounding personalities, such as Emperors of Ethiopia.

For most untravelled non-Africans, our continent remains on the outermost borders of their consciousness. For some, Africa lies still further beyond that – an unimaginable place. So long as a land lies outside a person's mental atlas, it cannot exist for that person – even when that person's ancestors inhabited that land for countless generations.

At age eleven, in Europe, I witnessed all the wonders of my *Caput Mundi* summarily banished to the black hole of my godmother's sense of outer space – beyond curiosity, beyond any trace of interest. From her and too many other non-Africans, I learned that my centre stage was not even a side-show. For much of the world's population, Africa, our ultimate motherland, is a meaningless non-existence, let alone its component places and people. These pages protest, but also represent a now elderly teacher's effort at answering Lydia's question. Along the way its pages, like any artefact, invite graphic diagrams of the ideas expressed. As an artist, an illuminator, I try to oblige.

Every species of animal or plant has a place (often an island, more often a continent) of origin. Thus, kangaroos originate and flourish in Australia, while armadillos, tomatillos and tinamous are all American.

How about humans? Only a few non-Africans have learned that they, like Lydia and house sparrows, are actually ex-Africans – the children of voyagers from a homeland that is actually and incontrovertibly shared by all humanity. Everywhere that is not Africa is territory colonised by Africans.

This is a wholesale inversion of the way much of the world view their origins. A continent of unparalleled wealth, combined with the fact that the most complex of all mammals mainly evolved there is, itself, a measure of Africa's supremacy. I first asserted this in *An Atlas of Evolution in Africa* and then, in more exploratory detail, in three books – *Self-made Man*, *Island Africa* and *Lowly Origin*: *where, when and why our ancestors first stood up.*

It was surprising to find myself author of the first substantial book that examined the origins of bipedalism in humans. It seemed such an obvious and fundamental topic to explore – had people just taken it for granted? Or had too many creationist doctrines undermined research and punished curiosity?

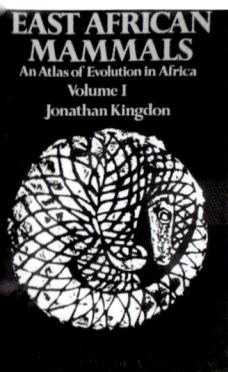

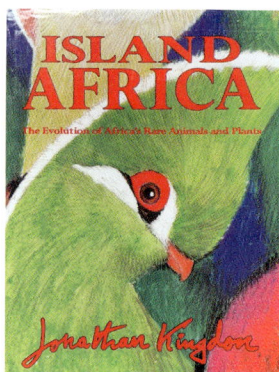

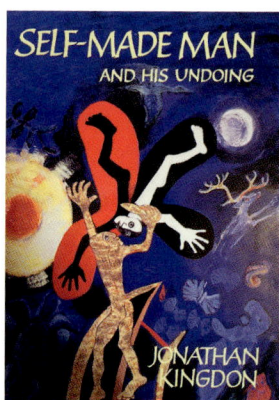

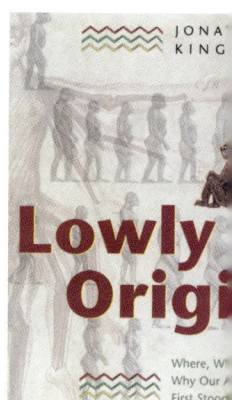

*Four book covers (*East African Mammals, Island Africa, Self-made Man *and* Lowly Origin*).*

Origin Africa is a more personal account of how, together with many other naturalists, I came to join the grand and ongoing enterprise of exploring how the magnificent natural history of our continent embraces our own evolution. I will show how we began as very minor primates and only quite recently replaced elephants as a world-defining 'keystone species'. Instead of this architectural prototype (coined by the brilliant ecologist Bob Paine) I prefer to describe our ecological role as premier 'thief of niches' – that concept too, will bear examination (see Chapter 18).

It is also a belated effort to provide a few facts and thoughts in answer to Lydia's difficult question, as well as augment those of her New York compatriot's fireside challenges.

Perhaps the plunder and rape that has dominated the written history of our continent has inhibited curiosity and fed a wider hubris among ex-Africans. Africa's very outline was mapped and named by pirates searching its coasts for booty – Slave Coast, Ivory Coast, Gold Coast, Coast of The Blacks, Regnum Melli, or Mali (Kingdom of the Blacks). For hundreds of years the thieving of human labour, ivory, minerals, timber and plantations has driven and continues to dominate our economic and political history. For centuries, elephants and humans, both richly fertile animals, have been magnets for predatory people from both within and without Africa. It has been a tragic and shameful history in which slaves and ivory became the prime 'commodities' that drew pirates to our shores – they keep coming, carrying away inanimate minerals and cadavers carved out of once animate beings.

An entire continent has been consistently misrepresented, partly because our coasts were first viewed by rapacious seamen who also happened to be compulsive map-makers.

Africa as mapped in 1554 by S. Munster.

I am seeking here a more natural history of humanity's ultimate homeland. Even so, all histories are human artefacts. They are translations subject to the limits of human minds, languages, their iconographies, technologies and concepts of time.

As with any child, those who care for you in earliest childhood remain inescapably integral to the adult you become. In my case three adults remain unforgettable for the different perspectives they invariably brought to any subject, as illustrated by the following fragment remembered from an Arusha garden.

From an armoured helmet, three long spikes protruded, rhinoceros-style. Behind it a narrow, saw-toothed back arched and heaved back and forth. Calipers on robotic arms clamped onto the limbs of a smaller being, quaking beneath him. Her equally leaf-shaped, equally leaf-coloured body bore all his weight while her long, raised tail, like a fifth limb, curled and uncurled as if signalling surrender. Meanwhile, beneath her more modestly spiked helmet, two turret-eyes swivelled about in a mechanical, meteoric motion that seemed to plead for relief from her robot burden.

I called out to Saidi, '*Njoo! Tazama! Wanafanyaje? Tusaidie yule chini?*' ('Come, look! What are they doing? Should we help?')

'Eh! *Ndugu ndogo*, little brother, we are very, very lucky to witness this act. These are messengers from ancient days. They say, "One day your skin will wrinkle all over, like us, you will move very, very slowly, like us, but you and your *watoto* will live long, like us."' Then, with a side glance and a conspiratorial grin, he added, 'But neither you nor me will ever change our colours like them.'

'But what are they doing?' I persisted.

'*Kinyonga Faru* is making *watoto*, babies, and see, she does not cry. From that hole he is filling with his semen, baby *kinyongas* will, one day, drop out inside envelopes, as if they were messages from our grandmother, Bibi Minnie. It is our great good fortune to witness what few others have ever been allowed or will ever get to see. Never forget this day.'

I ran back to the rondavel; Dorothy, my mother, must share the privilege of witnessing this living portent. By the time she arrived, the chameleons had uncoupled, the female was hiding while the male was rolling his eye-turrets, searching for an escape route. 'What an extraordinary and marvellous creature! Just like a miniature horned dinosaur – I must draw him!' She hurried off to fetch her sketch-book.

At bedtime, Teddy listened to our accounts. 'Hmm, sound to me like real rocking reptiles' (*see* p. 30). Shown Dorothy's drawing, he exclaimed, 'Aha! That's Jackson's chameleon, a lizard of the highlands named after Sir Frederick Jackson,

LEFT: *Jackson's horned chameleon (*Trioceros sp.*) (etching by Thomas Zziwa).*
MIDDLE: *Dorothy's sketch.* RIGHT: *Cape chameleon by the author.*

former Governor of Uganda – friend of Uncle Harry and a keen naturalist. Sleep tight, good night.' Ever the civil servant (or, for that matter, the busy schoolmaster). Teddy collated, classified, noted, filed and moved on.

All children rely on families and friends to make sense of whatever world they are born into. Families, in turn, inherit languages, customs and beliefs that have sustained still more generations. These three adult authorities, Saidi, Dorothy and Teddy, presided over the first seven years of my life. Each interpreted almost every event of our lives in fundamentally different words and ways. From them I learned, very early on, that everything, especially anything natural, could be seen in more than one way. It was not just a matter of switching from Swahili into English or French, or converting ideas about things into pictures. Making sense of the world needed translation into mind-models as much as into language. To my mind it also requires that self-education must become a lifelong pursuit.

Teddy's response named and domesticated its subjects, attached a contemporary and familiar name-tag, supposedly lightened up with a heavy-handed joke. He was an exceptionally conscientious teacher/public servant, a District Officer in Tanganyika, the vast territory that lay between the Great Lakes and our Indian Ocean seashore. He owed his social standing to as excellent an education as was then available in Europe, yet schoolboys of his generation were groomed for war, to be riflemen on civil war killing fields. For his ease with blood sports and weapons he was also beholden to his godfather, a county squire.

Visiting Egypt and its wonders as an eighteen-year-old, Teddy liked to quote a sentence that had survived more than 300 years, because it hinted at all the unknowns, even unknowables, of Africa and of ourselves. As Thomas Browne wrote in 1635, 'We carry with us the wonders we seek without us – there is all Africa and her prodigies in us.' This insight was published some three centuries before modern genetics and the discovery of DNA.

Acknowledging his debt to education, Teddy never lost the futuristic outlook and motivation of his first vocation as school teacher. He worked with the future in mind, while maintaining some measure of disciplined detachment and the embryo of a scientific take on life. He took pride in building schools.

Dorothy was also a teacher and an exceptionally talented professional artist, impelled to MAKE something of every experience. She was physically tough and stoic, but she had sensitive manual skills and an emotional, open-eyed, curious mind. She became remarkably self-sufficient because the patronage and feedbacks on which most Western artists depend were just not there in a territory leashed by penny-pinching bureaucrats twitching away up in Europe. There were no self-appointed disparagers to belittle work, so she could cultivate visual ideas that 'grew' on a canvas or sketchbook, no less than her vegetables in the kitchen garden. While acknowledging the limits imposed by her time and place, she only had to measure up to her own, never alien, ambitions. As a mother and wife she was loving, loyal to a fault, alternately sociable and reclusive, easily distracted and, when sick, could be quite irascible.

Teddy on a mule in Egypt, 1920.

Dorothy sketching, Kibondo, 1933.

Dorothy could translate any experience of things into immediate images, drawings, photographs – memory's visual aids, storable visual documents. She was a woman of her time, inconsistently both feminist and existentialist (she loved swing and spirituals). She had an understated self-assurance, immersing herself in drawing and writing every morning, at which time she brooked no interruption. In respecting her intense preoccupation, Saidi was an admiring observer rather than an obedient servant. He shared her inquisitive eye, while attributing her skills in representation as pure magic.

I have discovered that others can share my own pleasure in Dorothy's paintings, sketchbooks, photographs and maps. She lived for the present and carefully captioned her drawing of that horned chameleon 'Lake Duluti, Arusha – Dec. 1940'. For me, her drawing serves to anchor Saidi's injunction never to forget the wonder we had witnessed.

Perversely, perhaps, that copulating pair of chameleons is reinforced in my memory by the sight of a dozen or more newborn chameleons crawling through a garden shrub, each one announcing nature's imperative to procreate, while nature's cruel judgement was captured in the one thrashing out its last spasms in a shrike's hooked beak.

For Saidi Batale Abisa, my finding two tricorn lizards conjoined in reptilian ecstasy was as symbolic as Leonardo da Vinci's witty representations of a unicorn seeking solace, then sleep, within the arc of a maiden's warm lap. In my own, much cruder, version, the maiden tries to protect the unicorn from her brother, who symbolises archetypically masculine attitudes in which nature is there to be exploited for sport or to be eaten.

Akasimama juu ya kisuguu na kuumwaga unga hewani

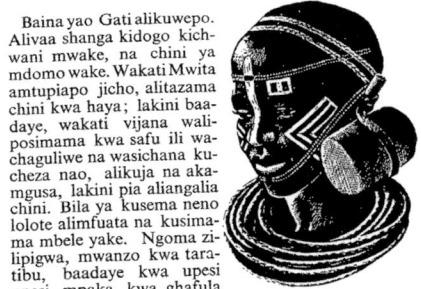

Baina yao Gati alikuwepo. Alivaa shanga kidogo kichwani mwake, na chini ya mdomo wake. Wakati Mwita amtupiapo jicho, alitazama chini kwa haya; lakini baadaye, wakati vijana waliposimama kwa safu ili wachaguliwe na wasichana kucheza nao, alikuja na akamgusa, lakini pia aliangalia chini. Bila ya kusema neno lolote alimfuata na kusimama mbele yake. Ngoma zilipigwa, mwanzo kwa taratibu, baadaye kwa upesi upesi, mpaka kwa ghafula zi lizimishwa.

Gati

Makao ya Kuria. Vibanda vinazunguka zizi la ng'ombe

Kikosi cha wapiga beni cha Simba Karata

Wanawake watatu walikuwa dunguni wakilinda ndege

Tegu and Goja

Msichana alimlisha na vidole vyake

Dorothy Kingdon's illustrations for African readers from 1930s and 1940s.

Saidi, my personal tutor, sentry and buddy, lived in a world of oracles but he was also an earnest observer and student of our parents – something we shared. At heart he was a hunter and surreal symbolist. He longed for the hunt and for past glories but he also longed for and lived to hear and move to music. Provoking social interaction, showing off and being shown off to was his idea of having fun. He was also the person I most wanted to be with – ever-present, endlessly attentive, more my beloved big brother than anything the grown-ups chose to call him. As a Nyamwezi, he was nobody's inferior, but he knew he, like me and anyone else, had much to learn. I had to wait until Lawrence Toynbee became my art teacher before I met Saidi's equal as a tutor. He was a model of sensory

ABOVE LEFT: Maiden and Unicorn *by Leonardo da Vinci.*

ABOVE RIGHT: *Maiden and unicorn as a metaphor for conservation and its enemies.*

Saidi's steadying hand – always there, 1939.

engagement, mental concentration and sometimes hilarious wit. Laughter never came and went with quite so little affectation, nor did affection flicker quite so easily from other eyes. Last seen waving farewell from the rail-tracks on Mwanza station, Saidi left me less unfailingly merry but just as prone to asking questions. The legacy of his influence has been lifelong.

My mother was a lot more independent-minded than her mother. Bibi Minnie was a protestant yeoman farmer's daughter and order in her life came directly from what she called the 'Book of the Generations'. This document (as tampered with by subsequent generations of scribes) can be read as a licence, a passport and a potted history as issued by Middle-Eastern seers. In Minnie's 'Book of the Generations' the awe of little boys for their powerful and, to them, extraordinarily versatile farmer fathers gets translated into metaphysics. Told that they will grow to be just like such an august parent, small boys happily accept being made 'after that likeness' with the added promise of inheriting bipedal Dad's elevated power over 'every creeping thing that creepeth upon the earth'. In these words, nature is but a slave, grovelling before its lord and oft-times executioner.

The pages of Minnie's book also scream guilt for the sins of the land-grabbing, soil-worshiping Neolithic, and the slaying of non-agricultural brothers. 'The voice of thy brother's blood crieth unto me from the ground'. In guilt and sorrow, in the sweat of tillers' faces, the sons and daughters of the Neolithic go forth to till the clay from which they were modelled. Like all of us, dust they are and unto dust do they return. Meanwhile, over some 10,000 years those nomadic human societies with the most realistic and intimate knowledge of nature have suffered and declined, as Cain's murder of Abel so vividly symbolised. That drama was still alive in the Africa of my childhood.

Culturally, farmer's daughter Minnie could be said to be a true heir of the Neolithic. Nevertheless, the wholesale poisoning of her parents' farm by a neighbouring copper smelter not only forced her family off the land, it was also one of countless tragedies inflicted by industry, which eventually demolished the Neolithic. Without Bibi Minnie knowing it, that Avon copper smelter was the herald of a new era: the 'Anthropocene' epoch had arrived.

Coined by Nobel prizewinner, Paul Crutzen, this new Anthropocene category of time asserts the inescapable fact that humans temporarily dominate Earth in a planet-dousing 'age of humans', much as giant reptiles once ruled the Jurassic 'age of dinosaurs'. Get used to this new name – it is here to stay (for as long as our time on the planet lasts).

The Neolithic's extension into industry and now into the global Anthropocene has been so fast that today there are few older adults who do not mourn those moments of childhood when a bird's song, a frog's leap or a rabbit's alert pause and twitching whiskers touches some pre-Neolithic sensitivity. An ever-larger number of people are examining pre-Neolithic Africa for hints of how humans got by with only the flimsiest shell of technology to shield their thin-skinned primate frames.

*Bandusya, a hunter and friend
from the Bwindi forest, 1969.*

*Charles Darwin's study at Down House, UK,
reconstructed to its original appearance and
arrangement of furniture (the author is reflected
on the left in the mirror on the mantelpiece).*

In my case, the pre-Neolithic took on the kindly faces of Ndorobo, Hadza, Batwa or Sandawe hunters, offering me a suck on the sac of a carpenter bee, a comb of honey, or the tang of orange fruit from a wild ebony.

Our hurtle into the Anthropocene has been so fast that many of us have had grandparents embedded in one or other Neolithic society. In Minnie's case, her 'Book of the Generations' stumbled out of Middle Eastern desert dust. Without gorillas or chimpanzees, without monkeys, without ancient fossils to provoke thought, the mysteries of life were subsumed in endlessly stressed tribal legends, convoluted allegories of family hierarchy, and endless squabbles over territory.

The most momentous break with all the contradictions of the Neolithic found its full expression in a single 1859 book – Darwin's *On the Origin of Species*.

We have been swept so fast into the Anthropocene that Charles Darwin's son, Leonard, was still alive when I was seven years old, and two of his great-grandsons, biologist Francis and artist Robin Vere, became personal friends. My professional homage to Darwinian Jim Watson, co-discoverer of DNA, took the shape of a painting I gave him at his 75th birthday party, accepted with an embrace.

In spite of Charles Darwin's proximity and his global celebrity, too many of my contemporaries and their children have been denied access to one of the most exciting,

Painting by Jonathan Kingdon donated to James Watson for his 75th birthday and the 50th anniversary of the discovery of DNA.

complex but all-too-slowly unfolding true stories of origin and ancestry. Charles Darwin has spawned a crop of blood-line descendants, but his finest legacy has been an ever-increasing number of scientists and students, all rootling after life's origins and expressions, as well as the ailments and frailties of a predominantly tropical bipedal primate from rain-blessed regions.

Scientists are contributing to new 'Books of Generations' that plumb greater depths than the shallows in which poor Bibi Minnie was baptised. In this respect Africa, guardian of humanity's origins, must put a priority on becoming the centre of excellence in all fields pertaining to human origins, and to the evolutionary processes that our very bodies and souls exemplify. We will express this in our research and in the revelation of life as it has been and will continue to be lived, on the richest continent on Earth.

From its beginning, my personal nuclear family was a nomad band, subject to Government-enforced peregrinations that moved us from one posting to another every two years or so, always interspersed with 'local leave' excursions or occasional 'sick leaves' to Dar es Salaam, Nairobi or Kampala. Dorothy's record of our peregrinations took the form of what she called 'quickies': sketches that just might get developed into paintings. Her example meant that sketching, particularly animals, became habitual for me. My subjects ranged from rhinos to reptiles.

It was by chance that Teddy's postings and leaves propelled us through all the major habitats of tropical Africa. We lived in little settlements that felt like rather squalid termitaries, sprouting within broad swathes of acacia savannah, each encircled by a halo of bare, over-grazed, over-hoed earth. We walked or rode through *miombo* woodland or over great grassy plains. We lived beside several

ABOVE: *Sketches of Grass Rhino* (Ceratotherium simum*) in Uganda, 1963.*

Grounded lianas.

tropical lakes and rivers and on the shores of the Indian Ocean. Sometimes we inhaled the dust of near-deserts, submitted to the ravages of locust swarms, then struggled to maintain gardens close to monkey-filled forests, whose roots fought wet-season floods to hold riverbanks together. On mountain slopes we found the ancient struggle of plants against axes and saws expressed in groves of trunk-thick lianas that, deprived of the tree-top trellises they once hung from, writhed over the ground, as if contorted in anger, like stray pythons.

For a young person new to the world, every detail of every habitat raised many more questions than the adults around me were equipped to answer.

The chapters and pictorial materials that follow represent excavations of fossil treasure chests and remembrances of safaris over cross sections of our continent. I have, like Dorothy, tried to grow ideas from diverse sources into compositions that are, and should be, open to more than one interpretation. Nonetheless, I have tried to convey something of the extraordinary fecundity and diversity of living things that have, over the eons, evolved ways of surviving in the world's richest continent. These words and images are also, more distantly, love-letters to my fellow-travelling mother. Her drawing of that horned chameleon is but one fragment of a pictorial archive that represents a talented

The Rufigi River in Flood. *Painting in mixed media by Dorothy Kingdon, 1947.*

and observant woman's response to her life in a little-known part of the world, during a forgotten era. She, Teddy and Saidi were by far the most vital people of my early life. I want to share my memories of them, as well as answer back to those who might dare to disparage the richest of continents, our lands or our peoples. I would like to change their minds.

Back then I was five, Saidi was twenty-five. I have lived a long life, learning much along the way and, as his words that day foretold, I now enter an ever-slower, ever more carunculated, lizard-like decrepitude. With some help from sculptor David Farrar, I once sculpted my memory of those two chameleons, mounting their enlarged copulation onto arched beams to become, literally, the rocking reptiles of my father's facetious remark.

On his death-bed 60 years later, Teddy said, 'We who are leaving the scene saw you blossom as a naturalist. Isn't it time to let others in on the story?'

'What story?'

'Where your mission for natural history came from. It certainly wasn't from me. I used to be as scornful of 'conchies' [conservationists] and all that as everyone else. You converted your dad – go convince others.' It was as a very late convert that Teddy could remember participating in the depths of hostility that naturalists, evolutionists and environmentalists had provoked among his peers,

Teddy's Rocking Reptiles.

himself. This hostility still unites hosts of social critics, post-modernists, 'red-necks', religious fundamentalists and ecological vandals.

The gaps between parental worlds and those of their offspring widen with every generation, but was this ex conchie-baiter, all of a sudden, inviting his son to become an eco-warrior?

His life had spanned nearly a century, closing with a millennium which ended with my mother close to dying of grief. Otherwise, on 1 January 2000 I was not alone in taking stock of my own and my family's share of the last 100 years, and in feeling apprehensive about the new century.

In its millennial edition, the journal *American Scientist* listed 'one hundred books that shaped a century of science'. Among them was my *East African Mammals: An Atlas of Evolution in Africa*, a seven-volume handbook of eastern Africa's mammalian fauna, including snatches of whatever evolutionary history was known or guessed at by 1970. In it I very deliberately opened my accounts with *Homo sapiens* as one of many primates, all with mostly hidden evolutionary histories within a much-neglected, much-misrepresented continent – our common home.

In researching, writing and illustrating those volumes, I was student, researcher, translator and teacher, all at once. Future, unknowable scholars with future, unknowable techniques will treat my earnest ambitions as premature, ill-equipped, even childish. They will be right. Wondering where you and your kin came from, even where you are headed, is indeed within the province of childish curiosity. Knowledge, scientific or otherwise, can be as ephemeral and fragile as the accidents of childhood, the generosity or perversity of parents.

Unwittingly, all my family conformed with some fundamentals of animal existence. They all selected and tried to induce behaviours, in me or themselves,

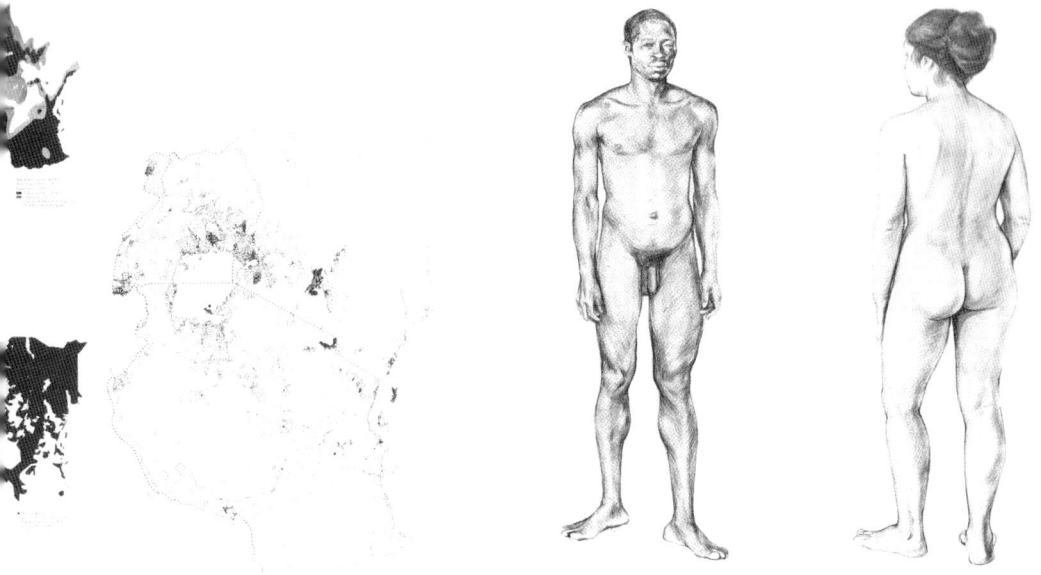

Opening pages of East African Mammals, *1971.*

that might reflect well upon us, singly or severally, in the eyes of others. Tradition-bound niceties are flouted by many, but each narcissistic one of us is a tiny lode-star for our very social species' voyage through time.

In attempting to fulfil Teddy's challenge and excavate the roots of my passion for Africa and natural history, I have had to burrow deep into strange soils. I have come to share something of my mother's sense of surprise, even bafflement, when she wrote of her four-year-old son, 'This cheerful, independent little person is sometimes quite difficult to understand and I wonder where he belongs. He can mimic a toad's belch or a bush-baby's bawl to perfection and already knows more Swahili than I'll ever learn.' More than 80 years later I hover between writing 'him' or 'me'. With such surprises hidden within the letters of my mother's bequest, I have become as much the chronicler of an inquisitional busybody's questions as an old, widely travelled man with a long memory.

Raised in Dorothy's very diurnal universe, bombarded with carnivals of light, colour and action, I watched her thrill to all sorts of visual phenomena that she saw as 'crying out to be pictured'. She made it easy to grasp the fact that individual human hands at the behest of human sensibilities can translate the most ordinary of optical experience into drawings, paintings, prints, poems, even photographs and films, all thrilling to the mysteries of landscapes and their inhabitants. Every vista swarmed with patterns, in which geology and its countless past vicissitudes provided the substrates for every expression of life. I was taught to recognise the intersecting wars of living things, from lichens to Leopards, all seeking survival and territory wherever chance and birth had deposited them. Teacher Dorothy schooled my senses, my brain, my hands, to respond to nature's prodigal magnificences. She shared the privileges of existence, of health and of education, through rare skills in representation and expression.

ABOVE: Cavalcade, Mbono Valley. BELOW: *Dorothy and 'Nippy' the Leopard.*

This was an infancy that resembled many others in taking place within a relatively closed circle. Apart from insects breaching our defences, our habitat remained well insulated from the larger world outside. When a big black buffalo bull blundered, lost, through our Tabora garden, searching for its herd, Teddy promptly shot it. The same fate was met by a wandering giant cobra behind the kitchen. Our neighbour had hand-raised a Leopard cub which was already three-quarters grown and free-ranging at the time she became very attached to Dorothy. This period coincided with my mother's pregnancy, which perhaps influenced this apparent bonding between females. Shortly after I was born, Dorothy watched nervously as the Leopard licked me as if I were her own newborn cub. In any event, the two she-mammals' affection was mutual and may have helped tame my mother's perceptions of the wilderness around her.

Incidents such as these might be seen as caricatures of colonial existence. Yet by the time I could walk and talk, the family had already swapped homes and communities several times, while various wild beings had come and gone from our quarters. The only stable factor was our nucleus of four individuals. Did we belong anywhere other than under a sun and moon?

More than 2,000 years ago a juvenile Lucretius, growing up on the slopes of Vesuvius, declared that he could not imagine a world beyond its towering maternal silhouette. No less, for me – a few long, dusty roads and a railway marked out a triangle between a still more gigantic, two-peaked volcano, Mount Kilimanjaro. Like any other child, I must have perceived myself as existing at the centre of the only universe I could possibly know. How can a birth-place be anything other than Centre of the World for the babe born there?

Any identifiable sense of place or self is quickly overlaid by a mass of behaviours and thoughts, copied from or induced by others. Lifetimes of learning are superimposed upon, or integrated into, whatever vestiges of earlier mind or memory survive.

What, then, of dreams? What substrate of the self do they come from? On waking, every infant, somehow, learns to reconcile their material, public

Xeno, our home farmhouse below Kilimanjaro.

Snows of Kilimanjaro through flowerbeds and Heliconia.

existence with the private magic of imaginings in which the most terrifying ordeals as well as the wildest extravagances become possible, including the ability to hover and swoop like a dragonfly.

Is it in moonlit privacy that lunatics, seers and prophets are born? For most growing children the adult, daylight world must progressively over-ride, contradict or forbid self-centredness, even imagination – even, perhaps, censoring dreams?

Infant egos seem capable of melting themselves into what they take to be their parental home habitat. Although my life as a messenger has taken me to every inhabited continent, I have retained an almost infantile attachment to East Africa. Peregrinations spread over an entire childhood familiarised me with East African shores, lakes and mountains, with plants and animals of all sorts, very varied local seasons and local climates, droughts, floods and famines. This nomadic, naturalistic childhood helped me bring direct experience to my grown-up explorations in biogeography and evolution. It also informs all of my work as an artist, and to the spirit in which I try to practise being an author, scientist and teacher. A childish sense of being born at the centre of the world may well have intensified those most persistent, even infantile, questions – how, where and when did people, my very own human people, begin?

By chance my birth deposited me into what, during my lifetime, has become revealed as birthplace of the human lineage. My generation of Africans has been the very first to know, with all the certainties bestowed by fossil skulls,

Jonathan and Teddy at Olduvai, 1939.

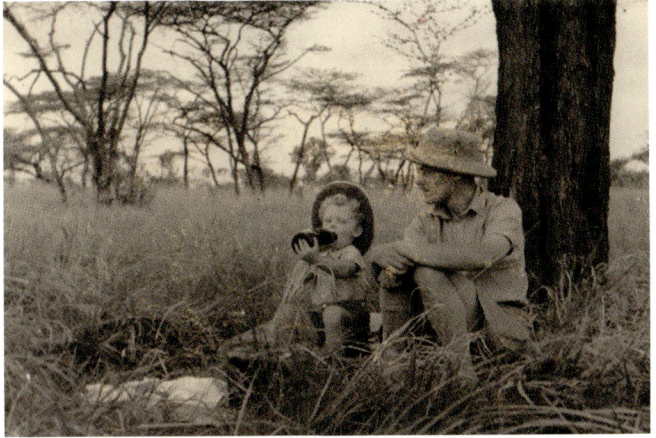

stone tools, dated soils and living genes, that we walk over the graves of our fore-fathers. Not only fathers of the last few centuries or mothers of many millennia but truly universal ancestors, taking our special primate lineage back tens of millions of years and more.

Skulls and stone axes have an enhanced reality for me, from my having braced the muscles of my arm to carry the weight of their stony heaviness on the palm of my hand. Some have been excavated, before my own eyes, from the soils of Olduvai Gorge, Olorgesailie, or the canyons and pinnacles of Isimila – all casual childhood picnic sites.

Pinnacles at Isimila.

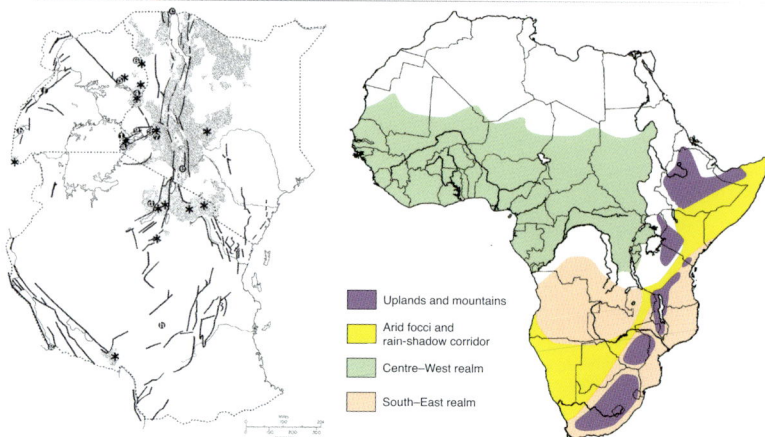

FAR LEFT: *Map of East African fossil sites, volcanoes, rifts and the 'Victoria' microplate.*

LEFT: *Biogeographic realms within Africa. A novel way of mapping Africa.*

Legend:
- Uplands and mountains
- Arid focci and rain-shadow corridor
- Centre–West realm
- South–East realm

At every stage, from hominin to *Homo*, the eastern and southern half of Africa can be plausibly identified as their centre of origin. It was certainly my personal centre and I was one teenager, as serious as my role model, Louis S. B. Leakey, about its claim for the cradle of humanity, even before anthropologists outside Africa, many immured in medical schools, began to look our way for illumination.

Listening in on the grown-ups, I once asked, 'What's a Riffallee?'

Rift valleys challenge anyone's imagination, let alone that of a child. We now know that land-masses are thick plates of relatively light stuff riding high above or next to thinner but dense plates covered by ocean. Beyond such generalities, our continent's geology is actually unlike that of any other. Creased and buckled with innumerable cracks, splits and faults (the most extensive named 'rifts'), volcanoes blast through above hot-spots or rip-apart pressure points, as vast rafts or 'microplates' jostle or grate against each other. Africa is also ridged and domed, Precambrian substrates having been pushed up from below by mysterious interior currents in the magma. Sustained erosion from these domed regions has deprived our thin Precambrian soils of essential nutrients, but wherever there are volcanoes, there will be thin skirts of richer soil, which help sustain uniquely local life.

Eastern Africa is defined by an incipient ocean along its western, central African borders. Here, the Great Rift Valley is a deep crack across more than 2,000 km of uplifted continent. Its northern end splits Ethiopia. Its southern end helps drain the Zambezi. In Kigoma we looked down on Lake Tanganyika and I was told that these shimmering waters filled the longest, deepest and oldest of Africa's rift valley lakes.

Driving up the steep escarpment road above Lake Manyara, these were the sheer walls of a second, lesser, but still continent-cracking split, the Eastern Rift Valley.

The land outlined by these two inland rifts is called the Victoria Microplate or VM, its western margins grating against the great continental mass that geologists

Kichwamba: a landscape of explosion craters.

call Nubia. The VM's upper crust varies in thickness from about 25 km at its thinnest down to about 120 km toward the southern end of its 'keel', which floats, like a jostling boat in a marina, above Earth's molten core some 300 km down.

The basin between these two rifts is currently partially flooded by the vast, shallow pond that is Lake Victoria, but should its northern margins buckle down a fraction, like the crumbling of a dam wall, the whole lake would empty and run down the Nile valley, as it has done at least twice in the past. Raise those margins just a fraction and the lake spreads far south, tripling in size.

Cranes are ancient birds so the Grey Crowned Crane (or 'Kavirondo crane') may well have seen all these comings and goings – its bugling and the geometry of its plumage accompanied my entire childhood over and over again. These magnificent

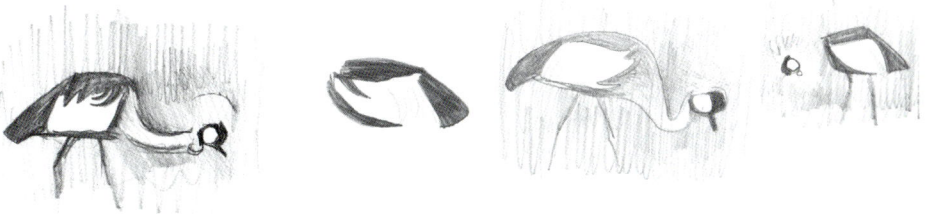

Crane sketches.

ABOVE: *Crane as emblem.* ABOVE RIGHT: *Crane tailoring.* RIGHT. *Crane composition.*

birds have challenged me to explore their beauty and pay them homage in draw-ings and paintings, even coats-of-arms. This crane is the lake's herald and emblem.

My birth-place, Tabora, is at the centre of that theoretical extension of Lake Victoria. Wide swamps and flood-plains lie west, south and east of Tabora, which occupies higher ground. At the height of a prolonged and heavy rainy season, the flooded lands around do indeed feel and look like a poorly drained lake bed, especially when great flocks of cranes, storks and waterfowl suddenly appear. Tabora, with the giant Camel Rock near its centre, resembles a citadel on what would become an island, were our game to become real.

Emptying the Victoria basin of its water was a historical reality for our orni-thologist friends Robin Fuggles-Couchman and Hugh Elliot. It helped explain

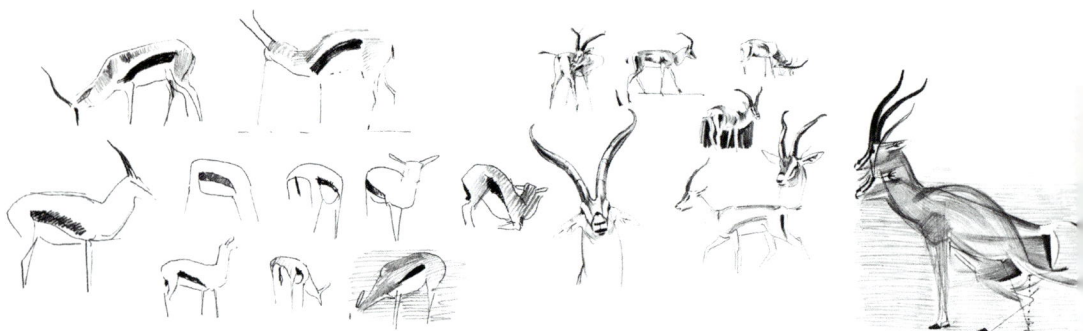

what the desert-adapted Northern Anteater Chat and the Sahelian-dwelling Singing Bushlark were doing just east of the lake shore. And why Striped Hyenas, Golden Jackals and two species of gazelle – all four of these being true desert animals – were at home in the Serengeti. Contemporary borders between related species of plants and animals present challenging questions for naturalists. Since there are many more borders than there are species, and most borders are quite labile, there's plenty to think about, plenty to argue about, and plenty to name.

Questions jump out of the grass, speed over the plains, warble from the trees or hide underground. Two flocks of go-away birds yip and gobble at one another, along territorial frontiers that are ecologically a lot more logical than those drawn by imperial schemers around a table in a Prussian Palace.

Underlying such unmarked borders are further questions as to why and how species exist. We are arriving at an understanding that, although Africa is pretty monolithic, its varied and distinctive ecologies are densely inhabited 'islands', territories with histories no less real than the ocean-separated Galapagos Islands.

ABOVE: *Go-away birds* (Crinifer *sp.) in territorial dispute.*
OPPOSITE: *Gazelle sketches.*

If the contemporary fauna and flora of tropical Africa include some of the earliest inhabitants of our small blue planet, they are but yesterday's children compared with its fabric of minerals, rocks and soils.

This was impressed upon me by our family friend, geologist 'Great Dane' Max Coster. Singida town, where my father was District Commissioner, sits between two soda lakes, and its buildings, roads, trees and gardens fit in among great outcroppings of rock. Shortly after his arrival I found myself sharing the top of a giant boulder with Max as we watched a flight of flamingos cross a near-setting sun. Our dusk perch was one of many eroded rock formations on the outskirts of the town and we were close to the promise of sundowners at the *Boma*, the massive German-built fort that our family called home. 'The parent gneiss for the rock we're sitting on could go down, say, 40,000 feet beneath us, and it became rock some two billion years ago,' Max explained. The fact that our sandals and trousered butts rested on such venerable material was a routine observation for Max, but a big imaginative challenge for me.

The one constant is that our planet has always had a relatively warmer atmospheric girdle, between two cold poles where the regular, periodic absence of sunshine returns each polar surface to something like Earth's lifeless beginnings. Meanwhile, Africa's equatorial belt harvests evaporation from two oceans, resulting in rain that falls all year, or in two wet seasons in quick succession. Max invoked the example of a boiling kettle that evaporates a lot more water than a cold one. Africa was dry and cold during the Ice Ages and hot and wet during thermal maxima, while over the last few millennia it has been about as ideally suited to humans as it ever could be – something that should never be taken for granted or assumed to be permanent.

The slow wars of jostling continents have spilt magma at weak points, on land or over the sea floor. Wounds that begin white-hot, quickly cool. They may form conical pustules which come and go as storms, waves and deluges wash them away. These are volcanoes – ecological islands on land, physical islands in the sea (remember that among the tens of thousands of marine mountains, at least one rises 9 km above the sea floor, a lot taller than Everest).

In East Africa the familiar cones of Kilimanjaro, Rungwe, Kenya, Elgon and Bufumbira all invite questions about geological history and the hidden forces that generated our pimpled, rifted landscapes. Inhabiting every mountain top to every shoreline are floras and faunas of countless entities. Their interactions with one another and with their surroundings hold inconceivable complexity. Their histories are yet to be untangled, let alone understood.

Rock-perched musings with Max aside, huge events like global warming and freezing, tectonics, volcanic eruptions, asteroids and their impact on the history of life on Earth were not part of my formal education – I wish they had been. Even so, subsequent expeditions to the heights of Kilimanjaro, Kenya, Rwenzori and Rungwe challenged my senses and sensibilities – every rock has a story to tell. Today, sundry multitudes of scientists are bringing the unique insights of

Baobabs in Kunduchi Bay.

evolution, genetics, biogeography, geology and astronomy to bear on human adaptations and prehistory. We now examine islands, mountains, landscapes and climatic periods, asking what set of circumstances could possibly have given rise to the most interesting and complex of all animals – mammalian primate humans.

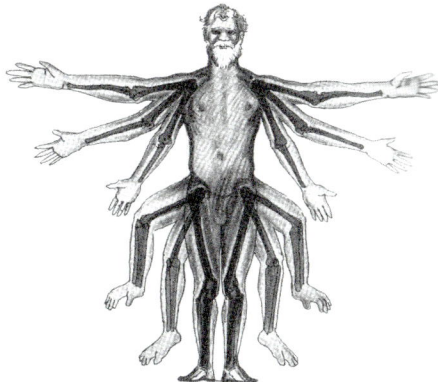

Ape and human limb proportions compared.

Kilimanjaro, seen from Mweka, and imagined erupting.

GONDWANA'S CORE

In which frogs, with some help from geologists, gift us a truer and older history of continents. Their legacy, with lungfish and scorpions as guides. Native of a microplate.

In 1963 my dear friend and fellow biogeographer, Wilma George ('Mama-Gundi'), invoked land bridges between fixed land masses to explain some puzzling animal distribution patterns. Given that many offshore islands and sandbanks had long been shown to have had former dry land connections, her speculations were forgivable. Furthermore, her most powerful and influential contemporaries, pundits in the USA and commissars in Russian petroleum industries, all envisaged fixed land and marine surfaces just rising and falling, not slewing and sliding about, *en masse*.

They and Wilma were wrong.

Amid the many mysteries of our planet, the true story of the formation and movement of continents has been among the most recent to be discovered. We owe our knowledge to one of the great heroes of science, Alfred Wegener – geologist, astronomer, meteorologist, explorer, non-stop smoker and a martyr to science. He first presented his research at a 1912 meeting in one of the world's most splendid (and my favourite) of museums, the Senckenberg, in Frankfurt. However, his discovery of what he called 'continental drift' (now more precisely the science of plate tectonics) only found its full acceptance during my lifetime (after a half-century of ferocious rejection by most self-appointed authorities). In 1944 a humble geologist, Arthur Holmes, who had first tried out his field skills in eastern Africa and Borneo, was the first to document the spread of mid-ocean ridges and confirm that continents slither, like fragments of loose peel over the surface of our global tangerine. Wegener, Holmes and the science of plate tectonics explain the eruption of volcanoes far out in the mid-ocean, like the Galápagos, Hawaiian, St Helena and Mauritian islands. Most important of all, evidence from many disciplines has now been brought together to allow a reconstruction of Pangaea, the supercraton mother of all continents. Even then, her vast amalgam of land and rock took up less than a third of Earth's surface (Mother Earth for us primates, but Mother Ocean for whales and extra-terrestrials). Estimated to have held together for some 150 million years, Pangaea's equatorial waistline then arched obliquely over today's Sahara, but mainly it warmed a much more extensive belt of sea.

Pangaea's fracture into northern Laurasia and southern Gondwana is thought to have begun about 300 million years ago when a rift valley first split Morocco's Atlas Mountains away from New England's Adirondack Mountains. An

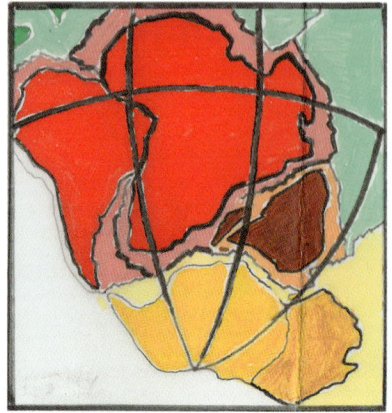

Schematic maps of Pangaea and Gondwana's breakup.

embryonic 'sea of Tethys', the future North Atlantic, swept into the cleft. Then the sea's floor steadily widened, just as Arthur Holmes was the first to show.

The mental challenge of imagining events such as oceans cascading into rift valleys or over a Gibraltar waterfall brings with it a peculiar sort of thrill. Authors of the Old Testament understood this well enough when they concocted the scene of Noah's ark floating off on a Mesopotamian flood (real enough for some likely originators of the story). Moses striking his magic staff to part the waters of the Red Sea and let the Israelites pass over could plausibly have had its origins in some upstream landslide creating a dam, or a super-drought somewhere in Mesopotamia. The big difference is that Wegener and Holmes dedicated their lives to extracting true stories, backed up by their own and their colleagues' painstaking research into natural processes. However, contemporary idolisers of 'sacred books' tend to bring literal minds to bear upon symbolic stories that invite interpretation (perhaps designedly so) in more than one way. Expulsion from the garden of Eden was once understood as the end of childhood, while Cain's murder of brother Abel was allegory for the farmers' displacement of nomadic foragers – both tales referencing the personal experiences of the Middle-Eastern authors, and other early converts to these fledgling faiths. For all their poesy, here are ambiguous and misleading materials for mullahs, mothers and grandmothers to tell bedtime stories. Richard Dawkins and Dave McKean have explored this dichotomy brilliantly in their very beautiful book *The Magic of Reality*, which should be in the library of every school, worldwide.

Had I been raised by such mentors with such school-books, and innocent of an over-arching, global civil war, I might have seen my environment very differently. Luckily, I can go on learning and discover that I was born close to the centre of a grand and fertile stretch of territory – the VM or 'Victoria Microplate'. Much of my life has been spent exploring fellow animal and plant inhabitants right out to the lakes, mountains and rift valleys that demarcate the VM's far-flung and often heart-quickeningly beautiful borders.

Living on the fertile slopes of Kilimanjaro, watching a setting sun paint those beloved hillsides orange, brief as flame, conjured biblical visions in

Afro-Arabian continent about 62 million years ago (below) and today (right).

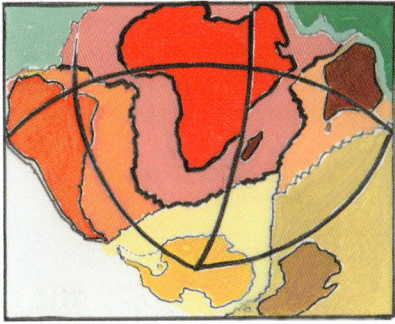

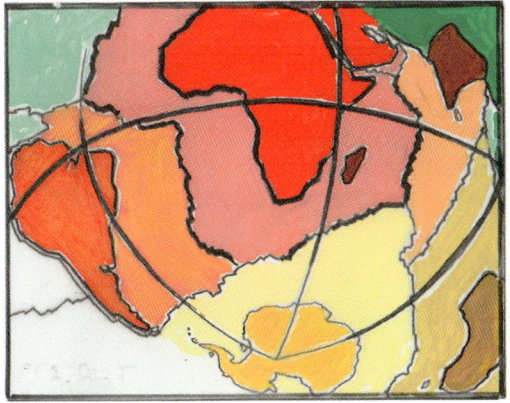

which fountains of magma built the mountain, the very slopes we now inhabited. Humans may not have witnessed Kili's most recent eruptions, but will a day come when people, our people, see the mountain wake and come alive again?

This native of a microplate has lived long enough to expand his habitat and describe it as a sort of 'life on central Gondwana'. These terminologies are very distant from those that territorial colonialism, nationalism and religions have imposed upon us. They are a part of embracing the deep past and its consequences – a necessary part of the emancipations and joys of becoming a naturalist. Gondwana was, perhaps, 150 km thick, which did not stop it from fracturing. Africa became the central supercraton of four – Antarctica, Australia and South America budding off (in that order).

There are Wegener fans who call Australia 'Eastern Gondwana' and South America 'Western Gondwana', while Africa occupies Central Gondwana (central, that is, to the pre-break-up mega-continent). This core is still actively breaking apart, having already sloughed off (about 165 million years ago) those big chunks of ex-Africa that we call Madagascar, India and Arabia, the latter creating the 2,200 km-long, narrow, but still widening Red Sea. Today, an extension of that ex-rift valley is opening as a continental crack that runs from the southern end of the Red Sea to the Zambezi delta, 3,000 km away. Like Arabia some 50 million years ago, all the land lying east of that eastern rift is today pulling away towards the Indian Ocean. The creation of new islands is a violent process. Mount Kenya's slopes are peppered with lava bombs (now cloaked in moss) beside the living grace of Afro-Alpine groundsels and senecios.

The deeply fractured trenches of the western Rift are more complicated. They demarcate the western margins of the VM, now known to be one of the most extensive chunks of thickened crust and mantle in the world. In spite of its thickness, this colossal microplate has been swept along by Africa's swivelling but mainly northward drift. Because it extends more than 100 km down, deep into the Earth's molten mantle, it is particularly 'sticky' along its northwestern hinge with Africa's main body (the Nubian plate). 'Hinge' is tool-language for the Rwenzori mountains, where the VM has crumpled up thousands of metres high

Stromatolite landscape.

(today eroded down to peaks of just over 5,100 m). This vast uplift is due to the sticky VM going into a counter-clockwise swivel against its fracturing junction with the main mass of Africa. The microplate's southern tip, lying beneath my family's little home town of Mbeya, grates against the Nubian plate to the west, while the rift that marks the VM's long eastern margins is pulling a 3,000 km-long strip of land eastward. Like Rwenzori at its northern end, this southern hinge is associated with volcanics and massive uplift. The natural history of the VM's extraordinarily beautiful landscapes, especially around its outer extremities, begged to be explored by my youthful self.

Before our time, it was inconceivable that the surfaces we walked or rode over were anything less than the bedrock of our grasp of reality. Scientific hero Wegener, for all the pedantic style imposed upon him by a provincial academic tradition, showed that we are mariners on rafts moving at rates, varying rates, that can now be calculated.

Before she crashed into the sandy shores of Asian Himalaya, India detached herself from Madagascar and broke records by speeding across the Indian Ocean at 15 cm per year, which is pretty fast for a migrating subcontinent. Australians, their males prone to growing stubble, like to say that *their* continent is moving more slowly, but at the same rate as hair growing on an Aussie chin.

Scientists have estimated that Earth formed from its constituent matter some 4.6 billion years ago and, very tentatively, that life began some 4 billion years ago (give or take quite a few million years).

To find anything resembling 'first life on Earth', visit the slopes of an active, sometimes smoking volcano in the East African Rift Valley. Its perfect cone, nearly 3,000 m high, will tower overhead. Below you, the sun or sky will reflect off the glassy, salty surface of Lake Natron. Standing on the banks of a pretty little stream that flows off Ol Donyo Gelai, look down into water that is disarmingly clear but actually resembles the sort of chemical soup that flowed over the cooling surfaces of planet Earth all those billions of years ago.

In those shallow waters you will see rounded mineral accretions, stromatolites that are the product of mineral-trapping bacteria, by ancestry as ancient as the chemical soup in which they live. In Australia's Shark Bay, stromatolites take the form of giant mineralised 'mushrooms', which look like a vast gathering of globular tents scattered through extensive lagoons of warm, shallow water. The upper surface of each 'mushroom' consists of a mat of living bacteria, and it is one of life's thrills to watch a fine film of oxygen balloons bubbling over that broad, bland surface and realise that you are witnessing a process that first evolved about 3 trillion years ago and that you owe it to that bubble-wrapped mushroom that you can breathe. Filling and emptying of lungs, steady, even while you sleep, or deep and gasping after a run, is taken for granted until you kneel, warm and wet, beside the altar of a stromatolite, mother of Earth's oxygen.

There are good reasons to feel reverence and gratitude here, because long before the emergence of Pangaea, in the shallow waters of future land-masses, various microbes were busy photosynthesising, and emitting deadly gases such as methane. For more than a billion years, Earth's atmosphere was pretty toxic. Among the photosynthesisers, a single form of bacterium evolved the ability to break a particularly robust bond – the H_2O of water. This oxygen synthesiser consisted of four manganese molecules that freed oxygen out of water, releasing oxygen as a free energy source. Recent research has identified the 'moment' when the oxygen released by bacteria overtook the toxic gases released by other bacteria. That moment, about 700 million years ago, is called The Great Oxidation Event (GOE). Thereafter, complex life and more sophisticated types of photosynthesis evolved but all began with bacteria, drawing life from the rays of a life-giving star – the sun.

Our planet's history is inscribed in rocks and bacteria but also in much more complex living things – take lungfish, old-timers with an ancient phobia for salt water. Africa boasts several lungfish species in the genus *Protopterus*, unambiguously Gondwanan in origin. Compare them with the Barramunda or Australian Lungfish, as well as with the South American Lungfish or *piramboia*, the only lungfish species found in the Americas. It is astonishing to find such close resemblances enduring since the break-up of Gondwana.

African lungfish
(Protopterus *sp.*).

Fossil lungfish were quite diverse 400 million years ago. Even then they tolerated extremes of drought by aestivating, yet (in spite of being distantly related to those 'living fossils' the coelacanths, deep-sea fishes that can live for more than 100 years) they are completely unable to tolerate salt water.

When I was a boy, the giant eel-like bodies of the African lungfish were often on offer on open market stalls. Chunks chopped up with a *panga* (cutlass) ended up in onion, brinjal and pepper stews. With biologically minded friends, we called them *Protopterus*, or 'gloppy-bloppy-opteruses' because they seemed to swim in slime. An individual *Protopterus* might arrive in market attached to a stout yoke and hanging as tall as the porter carrying it. Given their great size, it was not surprising to be told that they could live as long as a human.

One of my more enduring memories is of two youths crouched in the reeds close to the foreshore below our house in Mwanza. 'What are you digging for?' I asked as they cut and levered away with *panga*, and a sharpened stick. Could it be *Mmaamba* – a crocodile? I recoiled, but it was no crocodile that emerged from all their spading and probing (the same Swahili word is used for crocodile and fish, but the latter is discriminated by drawing out *mmaa*, while a short, blunt *mamba* signifies the reptile). Tearing away another clod, one youth plunged his arm down into the muddy hole he had excavated. Breaking up an envelope of brittle material resembling aerated plaster, he drew out a pale sausage folded over itself and covered in messy slime. The indeterminate creature, a *Protopterus*, writhed in slow motion as it was skewered behind its blunt head, and I too shuddered as I watched the young men set off with it to the open market in Mwanza town.

As its habitat dries out, the fish burrows by biting its way down, allowing mouthful after mouthful of soft mud to escape through gill arches. The passage of its efforts is marked by a hole through which the lungfish can continue to breathe, and it must have been this air-hole and its disturbed surroundings that gave the cocoon away to the young foragers. Once embedded deeply enough, the *Protopterus* foams up a frothy mix of gluey secretions and mud that harden into a cocoon around the ever-more-immobile, usually U-folded, fish.

Lungfish adapted to droughts some 400 million years ago and have survived by outliving the most punishing of global perturbations. Since lungfish have never been able to tolerate salt water, they are true relicts from terrestrial Gondwana.

Having a genome that is 36 times larger than that of a human and 360 times larger than fellow fish *fofu-nungu*, or Long-spine Porcupinefish raises all sorts of interesting and important questions. The answer to one, deduced by geneticists, is that the common ancestor of humans and *fofu-nungu* had 12 chromosomes. How lungfish came to acquire a genome with more than a billion base pairs remains a mystery, but selection for a way of life that has survived for 400 million years has to have some relevance, as well as testifying to the durability of lungfish under the most extreme of conditions.

Killifish from west to east.

This also raises questions about what it was about the southern continents that has allowed lungfishes to survive there but not in the north – a question to be examined in the next chapter.

The aestivation that has helped lungfish to survive the almost unimaginable vicissitudes of all those millions of years is paralleled in another group of extremely ancient salt-water-intolerant fish – the killifish (also known as rivulins and annuals). These tiny, slender fishlets probably evolved in Gondwana, and their survival over the ages has been enhanced by their spending most of those ages as embryos, not adults. Why? Because their fertilised eggs and embryos can survive where adults cannot, and can endure drought, heat and climate change, as well as avoid predators. To achieve this, embryos have thickened their outer membrane to resemble the impervious skin of some ingenious time capsule. Furthermore, that well-protected embryo can respond to other external challenges by delaying its own development at any of three embryonic stages. Just how lengthy such delays might have been in the past remains to be determined. The eruptions, earthquakes and heat-waves that surely accompanied the break-up of Gondwana must have severely tested these animals, among the most advanced forms of life on land at that time.

Male killifish court the dowdy females with some of the most glorious displays of colour and pattern ever to have evolved on this planet. Here, evolution has captured the entire spectrum of the sun's radiance in fish scales. Here, tiny male killifish in seasonal puddles (even rainwater in elephant footfalls) combine all the potentials of touch and vision to caress and impress females, the vessels for their very meaning, their brief moment as males.

Their flat, two-dimensional fins take bill-board advertisement to its extremes. Like many other organisms, segmented beginnings, genetic structures and the structural sensitivities of eyes favour repetition, such as a pattern of spots and stripes. Additionally, sensitivity to colour allows sides and fins to evolve elaborations of these repeated patterns. The outlines of fins alter easily, extensions acquire streaks and stripes, margins enlarge, spots become red, blue, yellow – all to what end? To out-compete other males in the beauty stakes? To seduce drab, functional females? The winning moment is to be there, sperm-ready, fins caressing her as she, visually mesmerised, expels her few and precious eggs.

Way out west, in Djallon, female killifish respond to zigzag movements and complex patterns. In Togo, perhaps the hypnotic stare of disembodied eyes is enough to convince them to release their eggs, but no, a female's girdling by an enormously enlarged male tail could just as well be seen as the fish equivalent of a peahen submitting to the inescapability of the male's quivering galaxy of eyes. In the species south of Benue, it is red and blue stripes that win over the females, while around Buta a jeweller's shop-window of rubies and diamonds set in silver holds the key. The male of one East African species throws the works at his intended. An opalescent eye stares out from a facial mask of fluorescent vermilion. This graduates into a succession of about 30 zebroid stripes of amber, then ruby, each alternately set in turquoise, each just one scale wide. This glowing procession of jewels ends in a tail-tip of blazing orange gold. Remember, a fish that lives for just weeks cannot become the year-round diet of any predator. Killifish males can afford to be as garish, as bling-worthy as wins the race for the longevity of your species, if not for your own month in the sun.

Lest we take too much comfort in our own supposed longevity, remember that male killifishes have been capturing rainbows perhaps for 170 million years. Here, what we call 'beauty' has been thrilling female killifish for all those years. Here, in the briefest of brief moments, we witness life's meaning, moments in which males can dazzle, moments in which male determination to impress meets its maker – the female principle, a male's only guarantor of continuity, the only hope of a future for his kind.

Here life's capacity to surprise found expression before there were human eyes and brains to explore their meaning; long before that same species evolved with a determination to make all nature serve its animal appetites.

Now, let us turn to frogs. Because they had evolved by at least 250 million years ago and cannot tolerate salt water, land-lubbing frogs are a gift to biogeographers. Each continent, even each former continental mass, even odd piles of debris such as the Seychelles and New Zealand, left behind from the safaris of wayward continents, has its own archaic families of frogs. The most ancient forms evolved while all the continents were one mass – Pangaea.

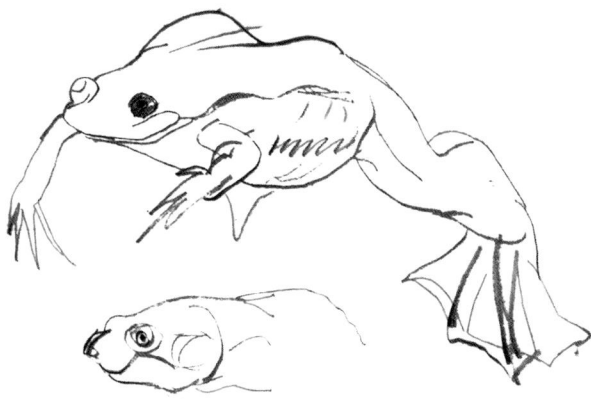

African Clawed Frog
*(*Xenopus laevis*).*

It is a near-miracle of survival that Pangaean frogs should have survived in New Zealand, but their survival there is a tribute to the durability of land-life-support in just such remote localities. The parting between Laurasia and Gond-wanaland, some 175 million years ago, stranded yet more primitive frogs in the southern mega-continent.

Lake Mutanda in the Rain.

In Mwanza our evenings were sometimes punctuated by noisy clicks made by an extremely primitive frog, *Xenopus laevis*, or the African Clawed Frog. *Xenopus* are air-breathers, adapted to burrow down into mud as their freshwater ponds dry out. They can aestivate for a year in a slimy cocoon, and live relatively long lives (15 years is long for a frog).

South America hosts a closely related species, the Sabana Surinam Toad or *Pipa parva*, which has maintained a similar body plan and habits, even though the last common ancestor that it shared with *Xenopus* had lived before South American 'West Gondwana' broke away from Africa's 'Central Gondwana' some 140 million years ago. In eastern Africa, *Xenopus* are most in evidence during the wet season, when the silhouettes of their splayed, black-nailed limbs and bloated-looking, small-eyed bodies seemed to float, like corpses, in many a pond or river-bend. Misleadingly torpid, they dive out of sight and swim away surprisingly fast when disturbed, in spite of being incapable of leaping, only swimming and scrabbling. In their inability to jump or hop, and in their very primitive mouth parts, they are unlike other frogs. Because they lay large eggs and are easy to keep under laboratory or aquarium conditions, *Xenopus* are favourite experimental animals and thousands have been exported and established in labs all over the world. For many years, *Xenopus* was used to test pregnancy in women, because female frogs exposed to the urine of a pregnant woman responded by laying eggs. Today, these frogs, their eggs and tadpoles are employed in countless biology labs for an astonishing array of research topics. In genetics they are used to reveal the function of particular proteins, because the individual genes that control for particular proteins can be easily knocked down and thus reveal their function in the expression of mature structures.

In the mountainous regions of central Africa, high-altitude lakes have formed in valleys peppered with conical hills that are miniatures of neighbouring giants, some of them active volcanoes.

In some of these newly formed lakes *Xenopus* got there before fish, and multiplied enormously. Around Lake Bunyoni, in southwestern Uganda, both otters and the local people adapted to a diet of frogs. It was there that I sampled quite tasty *Grenouille Provençale à la Kigezi*, without knowing my dish's ancient

Table Mountain, South Africa.

ancestry. I wonder how my mental child-scape might have enlarged had I the capacity to imagine such knowledge at the time. With the wisdom of hindsight I can vouch for their tangible reality, because I have contemplated, handled, drawn and even eaten beings that were here in scarcely different form from long before there were dinosaurs.

Another pre-dinosaurian has survived in the shape of 'spookpaddas' or Table Mountain Ghost Frogs, *Heleophryne rosei*. Emerging as a distinct lineage some 140 million years ago, these South African endemics are the epitome of evolution's astonishing capacity to adapt to just about every imaginable vicissitude, to survive against all odds, yet be abruptly extinguished when some vital property of existence is withdrawn, or some novel disease penetrates an organism's defences.

South Atlantic swells have carved this last chilly outpost of Africa, sometimes encircling it as an inhospitable island, but Table Mountain's hard, weather-beaten granitic sandstone has endured, as have spookpaddas and, remarkably, their tadpoles. Chilly waters and a pauper's diet slow their development down to an entire year, yet these hardy little larvae have survived by broadening their lower jaw and adapting their throat skin into a strong adhesive suction-pad that can cling to slithery rocks, while their slipstreamed bodies withstand a year of survival under rushing torrents of near icy water. How could such persistence endure and evolve over so many millions of years, and why?

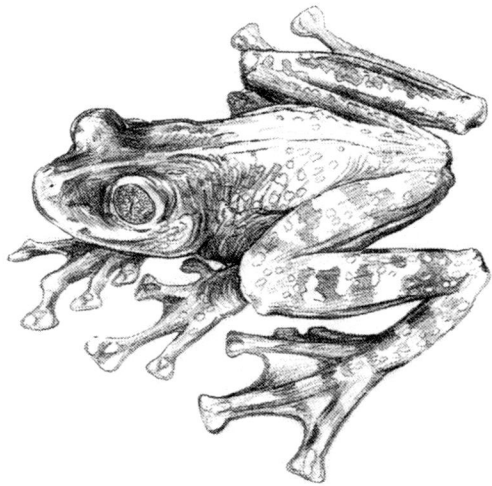

Spookpadda or Table Mountain Ghost Frog (Heleophryne rosei).

Their habitat's most durable, most reliable, least changeable but least nutritious resource has been the green algae that coat those slimy rocks, and this is the tadpole's only food, harvested by tongues armed with tooth-like rakes. Only after they have matured into (still rather flattened) adult frogs can they shift to a more nutritious diet of worms and invertebrates, but their webbed and spatulate digits signify that they remain champion clingers, as well as fast, strong swimmers and expert hiders in crevices, but hopeless hoppers. The steep, shady, once densely forested valley in which the ghost frogs were first found is called Skeleton Gorge, hence the frog's spooky nickname.

The city of Cape Town is built on lowlands lying east of Table Mountain. The town draws its waters from reservoirs built high up on Table Mountain, which capture and store an abundance of rain coming off the cold South Atlantic. Overflows still course down steep, once spookpadda-friendly canyons, but European colonisation brought several unprecedented changes. Reservoir water gets heated by the sun, so the overflows, even when sheltered from the sun, are warmer. I have also seen for myself how fast-growing commercial conifer and poplar plantations have replaced the indigenous forests that were felled long ago. Also introduced exotics, such as livestock and the goat-like Himalayan Tahr, churn up banks and streams, making once crystal-clear waters turbid and contaminated. Worldwide, the Cape is celebrated for its flora and for an abundance of supposedly lesser beings such as spiders and some spectacular insects. Notwithstanding all the vicissitudes that the Cape has suffered at the hands of vandals, generations of noble conservation-minded locals have done much to mitigate some of the horrors associated with pioneer urbanisation. Thanks to their efforts a splendid Cape Peninsula National Park now administers and protects those spookpadda habitats and other wildlife-rich areas that remain.

While almost every geological era has retained vestiges of history in the form of living frogs, India's detachment and cruise away from Madagascar and Africa allowed it to transport many plants and at least two distinct Gondwanan frogs, one survivor being the *Xenopus*-like, miniature-headed Purple Frog or Nasika Frog, *Nasikabatrachus sahyadrensis*, that just survives in one corner of the southwestern Ghats.

The Seychelles archipelago (fragments of granite left behind during India's cruise across the ocean) still hosts several species of Gondwanan palm frogs of the genus *Sooglossus*. These extraordinary miniature frogs develop from eggs deposited in a cluster on damp ground. A parent then guards this crèche until the eggs hatch into froglets, upon which they wriggle their way up onto the parent's back. Here, glued by some strange batrachian mucus, they grow to maturity. The vicissitudes of

Seychelles frog (Sooglossus sp.).

West-facing view from our home in Mwanza; oil painting by Dorothy Kingdon, 1941.

their islands' long oceanic isolation presumably included periods in which natural ponds dried out, favouring this strangely contrived ontogeny.

In a very real sense, lungfish, killifish and frogs are older than the continents and almost all the world's mountains, including the mighty Himalayas, which are but recent by comparison. In Mwanza I shared the lake shore with fellow beings, fish that were bigger and heavier than me, that had been around in little-changed form for all those many millions of years.

My personal good fortune was to be at home beside a short stretch of lake foreshore where gloppy-bloppy-opteruses, killifish, spiders, *ngo-mwenye-sumu* (scorpions) and an orchestra of frogs lived within the ambit of our verandah. By the time we left Mwanza it had been home for nearly half of my then life-time. Even now, among the many recollections of an already long life, those two and a half years beside the great lake still seem to occupy a disproportionately large portion of my memory. I am sure that many more of my perceptions were formed there than I can now begin to re-examine or unearth.

Robert Blackburn's map of airfields from London to Cape Town.

Had I known it at the time, perhaps some such explanation might have tamed some fearful moments that I can still remember as animal terror. My Mbeya School bed bouncing about as if it was being shaken by an amorphous angry dog – cracks opening up across the plastered walls of our dormitory – being peppered by bits of loosened thatch – very loud, rumbling growls from deep, deep below during the darkest hours of night.

An explanation of what was going on might have been calming – but for an eight-year-old animal on his own in a strange bed for the very first time in his short life, perhaps not. Anyway, I never forgot that first earthquake in Mbeya school, where some of us, children of World War II, could imagine we were bound for an abattoir where at least one of our teachers seemed to preside over an educational slaughter-house. That notwithstanding, I remember us as a cheerful little bunch.

Less than three decades before my incarceration in that school, a pioneer plane-builder called Robert Blackburn had an unlikely role in the creation of Mbeya township. During World War I he had turned his factories over to build a fleet of aeroplanes that fought and triumphed over those of the Kaiser's regime. After witnessing his fleet scuttled in 1918 and its remnants sold off to post-war Germany, he became one of the founders of Imperial Airways. Thereafter he battled a succession of British governments almost as incompetent as contemporary Brexiteers. Wanting to develop more versatile routes than the scatter of sea bays, lakes and broad rivers that Imperial's 'flying boats' or 'sea planes' had to splash down onto, he looked for an alternative. He saw that large planes with stout wheel carriages, able to land on *terra firma*, offered much more direct and profitable routes between Africa's capital cities. A brand new London to Cape Town service would link Nairobi and Harare (the latter a copy-cat 'Salisbury' at that time), but the two pioneer cities were nearly 2,000 km apart. Exactly halfway between them lay a broad, flat valley below a tiny Safwa hamlet called Mbeya. Here, Imperial Airways installed an airstrip, a post office, a fuel dump and a small hotel. The sudden, totally unexpected arrival of an airfield with international connections invested Mbeya like no other inland town with a new sense of being linked to the big, bad world outside. It also attracted Nazi party members, who set up a school there to indoctrinate German children. Such were the precursors for our earthquake-prone, ex-Nazi, Mbeya School.

Today, Mbeya town sprawls over the southern tip of the VM, a region of up-lifted rift walls (some of them sheer cliffs dropping into lake waters) and long ranges of hills and mountains. A few kilometres south of the town stands Rungwe mountain, a currently dormant volcano, rarely revealing its summit through the clouds that envelop it. The entire region is freckled with volcanic craters, hot springs and avalanches of pumice. I soon became as blasé as any other locals about earthquakes – they were that frequent. In Mbeya the rumble of Gondwana's fracturing feels as real as those so-ancient continental partings that stranded lungfish and clawed frogs on its floating fragments.

Mbeya valley, as seen from Crater Lake, Mbisi.

Lake Tanganyika, showing Bujumbura, Mpulungu, Tabora and the eastern Congo watershed.

Old Gondwana's break-up finds something of a replay in Lake Tanganyika, even though the rift it fills began to form a mere 12 million years ago. That rifting has been progressive, the lake deepening and extending south by stages. The lake actually comprises two distinct underwater basins, but its waters have risen and fallen many, many times.

Most fascinating of all, Wegener has taught us that the Earth's land surfaces have pitched and rolled, like floating lilos or the decks of sluggish catamarans caught in cross currents. In Africa it has been suggested that periodic overspills might, at different times, have sent Tanganyika's waters towards at least three points

Lava flow near Mzima springs.

Chyulu hills from Kilaguni.

of the compass – northwards into the Nile, south into Lake Rukwa and beyond, and its present intermittent overflow westward into the Atlantic, via the Congo basin. The rifting margins of slumping valley bottoms often rise up, tilting up former flats until they become near-vertical stratified hills or mountains such as the Rwenzoris. Upstart 'nations' or 'peoples' often use such margins and rifts as natural territorial boundaries. These ancient ranges, especially those that encircle the VM, have served as refuges that harbor interesting endemic biota such as relictual spiders, proteas, *Podocarpus* trees and worm-like amphibian caecilians. I visit all of these in later chapters.

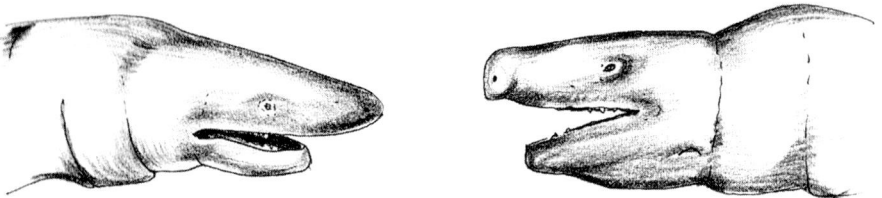

Two caecilians from the Seychelles: (left) Hypogeophis rostratus; *(right)* Grandisonia seychellensis.

DAWN AND CONSEQUENCES FROM CHANCE

3

In which the dawn of Earth as we know it today is shown to be the aftermath of a cosmic car-crash. The Chicxulub Crater. More frogs and more survivors in The Cape. The Mouse that ate the Laws.

On 20 July 1944, chittering monkeys hugged one another while dogs howled and spurfowl clattered out their evensong. Believers in The Day of Judgement prostrated themselves in terrified prayer. For those in East Africa and India who were unprepared for the sun's eclipse, it was a frightening experience. For those already alerted, pieces of exposed film were held up to witness the moon glide between viewers and the sun, its traverse an ever-larger nibble of The Big Orange until, for a brief moment, a perfect disc appeared, its halo blazing. Then, as the moon moved relentlessly on, sunlight returned to the Earth sliver by sliver.

Was it the sun that moved? Or the moon? Or the Earth beneath us? The ability to predict that eclipse was ultimately a byproduct of countless curious minds, over lifetimes of seasons, all registering links between sky-gazing, shadow-watching and ground-living. These were minds that put a value on close observation, the sharing of facts via language and print, devoted to keeping records and to explaining the sun's prodigality in the tropics – its annual withdrawal at the poles. Sunbeams fathering sunflowers. What we learn from science helps us reassure the little girl who thinks her shadow wants to bite her heels.

Seeing the sun climb, seemingly all wet, out of the sea, then, as it soared slowly overhead, watching crisp shadows glide from pointing west to pointing east until the sun hid behind the Usambara mountains – all of this was physical sensation. At dusk I felt robbed of all that hot light. Mosquitoes arrived and I was packed off to bed under a mosquito net. To interpret and translate such mysteries, a hungry mind depends upon what it is fed by parents, priests, teachers or scientists.

My father's efforts to teach me how Earth circles the sun and how the moon circles Earth were early and memorable. With an orange for the sun in one outstretched hand and a physalis berry for the Earth in the other, he slowly pirouetted himself and the berry around the orange as if he were a ballet dancer –'A full circle is one year.' Then he spun the berry between thumb and digits – 'While the berry faces the orange that's a day, and because it spins, like your top, half of its spin is in the dark – that's a night.' Then he made a much tighter, faster manoeuvre with a pea, making it whizz around the berry. 'That's the moon, but notice that when it's fully lit by the sun but we are in our night-time cycle we

OPPOSITE: *The sun as furnace of life on Earth.*

call it a full moon but while the Earth cuts out more and more of the sun's rays the lit portion of the moon declines, only to enlarge again while what we call a lunar month goes by. When the moon comes between us and the sun, the sun's light gets blocked out for some people somewhere on Earth, and that we call an eclipse.' Such intimate modelling served to reduce the unimaginable vastness of the Universe, let alone the sun, moon and weather, and translate it down to a scale that a child can imagine or visualise.

People have long tried to explain the movements of planets, comets and the starry, moonlit Universe above them in terms of their own little patch of terrestrial territory.

A most surprising and entirely unforeseen connection to ancient Africa and its inhabitants appeared above me during a visit to Egypt, dominated by museums, pyramids and a launch up the Nile. My mind and imagination seethed with exquisite images of Thoth, Horus, Seth and beautiful queens served by regiments of commoners, all depicted in sculptures, bas-reliefs and paintings from Africa's greatest and longest-lived civilisation, all built on the banks and delta of our longest river.

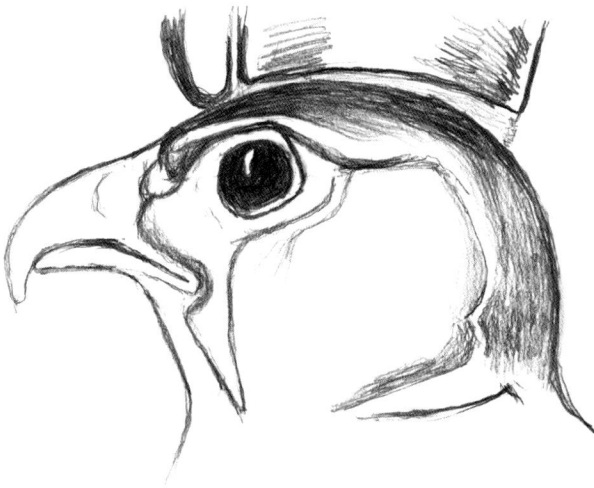

Gold head of Horus with polished onyx eye, Antiquity Museum, Cairo. Extraordinary attention to detail and proportion suggests it was modelled from specimens, alive or mummified, or both.

To avoid a scorching midday we were on a dawn outing to a mosque built on Cairo's not-very-high heights. Emerging straight out of a freshly risen sun, a Peregrine Falcon came stooping down to strike one out of a panicked flock of feral pigeons. Here, right before my own eyes, was Horus, Sun God and falcon-god, announcing his arrival to metropolitan inhabitants of the World's Centre, a city of plump pigeons as well as donkeys and people.

The cliffs of Jebel Iweybid were too far east for ancient Cairenes to learn that Horus and his predecessors spent their nights roosting there. Even so, I cannot be the only non-Egyptian to feel moved by the exquisite crafting that was lavished upon images, even mummies, of this bird – a smallish raptor that made manifest the abstract wonder of dawn breaking over this precious globe of ours as we all hurtle, spinning as we go, through space.

My father's astronomy lessons were augmented by a Royal Navy-issued telescope that he had inherited from his own father. Perhaps it was my grandfather's subordination to naval discipline and his imposition of it upon his offspring that influenced Teddy's stress on order, in both society and in the known Universe.

If so, our sessions gazing at the moon through his telescope forced him to concede that her plenitude of impact craters showed that the moon had been peppered by almost countless comets, meteors and asteroids. Was it only the moon that was some sort of a target on a shooting range? Or were bits and pieces of cosmic rubbish a perennial hazard out there?

Stained-glass window of moons and moonflowers, Rondo Chapel, south Tanzania.

Teddy was forced to admit that disorder could rupture his predictable fruit salad universe, as he knew at first hand.

In 1930 Teddy was in Mwanza when a bolide was seen to burn up in the atmosphere, scattering fragments all over the Sukuma hamlet of Malampaka. He also described the night skies of 1933 being lit up with frequent showers of meteors, burning up as they entered the atmosphere.

On several occasions Teddy drove the family out to Mbozi where we picnicked beside 'Kimondo', a then barely exposed meteorite. This chunk of iron and nickel has now been calculated to weigh 25 metric tons. Eating sandwiches, we sat on a surface which, with some imagination, could be said to resemble the back of some prehistoric reptile. Yet no-one has yet devised a date for this metallic bullet's ballistic impact. All traces of its crater were overlaid by later geological upheavals (which have been very substantial all around this disjunction between two rift valley lakes). The oldest craters are now invisible on Earth's surface, but meteors can leave a magnetic record deep underground. Thus, the misbehaviour of compasses in central Africa long ago revealed what has come to be called the 'Bangui Magnetic Anomaly'. Some geologists think this might betray the impact of a meteor strike well over 540 million years ago. Other scientists suggest that perhaps early basalts go down unusually deep below central Africa.

Earth has been hit by random comets many times (more than 300 impacts are on record).

One particularly well-known crater (dated to a mere one million years ago) is hugely popular in Ghana. Close to Kumasi City, this crater is called Lake Bosomtwe. Formed by the vertical strike of an iron meteorite not less than 500 m wide, it created a 10 km-wide, nearly circular crater with its central bore-hole about 8 km wide and 750 m deep. That the levels of the lake's waters have fluctuated wildly is betrayed by fossil fish being found near the peaks of surrounding hills, and by divers finding drowned tree stumps on the lake's floor.

Known to have shrunk to a puddle during past periods of drought, this lake figures in a telling folk-tale from long before the Ashanti became the populous people they are today. Pursuing what was probably the water-loving Sitatunga antelope, a legendary hunter lost all sight of his quarry when it submerged itself into what was then a vestigial lake lining the then forested crater's depths. Impressed by what he took to be the intervention of a god with a special fondness for antelopes, the legendary hunter and his audience named the lake after that 'God of the antelope' – Bosomtwe.

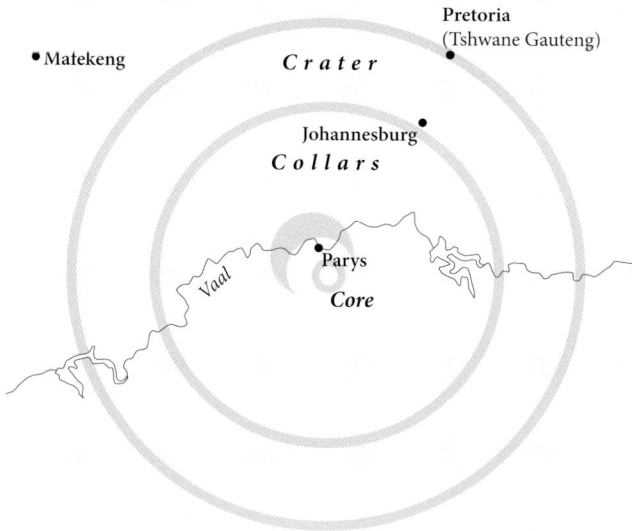

The Vredefort crater, also known as the Parys crater. BELOW: An 8,000-year-old image of a hippopotamus is engraved on one of the crater walls.

For the most pictorial visual effect, a perfect circle of cliffs in the Kalahari, some 70 km across, is perhaps the closest Earth gets to our equivalent of a lunar impact crater. Morokweng crater (with the charmingly named hamlet of Pom-Pom perched upon its rim) is dwarfed by its near neighbour, the 2 billion-year-old Vredefort crater which, with a diameter of 300 km, is the largest known crater on our planet. The meteor's effects are particularly well known because its impact liquefied the Earth's surface, melting a pond of minerals that included a lot of gold and uranium. The margins of this pond solidified into a series of very hard rings, known today as the Witwatersrand reef, and 100 years of gold extraction have left a landscape of mine tailings that eerily mimic the mounds of harvester ants, on a gigantic scale. Seen from the seat of an airliner, one rim of the inner crater is still just visible. Billions of years of erosion and the more recent cut of the Vaal river have blurred the rest of the imprint of a freakish bolt from outer space.

Was this the most grievous wound ever inflicted upon a world that just happened, at that very moment, to be in the path of a stray asteroid?

For the random event that has shaped our very existence, we must now turn to one fateful April moment 66 million years ago when a meteor, like many before it, nearly shot past the Earth. A matter of minutes, earlier or later, and gigantic reptiles might still be the dominant form of life on Earth. Instead, there was a cosmic car crash, as much an 'accident' as any collision on an autobahn.

We owe the story of that collision to a great father/son scientific duo: Luis and Walter Alvarez. It was their pursuit of an inexplicably rare mineral, iridium, that led them to be the first to describe the many dreadful consequences of a 10 km-wide meteor slanting in from the south to slam into today's Caribbean Sea, at about 25 km per second. The impact ignited a fireball reaching temperatures of 1,300°C and gouged a crater 14 km deep and 180 km wide. The whole site, sometimes labelled 'Cemetery of the Dinosaurs', is now under about 1 km of limestone, a tombstone that stromatolites and corals have helped to thicken with each passing year.

The closest contemporary on-land settlement to the point of impact is a Mayan village on the Yucatan peninsula, called Chicxulub. Get used to this tongue-twister because the Alvarez team gave this name to their giant, now underground, under-seafloor crater, and I will use it throughout this book for my single most important temporal point of reference.

The Chicxulub meteorite (just as correctly called an asteroid or a bolide) came in loaded with extraterrestrial iridium. When the meteorite and the walls of its crater vaporised, iridium, together with the tiniest of glass spheres, shocked quartz and traces of new, shock-induced minerals got blown through Earth's atmosphere and stratosphere. Some of this iridium-laced cloud of material continued onwards out into space. Eventually falling back to Earth, it left its momentary autograph all over our planet, with its heaviest particles falling closest to the crater, along the path of its blast and in the tsunamis and fireballs it engendered. The 'iridium layer' has become the geological and chemical marker that lies between the C(K)retaceous (K) and then Tertiary (T) periods of geological history. In Italy, that iridium layer can be seen with the naked eye as a very thin black stratum, separating the once life-filled K and near-lifeless early T. It is called the KT boundary and the catastrophe it marks is known as the 'KT Event' (often revised into 'C-Pg', but the more euphonius KT remains more popular).

The lynx's eyes.

We owe much of this vocabulary to Papa and Son Alvarez and their many collaborators, some of them Italian. Because Italy has a long and illustrious tradition in both science and in ceramics, her soils have long been explored and analysed by thoughtful minds and attentive eyes. The phrase 'eyes of a lynx' was appropriated by a brotherhood or academy of scientists founded by an 18-year-old in 1603 and based in secular Rome – this body is still called 'The Lynxes' or 'Academia dei Lyncei'. It was natural for Luis and Walter Alvarez to turn to Italian sources to lead them to a rugged gorge close to the hill-town of Gubbio, famous for its Maijolica pottery and a historic source of knowledge about local minerals. There, for every passer-by to see, is the KT Event writ large in coloured rocks.

Jonathan and Laura straddling the KT boundary at Gubbio, Italy.

Inspired by the Alvarezs' thrilling adventures in collaborative research across multiple disciplines, I and my wife, Laura, visited the gorge where their team had studied this particularly graphic exposure of the KT boundary. It was a thrill, perhaps a perverse one, to stand (and even be photographed) with one foot in the Cretaceous, the other in the Tertiary, each separated by that thin but portentous black line.

The prime victim of the KT Event was North America, where extermination of life was almost total. Europe and northern Asia fared little better. Australia, southeastern Asia and southern South America were all devastated, but their hardiest organisms, including particularly sturdy plants and some small animals, survived to inherit the Earth. Most of Africa was caught among three global catastrophes, all, so far as is known, deriving from the colossal jolt of impact.

At the same time as Chicxulub, a detached fragment known as 'Nadir' hit the continental shelf off Guinea. In a third, simultaneous cataclysm, a series of enormous outpourings of super-heated magma belched forth from the western edge of India (at that time a vast migratory island) – these lava flows are known as the Deccan Traps.

'Hot-spots' drive mid-ocean spread, so it is possible that Chicxulub may have jolted or influenced the opening-up of an already weak furrow cut by India's deep actively moving western margin, thus exacerbating this huge eruption of lava.

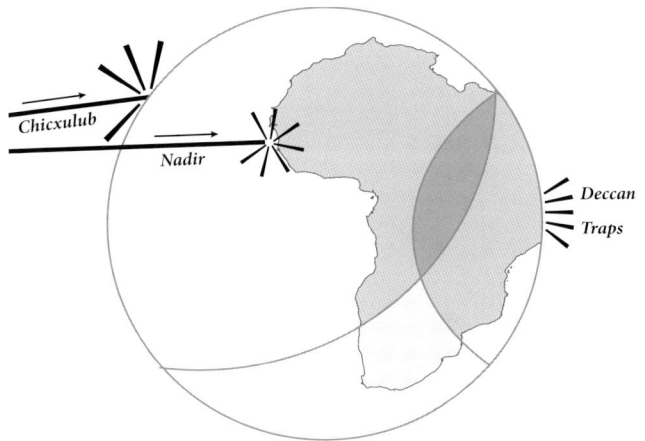

Combined impact of Chicxulub, Nadir and the Deccan Traps on Afro-Arabia 66 million years ago.

For most of Africa, the effects were dire. At the time of Chicxulub, Africa's northern reaches included today's Arabia, and all of this vast territory was caught between the three major sources of global destruction.

Southern Africa seems to have presented a brighter prospect. The previous chapter discussed the many resemblances between the biota of South Africa, southern Australia and southernmost South America, in terms of their including some survivors from ancient Gondwana. That may apply to some families of ancient vertebrates, invertebrates and plants, but by the time of Chicxulub many tens of millions of years had passed since the break-up of Gondwana.

We now know that the comet hit in the southern autumn – time to hibernate.

This potential for survival from Chicxulub's devastation can be illustrated by a lineage that has given rise to two frog genera that reinforce the likelihood that underground hibernation favoured survival after Chicxulub. Shovelnose frogs, *Hemisus*, and short-headed rain frogs, *Breviceps*, are deep burrowers, able to go

Two drawings of Isinana, *the Spotted Shovelnose Frog (*Hemisus guttatus*).*

into long-lasting torpor deep underground, or in crevices. Such a strategy could have allowed frogs or other animals to outlive the KT catastrophe, whether in adult, larval or egg stages.

As a small boy in what is now Kwazulu-Natal, I was shown a Spotted Shovel-nose Frog, *Hemisus guttatus*, locally known as *Isinana*. This burrower can survive long periods deep underground with all obvious life-signs suspended. In the rains it emerges to copulate and to release developing tadpoles or froglets, and to perhaps enjoy a brief aquatic existence. Today *Isinana* only survives on flat flood-plains beside a very few river basins in Kwazulu-Natal. It has a horny-tipped nose, fingertips and heels. Where muscular hind legs serve sudden and surprising leaps in other frogs, in shovelnose frogs they power forceful, persistent rowing through sandy soil. Females guard a jelly-cushioned underground nest containing some 200 eggs.

Their survival is consistent with the south being the only, albeit precarious, refuge from Chicxulub in Africa, but their mole-like behaviour has allowed closely related but somewhat more versatile species to expand their range. Among them are the rain frogs; family Brevicipitidae.

I remember my father charging me as a teenager with excavating three *hafirs* or reservoirs to irrigate his coffee shrubs. More than 2 m underground I encountered a fat, cappuccino-coloured frog, encased and immobile within its cramped little follicle. How this Rungwe Rain Frog, *Probreviceps rungwensis*, had managed to insinuate itself so deep underground was a mystery only just explicable while deeply cracked soils were so sodden that this apparently feeble frog could

Rungwe Rain Frog
(Probreviceps rungwensis*).*

Goliath Frog
(Conraua goliath*).*

half-dig, half-swim its way down many hundreds of times its own length through soils that become impermeable in the dry season. When attacked, both rain frogs and shovelnose frogs can inflate their bodies to almost spherical proportions, while sweating out noxious, glue-like exudates that deter most potential predators.

Another lineage that gave rise to many other African frogs includes the Goliath Frog, *Conraua goliath*. This species begins life conventionally enough, its eggs and tadpoles resembling those of related torrent frogs (among frogs, 'close' cousins can be more than 60 million years apart!). Goliaths earn their name by growing, growing and growing, until they reach up to 3.25 kg, with legs as long as a child's. As builders and sentries of stony, weedy frogs' nests beside rushing torrents in a few West African rivers, these, the largest frogs in the world, are doomed by collectors, zoos, road-side gourmands and most of the many afflictions that threaten frogs today.

In contrast to weak frog forefeet, many animals, especially the digging mammals, have hardened claws on tough, muscle-bound forelimbs. These

*Three Cape tortoises: (left) Geometric Tortoise (*Psammobates geometricus*); (middle) Speckled Cape Tortoise (*Chersobius signatus*); (right) Angulate Tortoise (*Chersina angulatus*).*

adaptations for subterranean life are undoubtedly ancient, not recent contrivances. Whether the ancestors of two deep-digging mammal lineages, the Aardvark and the golden moles, had parted ways by Chicxulub is still debated, but their divergence seems most likely to date from Chicxulub. Aestivating tortoises were other likely survivors.

Contrary to popular belief, ancestral mammals have been around just as long as ancestral reptiles. Less certain is just how far the radiation of placental mammals had proceeded before Chicxulub. Scientists working with molecular clocks reckon that many ancestral stocks had already evolved. Clearly, Chicxulub was the main 'releaser' for the radiation of mammals, but the rarity of their fossils before 66 million years ago tells us that most mammal lineages (as well as most of those of birds, and even frogs) radiated after KT.

To further complicate matters, these questions interact with controversies over *where* placental mammals first evolved. Mark Springer, at the University of California at Riverside, argues that the most primitive of surviving placental mammals are found in Africa, and that an early freak transport took one lineage to South America where sloths and armadillos are among their descendants. Nonetheless, neither Africa nor South America is the mammals' Eden – that distinction belongs to Asia.

Among Africa's survivors (first named 'afrotheres' by my dear friend Alan Walker) are Aardvarks and golden moles. These burrowing afrotheres are especially interesting in the context of Chicxulub because (in common with armadillos) they were perfectly adapted to survive the devastation by living underground. Even today, golden moles scarcely exist north of the equator and they are among the many groups of animals and plants that have found their main refuge in South Africa (where promoters of agro-forestry, sugar and other industrial crops now seem dedicated to their extermination). Burrowing afrotheres are interesting not only as survivors of Chicxulub, but also as models to exhibit some of the advantages mammalian hot-bloods had evolved over reptilian cool-bloods long, long before the meteor struck. In Chicxulub's wake, Earth became a sort of planetary bomb-site. On a blasted continent that was effectively a vast oceanic island, its southern extremities became the main source of survivors.

ABOVE: *Silver ghost moths (*Letovenus *sp.) mating.*
LEFT: *Three moths also endemic to the Cape: (top to bottom)*
Spiromyopsis, Brephos *and* Estigmene.

Emerging from whatever refuge had sheltered them, these survivors could subsist on other surviving animals and plants. An expanding profusion of plants fed herbivores of all sorts, but Chicxulub's total elimination of all large animals meant that all the earliest survivors were small. In what would have resembled a plant paradise, we can envisage huge swarms of small plant-eaters, especially invertebrates, dominating many post-Chicxulub ecosystems – locusts of the Palaeocene.

In common with many other schoolboys, I examined and made analytical drawings of animal skulls, both living and extinct. Some eventually got published

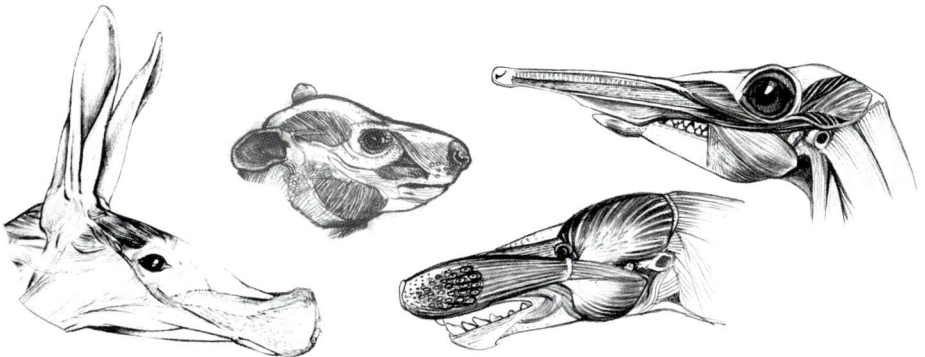

Four afrothere heads, their myology exposed – all from lineages that survived Chicxulub.

*Drawings of a giant sengi (*Rhynchocyon *sp.).*

as illustrations of mammalian weaponry. What struck me, even then, was the many ways in which the bony head of quite ordinary, generalised animals (or their teeth or claws) can evolve overriding elaborations, to serve ever-evolving diets or behaviours.

'Economies of scale' or, more accurately, selection for enlarged bodies, bellies and all the apparatus of harvesting and processing masses of vegetation, can be documented among many mammal lineages throughout the world. However, in Africa, following Chicxulub, the available mammals were all afrotheres. Fossils confirm this, but more interesting is the multitude of ways in which selection for ever larger bodies had to accommodate to the dictates of behaviour.

Six weaponised heads, from left to right: Hyracotherium *(Eocene perissodactyl);* Moeritherium *(Eocene proboscid);* Phachoerus *(warthog);* Ceratotherium *(grass rhino);* Giraffa *(giraffe);* Loxodonta *(African elephant).*

One such accommodation was competition with other members of the same species. Such contests (over mates or territory) help explain selection of enlarged tusks, teeth, horns, antlers or claws, right across the animal kingdom. In post-Chicxulub Africa, just such transformations appear among afrotheres. The tusks of elephants and sea-cows, the forehead spines of some head-butting tenrecs, the sharp fore-teeth of hyraxes and sengis – all of these are evolved weaponry that could be used against competitors of the same species. Most extraordinary of all were the paired nasal horns of *Arsinoitherium*, a rhino-sized herbivore first unearthed in Egypt. The discoverers of this giant showed a sense of humour in naming it after a local Ptolemaic Pharaoh. Arsinoe II, who married her brother, was a real bulldozer of a queen and she was sometimes represented in sculpture with two animal gods, Isis and Hathor, sprouting from her forehead, like horns.

The grotesquery of an *Arsinoitherium* head becomes more comprehensible when seen in the context of Africa's long isolation. After all, rhinoceroses have many resemblances with their afrothere predecessor, but *their* experiments in evolution had all of Eurasia (and eventually Africa too) in which to get horny.

While on the topic of animal extinction, I think it plausible that the relative biotic poverty of today's vast Saharan and Sahelian expanses might derive,

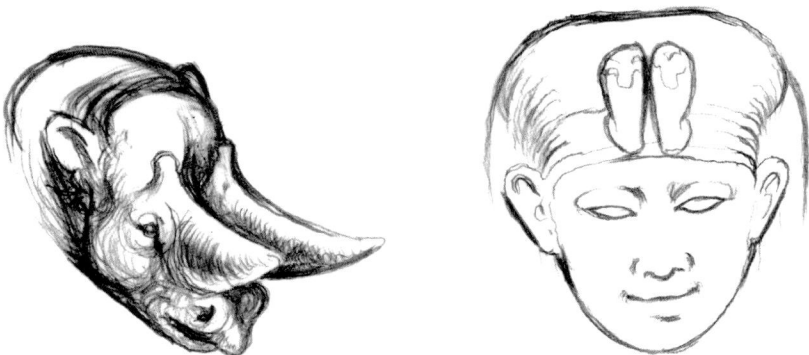

LEFT: Arsinoitherium. RIGHT: *Arsinoe II, Ptolemaic Pharaoh.*

even if distantly and only partially, from the time of Chicxulub, as well as from more recent droughts, floods, global climate changes and the depredations of humans.

At that time, Africa was several degrees further south than it is now. Our southern reaches have remained a hospitable refuge for innumerable ancient forms of life. Surviving Chicxulub was but one chapter in a long, long history which will preoccupy us in subsequent chapters. Once the skies had cleared and rains returned on a world without dinosaurs, it was mammals, birds, frogs and numerous plants that could evolve among relatively empty pastures. This was the dawn of a world where humans, elephants and Aardvarks could evolve.

Nothing pre-destined about an elephant.

There was nothing pre-destined about it, yet our own existence as advanced primates is just one of the many consequences of that accident. It is a major task for geologists, biologists, anthropologists and philosophers to plot our evolution as a set of responses to chance accidents, as well as the regular predictabilities of existence and the comings and goings of peoples, cultures and epidemics. All are deeply interwoven into our heritage.

Every one of us is the product of chance in the lives of our parents and in those of their parents before them. Extending the fruits of chance back into geological time should not present the difficulties that it evidently does for some.

In 1658 Thomas Browne, an arch-believer in Divine Providence, was forced to concede, 'I relate the history of my life, the occurrences of my days, the escapes, or dangers and hits of chance.' Science and civilisation are built upon recognition that predictabilities or 'laws' constantly interact with chance – that's life's central dynamic!

On the tiny stage of family life (in fact in our living room) my teenage self had a small lesson in life's potential for anarchy. My teacher was a mouse.

The *mvule* wood bookcase had been custom-made to hold a complete set of *The Laws of Tanganyika*. The dozen or so volumes spanned the full width of the bottom shelf. The upper shelf held a dictionary, Robert's *Birds of South Africa*,

some novels by Trollope, Buchan, Conrad and a small selection of French classical writers including Voltaire, Rimbaud, Molière and others, reflecting my
father's student years in Paris and Oxford, where he had studied French literature. Centre stage, on the polished top of the bookcase, was a very heavy copper
bangle, with a built-in, bladed axe-head, all hand-forged in Kibondo from melted-
down telegraph wires and obsolete German East African coins. Another axe,
Acheulean, possibly a million years old, lay behind it. On either side of the
bangle were two leather-bound photographs, one of my parents' wedding, the
other of my father standing, rather stiffly, on some steps and dressed in full
colonial uniform, white, be-buttoned, topee-hatted, with ornamental sword
on his hip. Like heraldic guardians, on either side of the sun-stained photographs sprouted two sky-pointing Marabou Storks, shallowly carved from
zebu horns.

In 1952 this horned bookcase stood in the corner of our lime-washed living
room inside a massive fort, built by the German government at the end of the
19th century and strategically placed close to some granite outcrops between
two soda lakes at Singida. At the time Singida was Teddy's 17th posting in a
25-year career as a colonial administrator and magistrate. The bookcase was a
piece of the paraphernalia that had accompanied us in our nomadism within
the vast territory that was called Tanganyika. I had never known a time without
that bookcase, which was more than an illustration of Teddy's personal history.
Ostensibly just a piece of furniture with a practical purpose, the bookcase and
its contents also stood as a secular icon or hearth altar, justifying my father's
existence as official representative of Law and Order. Respect for the law and
conscientious service to the imperial power of his time were central to his earn-
ing a living and to an almost familial sense of having been allocated a place, or
at least a role, in history. From that perspective, it was no accident that some of
the Laws of Tanganyika had been drafted under the supervision of his distant
cousin, Sir Donald Kingdon.

In 1952 I was home from school in Europe and sitting in the living room, reading, when I heard a grinding noise, punctuated by high-pitched squeaks, coming
from the horned bookcase. On my approach the noise stopped and I was about
to walk away when the squeaks began again and it was quite clear that they
were coming straight from *The Laws of Tanganyika*. I eased out Volume One, to
discover that the bottom right-hand corner was missing. Volume Two had the
same corner neatly perforated by a hole that ran right through, from front to
back. Each successive volume was perforated in the same way and I was just
about to pull out Volume Nine when a mouse jumped out. I soon cornered her,
impeded as she was by a minuscule pup that was attached to one of her many
nipples. Indeed, she was a multimammate mouse, and she had made her nest in
the last two volumes, lining it with shredded laws.

Because she was an agile jumper I put her in the white enamelled bath and sitting on the edge, did a series of quick sketches of her in watercolour. Some years

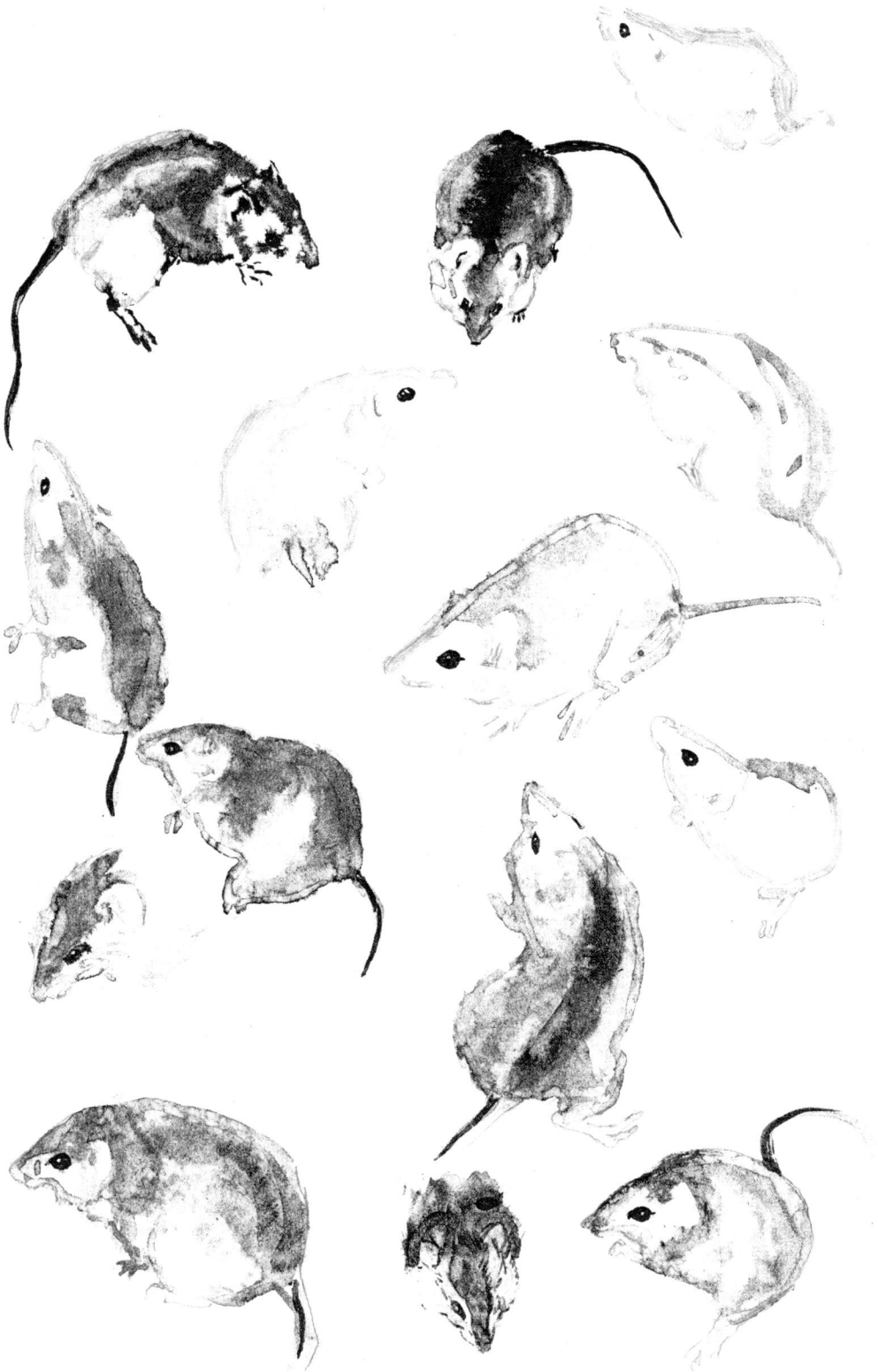

*Multimammate Mouse (*Mastomys natalensis*), 1952.*

ABOVE: *Multimammate Roman goddess of fertility.*

LEFT: Afrotheria. *Africa's sui generis mammals.*

later I published them, captioned with her scientific name, *Mastomys natalensis*. Like some Roman fertility goddess, this otherwise unremarkable mouse can have as many as 24 teats.

In my mind she remains 'the Mouse that ate the Laws' and in spite of her contempt of court she was released among the boulders to seek somewhere else to nest. For my teenaged self, her nest of laws seemed an apt reminder that everything man-made in Tanganyika, whether physical, institutional or conceptual, was destined to end up being gnawed or eaten away by mice, termites, fires or any one of the erosive forces that had shaped the landscape and moulded the granite boulders beside which we lived.

That such solemn and authoritative texts could be put to such a uniquely biological purpose simply confirmed that the laws that governed our lives were as trash beside the forces of nature. Nonetheless, they incorporated much of Teddy's *raison d'être*, and the formal trappings of a very recently colonised and forcefully united tract of central Africa.

Those few rather dated volumes in that bookcase spoke of a life that could do quite well without books, printed words, even laws. It was no meteor – it took no more than a mouse to make a mess of the rules that governed our lives. Today, it's a lowly virus (ultimately passed on to us from bats) that has brought civilisation to its knees.

Squared-up. *Checkerboard Wrasse (*Halichoeres hortulanus*)*.

4 AFLOAT ON PLANET OCEAN (ADRIFT AND HEADING NORTH)

In which homage is offered to Jurassic and other mothers. In which Afro-Arabia loses touch with the rest of the world. The consequences for natives.

My own most memorable encounter with the world of dinosaurs was triggered by some photographs and paintings by a German expeditioner, a few years before World War I. They showed a cadaver, being swarmed over by a veritable ant-hill of excavators. The hundreds of insect-like animals were actually human labourers, retrieving the fossil skeleton of one of the largest dinosaurs ever known, from the Tendaguru fossil beds in what is now Tanzania.

This specimen of *Tornieria africana* was about 26 m long and, while alive, weighed in at about 23 metric tons. The photographs resembled those taken during the building of the Panama Canal, showing thousands of workers dismantling entire hillsides. Today, the mobilisation of such an army of recruits for such a non-commercial, non-military project is unimaginable. Those historic pictures provided one incentive to visit the site of that extraordinary grave-robbing. My occupation of an expedition tent a mere 100 km or so away provided an opportunity to visit the Tendaguru dino-grave.

The site had reverted to open woodland. Nonetheless, I found a small fragment of fossilised dinosaur that had weathered out just below where all that feverish activity had taken place some 80 years earlier. A bigger challenge to imagination was to time-travel back 66 million years, to the cemetery of every big animal and plant on Earth. For me there was a nice symbolism in the descendants of some previously insignificant mammal resurrecting the skeleton of Earth's one-time biggest and best.

After Chicxulub, dinos had had their day. Now it was the mammals' turn and some would even out-do dinosaurs in size (for example, present-day blue whales can reach a length of 30 metres and

My own pasted-up remembrance of a long-lost photo of the Tendaguru excavation (possibly a bit exaggerated).

Whale mother and foetus about to be dismembered on a bloody flensing deck.

weigh 200 tons) but 66 million years ago whale ancestors probably weighed less than 1 kg or so, and were probably less than 30 cm long. Their transformation was the work of natural selection and time.

Once Chicxulub-darkened skies had cleared, the survivors found themselves in environments where the sudden absence of dinosaurs over entire continents changed the scale of existence for almost every living thing.

Previously, most opportunities for large-scale, above-ground existence had been closed off by every size and every style of dinosaur. Countless niches, especially the diurnal ones preferred by reptiles, and especially big reptiles, now opened up for survivors. Here was the ultimate classic of island biogeography where any colonist, endowed with one or more special skills, could find an empty or part-inhabited island wherein it could go forth and multiply (in both numbers and species).

In this case Afro-Arabia, Earth's second-largest continent after Asia, its northern reaches blasted from west and east, was that island, a real island, surrounded by sea. Surviving biota from the far south were best placed to spread and flourish over great distances but they (and any chance incomers) were spoilt for space and a super-abundance of potential niches. Some of these niches involved other, more venerable, animal lineages, such as termites, the ultimate survivors. These insects long preceded whole dynasties of plants and animals. Their cities, termitaries, have been background landscape from dinosaur to elephant and antelope eons.

Before exploring this post-disaster situation, I confess that the KT survivors that have interested me most have been today's African mammals, because I

Juvenile obsession with natural history. 1954 paintings of an agama lizard (Agama sp.) and a Nsenene bush-cricket (Ruspolia differens).

have had the privilege and opportunities to study them. However, birds, insects, fish and reptiles once preoccupied my many childhood enthusiasms – all subjects for early portraits in newly developed skills in watercolour and pencil.

For now, my first perspective must be prehistoric, terrestrial and geographic, remembering that Afro-Arabia had already been separate from other continents for some 70 million years before Chicxulub and remained more or less so for about 16 million years afterwards.

Even so, I pause to plead for thought, feeling and some empathy for all the crucial characteristics we share with all placental mammals. My sculpture of 'Mammalian Motherhood' was made in homage to the female principle, and in gratitude to my own mother Dorothy and all of her female, mammalian sisterhood.

Let me be your alter ego, beginning as an embryo formed by the fusion of my parents' two gametes. In the process of this embryo attaching itself to Dorothy's uterus, a placenta developed between the two of us. That placenta helped balance out our two separate interests as two soon-to-be separate beings. Dorothy's placenta inhibited her immune system from rejecting me as a foreign body, and it helped balance out her nutritional needs and my inherent foetal greed. That placenta helped prolong our shared period of pregnancy, allowing more time and resources to get invested in my brain, and in Dorothy's modified sweat glands, which would later secrete her milk. I share that individual history with you and with every mammal that has ever existed on Earth.

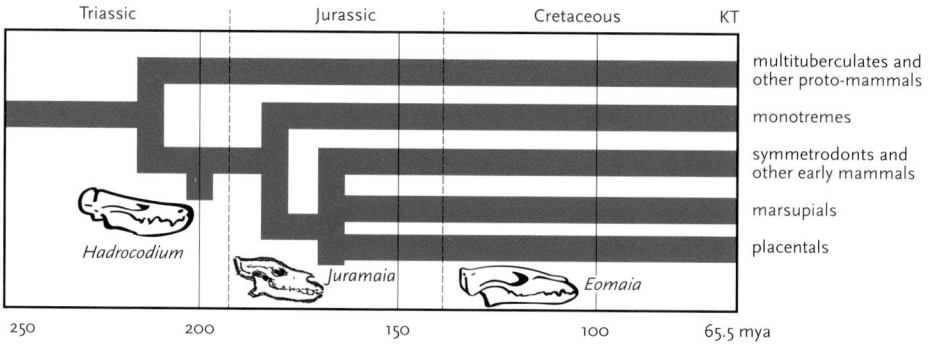

Triassic   Jurassic   Cretaceous   KT

multituberculates and
other proto-mammals

monotremes

symmetrodonts and
other early mammals

marsupials

placentals

Hadrocodium

Juramaia

Eomaia

250   200   150   100   65.5 mya

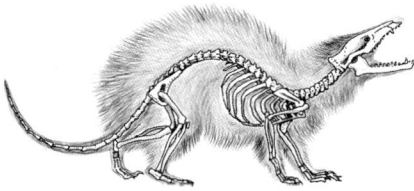

ABOVE: *The age of mammals.*

LEFT: *A later 'mother'* Eomaia scansoria *(a reconstruction).*

When did such a complex set of interdependencies develop, and what has it got to do with a natural history of Africa?

As recently as 2011, a team of biologists and geologists led by Dr Zhe-Xi Luo paid homage to the female principle by naming a new and super-significant fossil 'Jurassic Mother', or *Juramaia.* That 160-million-year-old Jurassic mother is currently the earliest fossil of a placental mammal known to science. Very small, furry, arboreal and probably nocturnal, this Chinese discovery was an immediate sensation because it proved that marsupial and placental mammals had already diverged by the mid Jurassic, an unexpectedly early date. *Juramaia* also confirmed that Eurasia was by far the most likely place of origin for placentals. Today, Africa is home to the greatest diversity of mammals on Earth but for most lineages it is not their original source. This apparent paradox is an important theme in what follows, especially in Chapter 14. Meanwhile consider the placentals' lot in a world dominated by dinosaurs – how did they survive, diversify and co-evolve over some 160 million years?

Dinosaurs were mostly very large, sight-dependent and diurnal, and at a disadvantage during very cold periods and at cooler latitudes. Several species did seem to have tolerated cold quite well, but that was certainly not the norm. Placentals, though, were mostly small with well-developed senses of hearing, smell and night vision. They were nocturnal, probably arboreal and best adapted to cooler regions.

Those mammal lineages that could aestivate over long periods, often deep underground or in caves, surely survived winters that way, as well as many a local catastrophe. Some nocturnal mammals had probably adapted to survive long winters, escape prolonged above-ground heat or drought – precisely the right recipe to survive the Chicxulub meteor's devastation.

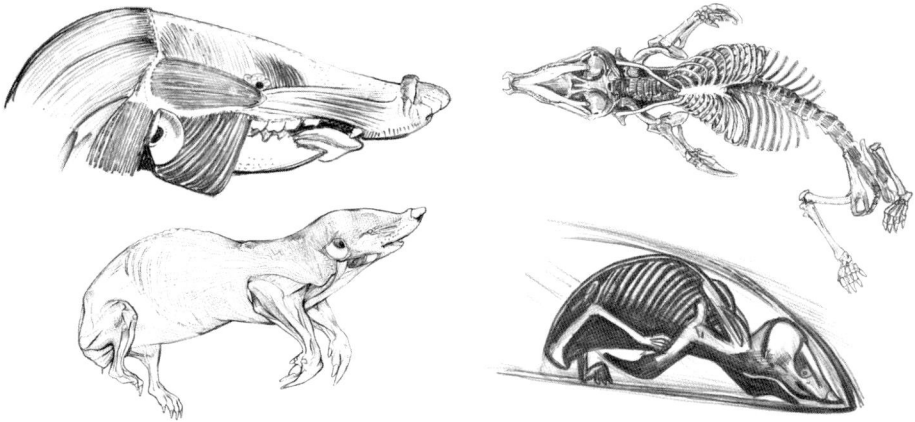

Looking in depth, digging in depth: the anatomy of a golden mole (Chrysochloridae).

Among many other key adaptations linked with underground existence was a diet of subterranean invertebrates, detected by highly developed senses of smell and hearing and excavated with sharp claws and muscular forelimbs. Even today, termites comprise the largest animal biomass in many habitats and they are a critical diet for many mammals, birds and modern reptiles (such as snakes and lizards). It is possible that termites are still the single most abundant animal in the world. These miniature colonial cockroaches long predated and demonstrably survived Chicxulub, and have remained the main diet for many sorts of animals.

Termites almost certainly sustained some of Africa's survivors of Chicxulub, but before exploring details there are still larger questions about how and when a Eurasian placental reached Africa in the first place. Most likely is that some time between 80–90 million years ago a very primitive form of early placental, possibly a single species, succeeded in crossing marine narrows towards the western end of the Tethys Sea (perhaps in the vicinity of today's Strait of Gibraltar).

This immigrant, the first placental colonist of Africa, then multiplied and diversified. Some 10 million years later, one descendant species crossed the still quite narrow gulf that then separated Africa from South America. It was probably a shorter and less perilous crossing than that made by its ancestors when they reached Africa from Eurasia, but that South American colonist, its descendants now called xenarthrans, retained a genome that reads as the most primitive of the three placental founders. The African branch, now called afrotheres may have once included xenarthran-like genes in its genome but has become more diverse and less primitive than surviving xenarthrans. The third, foundational placentals, the boreoeutheres, or northern placentals, are by far the most diverse and highly evolved of those three primary divisions.

There are very, very few fossils available to reconstruct what afrotheres were like before 66 million years ago but, taking their molecular phylogenies and molecular clocks into account, a tentative picture of some Chicxulub survivors can be extrapolated from their descendants. So, how did those earliest placentals respond to the dawn of a new world?

More than one afrothere lineage might have survived, each hinting at what it took to weather the terrible aftermath of the Chicxulub meteor's impact. Each suggests a distinct trajectory that has culminated in modern species, occupying one or more modern ecological niches. My friend and colleague, Erik Sieffert, has constructed tentatively time-tabled genealogies for extinct and extant afrotheres. Some now struggle to survive within the vestiges of once well-defined ecological niches, but each has built on a distinct way of living, some possibly pioneered during pre-Chicxulub times.

Aardvark (Orycteropus afer) *sketches.*

The oldest separate lineage within Afrotheria might be represented by the living Aardvark. Aardvarks dig as deep as 6 m and can tear the hardest termitary open with spade-like nails. Their long, pointed tongues are adapted to whip in and out of their soft, snuffling snout at a phenomenal rate.

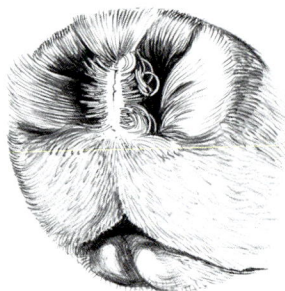

Details of head and snout of Aardvark.

Million and her servant at dusk.

Once, while I was feeding a habituated captive Aardvark from a jar held close to my face, I was startled to have my nostril probed by its extraordinary tongue. Another tame but free-living Aardvark was raised and nurtured by my friends Joan and Alan Root. Alan, with his taste for satire, named her 'Million', in this case triggered by the appalling sentimentality of Al Jolson singing in his New York accent, 'Maamie! Maamie! Aard vark a million miles, jess fir wun er yor smyles.'

Million the Aardvark had a personal servant, the mute and mentally challenged son of a neighbor whose job it was to fill in the trenches dug out all over the Root estate. Every evening Million set off in search of termite antipasti, before being served her main dish of burger paste slathered with formic acid sauce. In one of my more surreal memories, a fat 75 kg Aardvark swaggers into the sunset in a cloud of self-generated dust. Her biped attendant shuffles along behind her, spade in hand, sometimes irritating Million with an explosive sneeze.

Two studies of a very young giant sengi (Rhynchocyon sp.).

In South America, the armadillos resemble the Aardvark lineage and suggest that deep burrowing was one escape route from extinction. The next afrothere group, the Aardvark's closest relatives (that is, if a parting estimated by some at 78 million years ago can be called 'close'), are called sengis or 'elephant shrews'. Sengis have a very complex evolutionary history, their lineage embracing both omnivore and herbivore, and including miniature and middle-sized forms. Exceptional capabilities to digest all manner of toxin-laden foods in some miniature ancestor might have helped their ancestral escape. Today, most sengis subsist on ants, termites and other invertebrates.

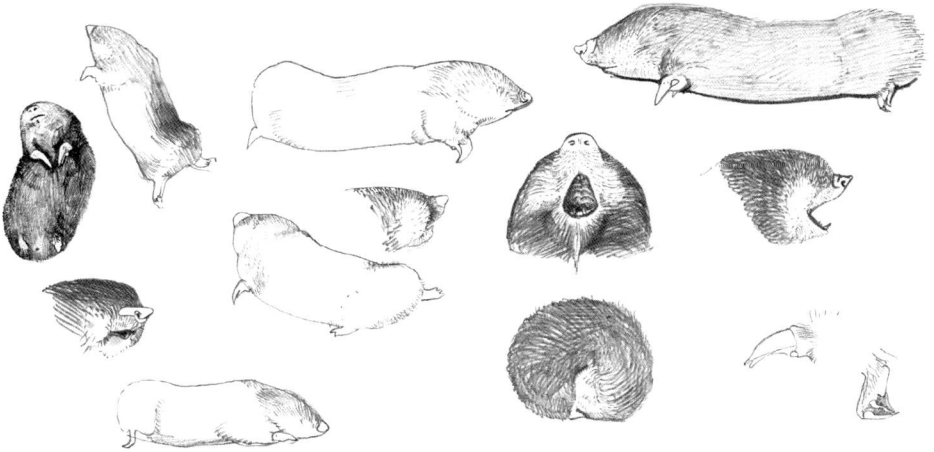

Drawings of a living golden mole (Chrysochloridae).

Even today, the massively clawed golden moles include miniature species (one is a mere 80 mm long). These afrotheres combine rapid and potentially deep digging with minuscule size, as if they were tiny Aardvarks. All species only occur south of the equator and most are concentrated around the Cape, which suggests that their single ancestor (molecular clocks suggest that all living species of golden moles share a single pre-Chicxulub ancestor) survived in optimum corners of the far, far south.

That southern distribution was clearly essential for survival of the tenrecs and potamogales or 'otter shrews'. These shrew- and hedgehog-like afrotheres have a poor fossil history, but the fact that three species are aquatic in Africa (along with a fourth, convergent, one within Madagascar) reinforces this group's strong association with water. Perhaps most significant of all, several species can aestivate for up to a year. In the context of outliving the Chicxulub catastrophe, here lies another avenue of escape. Pockets of uncontaminated water clearly lasted long enough for a wide range of freshwater organisms to survive as well, so afrothere dietary needs were probably assured.

*Drawings of the Giant Otter Shrew (*Potamogale velox*).*

These were not the only water-loving early afrotheres. Fossils reveal that one aquatic lineage of herbivores got substantially bigger very soon after Chicxulub. The common ancestry of sea-cows (manatees and Dugong) and elephants is recognised with the taxonomic category that unites them: 'Paenungulates'. With the demise of the dinosaurs and regrowth of a still-rich flora, there were enormous opportunities for plant-eaters. These opportunities were so great, both on land and in sea and freshwater shallows, that molecular clocks have correlated this paenungulate split with the immediate post-Chicxulub scene within Africa. Enlargement to elephantine proportions came by fits and starts in both aquatic and terrestrial lineages.

I have had several small but significant lessons in how both paenungulate lineages remain 'wed to wet'. In one dry-season incident I was overlooking a small lagoon when a herd of about 20 very thirsty elephants arrived on the far bank and rushed in to slake their thirst. Having imbibed many gallons of water, mothers with young calves were content to luxuriate in squirting trunk-loads of water over themselves, making sure their offspring kept to the shallows. The teenagers were under no such constraints. Once they had drunk enough of it, for them the water then seemed to morph into something in between the contents of a play-pool and the rediscovery of their one-time all-enveloping environment. Some waded into deeper water. Here they rolled over, using their trunks as breathing tubes. Once or twice an individual let the soft bottom mud support their weight while they waved all four legs above the surface, like some snorkelled cyclist pedalling away, but upside-down.

Dugongs and manatees normally graze sea grass in company. They leave long muddy trails across the sea floor as if exceptionally untidy lawnmowers have gone by. Dugong heads even look like mechanised cereal harvesters, elegantly styled by a designer called 'Nat Select'. This is because sea-dwelling

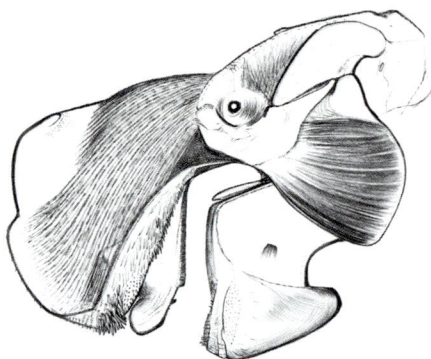

*Dugong (*Dugong dugon*) heads, exposed.*

Dugongs restrict their feeding almost entirely to vegetation growing on shallow sea floors. Manatees, though, have had to adapt to huge seasonal fluctuations in the tropical rivers and lagoons they inhabit. When the rains peak, flooded banks are lush with grass and sedges, while even tree canopies are awash with turbid floodwater. At just such a time I was astonished to see banks closely cropped and bankside sedges and mangroves browsed by manatees. It was as though these so-called sea-cows had hauled themselves up just enough to assert their land-lubber origins and an all-too-ephemeral affinity with browsing cows.

Hyraxes are the last afrotheres to hint at their ancestors' survival of Chicxulub. At that time this branch of afrotheres probably resembled paenungulates in having long been herbivorous, but there is nothing about their history to suggest any close association with water. The surviving hyrax species may bear little resemblance to their triumphant ancestor but their dormitories and latrines are often deep among great rock piles that Dorothy fancied as 'giants' playgrounds'. Hyracoids were the prime inheritors of the vast pastures opened up by the die-off of dinosaurs. They evolved numerous adaptive types, some resembling horses, others as big as rhinos. Millions of years later, bovids and the ancestors of horses arrived from Eurasia, and replaced them.

Seas surround our continent but, having shed entire continents off all its margins, Africa is left with narrow continental shelves with deep dark depths beyond. As a result, marine communities around Africa bear no comparison with those occupying much more extensive marine shallows. Southeast Asian waters had the added advantage of being the area most distant from Chicxulub. This may well help explain why Indonesia and the Coral Sea boast the greatest concentration of marine biodiversity on Earth.

OPPOSITE: *Two pictures of tropical fish patterns: (top) a rendering of Harlequin Fish (*Othos dentex*); (bottom) cigar wrasse (*Cheilio *sp.*).*

Transformations during a lifetime: two species of angelfish (Pomacanthidae), demonstrating how one pattern can morph from one design to another within a short fish lifetime, one generating vertical stripes, the other horizontal. This sort of ontogeny mimics evolutionary change through time.

Fish, wherever they swim, have evolved by natural selection, no less so than has land-life. For youthful me, the brilliant patterns and colours of reef fish seen or caught on all-too-brief excursions along our shores triggered a longing for explanation, for a grammar of comprehension for all that piscine beauty. A few notebook pages and coloured sketches fail to alleviate that longing. For generations yet to be born, with mental frameworks and equipment yet to be invented, we can hope that the processes and lives of marine ecologies, so challenging yet so abused today, will be integrated into the knowledge and culture of our descendants.

Overleaf is a panel depicting undulating reef life, painted in homage to Steve Wainwright, a friend who introduced me to marine biology and its unfathomable mysteries and beauties.

OPPOSITE: *Sketches of clownfish (Amphiprioninae).* ABOVE: *A visual lexicon derived from tropical reef fish.*

*Mimicry among reef fish. Blacksaddle Filefish (*Paraluteres prionurus*), a harmless mimic of the poisonous pufferfish above.*

ABOVE: *Leopard Moray Eel (*Enchelycore pardalis*).*

Undulations Below the Waves: *a celebration of tropical reef life. Duke University Biology Department.*

Fin-flasher fish, Gobioidea.

ABOVE: *Glass mosaic lunette based on gouache painting of squirrelfish.* BELOW LEFT: *Gouache painting of squirrelfish.* BELOW RIGHT: *Design for a dish, based on the same painting.*

To return to afrotheres and the mysteries of their histories, the colonisation of Madagascar by tenrecoids is a large and long story best left for Madagascans to tell, but here I indulge in a brief vignette of birds that illustrates the linked fates of Africa and Madagascar.

No-one can easily forget being greeted by a passing flock of helmetshrikes in the Serengeti. Loudly conversational, like school children let out of class, these pied birds always travel in parties of a dozen or more, all of them endearingly tame, always ostentatiously inspecting the ground or branches and trunks of trees, always on the move and sometimes following a routine course across the

*Grey-crested Helmetshrike (*Prionops poliolophus*).*

landscape on successive days. It is a memory that threatens to become actual history, because if ever there was a species heading for extinction it is the vivacious, fan-flirting exhibitionist, the Grey-crested Helmetshrike.

Near Tarime we saw them most often in the tall fever-trees that bordered *karongas* (shallow riverine gullies) or flitting in a noisy scatter from one whistling thorn to another out on a cotton-soil *mbuga* (flood-plain). They were unusual in being active around midday, possibly to mitigate the risks of being swarmed over by the fierce acacia ants that protect the thorn trees at other times.

What were they feeding on? Omnivores with a strong preference for insects, they were seldom short of food. For some measure of the abundance of insects on acacia trees, a few square metres of acacia foliage can support up to 650 individual insects, belonging to up to 500 species.

Helmetshrikes are winged fan-dancers in crisply precise pied costumes. Each yellow iris stares out from its own spiral spectacle of white feathers that swirls forward to converge into a single bladed fan above an ever-chattering, ever-snapping black beak. These tracts of feathers can shift or tilt to create a variety of 'helmet' shapes, alterations that presumably signal changes in mood or intention. When family parties meet they face-off from perches on the tops of trees and bushes. Every bird stretches tall and vertical, jerks its helmeted head from left to right, over and over, clatters its beak and fires off sharp, explosive volleys of sound, sometimes interrupted by some dive-bombing of the opposing flock. But there are quieter, more intimate moments. Each party has its own matriarch queen and her consort, and there are many small gestures signifying rank or subordination; flutterings of wings that almost seem to simper. Like costumed courtiers, helmetshrikes bow and nod with tilts and flicks of fans and eyes no less subtle, no less communicative than 18th-century French flirts semaphoring 'come hither' or 'get thither' with fancy fans.

When ornithologists first started collecting, naming and recording birds, these Grey-crested Helmetshrikes were recorded from a great arc of acacia country from the eastern shores of Lake Tanganyika along the southern shores of Lake Victoria and eastwards to the Rift Valley lakes in Kenya but this range seems to have been shrinking ever since, from some unknowable original extent.

Helmetshrike closer up.

Conspicuous, trusting and vulnerable to boys with catapults and to the char-coaling of acacia woodlands, these beleaguered birds now face an even more insidious rival. The range of this species is entirely encircled by a closely related and very similar cousin, the White Helmetshrike. The most obvious difference between the two species is the latter's broad fleshy wattle that encircles each eye and matches its yellow iris exactly. It is almost as though a fashion for fans is giving way to one for lemon-tinted lorgnettes.

These plumed birds are but two of eight African helmetshrike species. To evolve so many species, and their own system of co-operative child-care and education, helmetshrikes have clearly been around quite a while. How long?

Very recent genetic research has shown that their closest relatives are most probably the vangas, a group of shrikes exclusive to Madagascar, and that their last common ancestry was about 25 million years ago when some flock of proto-helmetshrikes/proto-vangas got blown across the Mozambique Channel to Madagascar. Once on their own island, these colonists radiated into one of the most spectacular of all island radiations – much more diverse than Darwin's celebrated finches and mockingbirds on the Galapagos. At least one vanga species still has some resemblance to its African helmeted cousins.

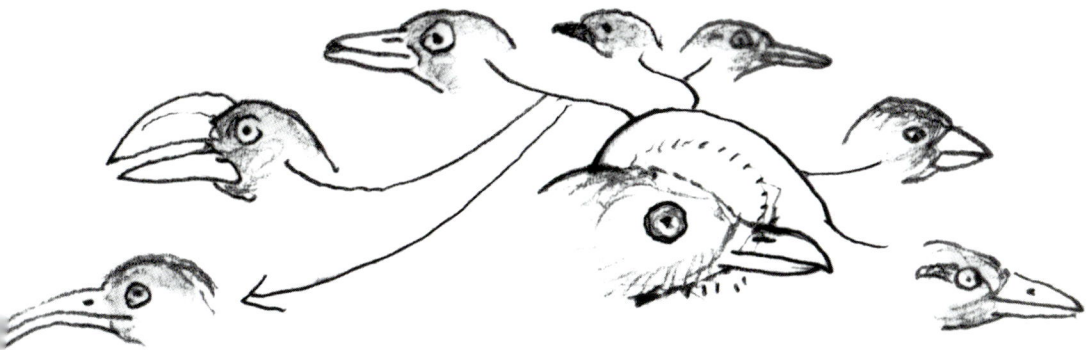

The vanga radiation generated an astonishing diversity of beaks.

From a distance, helmetshrikes resemble waving handkerchiefs. Among the farewells they wave are messages from tens of millions of years ago. These are the unexpected details that make African natural history so thrilling and its current, systematic extermination by industrial agriculture so unbearably tragic, and this sad story is playing out not just on land, but under the seas as well.

While Africa, rather than its surrounding seas, is the subject of these pages, our continent shares the larger fate of Planet Ocean. We now know that industries, industrial fishing and naval nations are melting, poisoning, blasting and devastating all of Earth's seas, even the most remote. Entire reefs are dying, entire coasts are being submerged and shorelines are shifting, among them our own. In spite of being the least industrialised and least sea-girt of habitable continents, we demand a say in how our seas and our climates are being abused by others. As scientists we are as alarmed as any of our now almost countless colleagues, worldwide. Together we have it in our hands to discipline rogue governments, rogue states and rogue industries, by imposing any or all of an arsenal of sanctions.

Above all, we have the responsibility to nurture the education of our citizenry. Education that joins the ceaseless search after our history and our ancestry. Education and research that finds inspiration in study and celebration of the many processes that nurture life as it is lived, right here, right now.

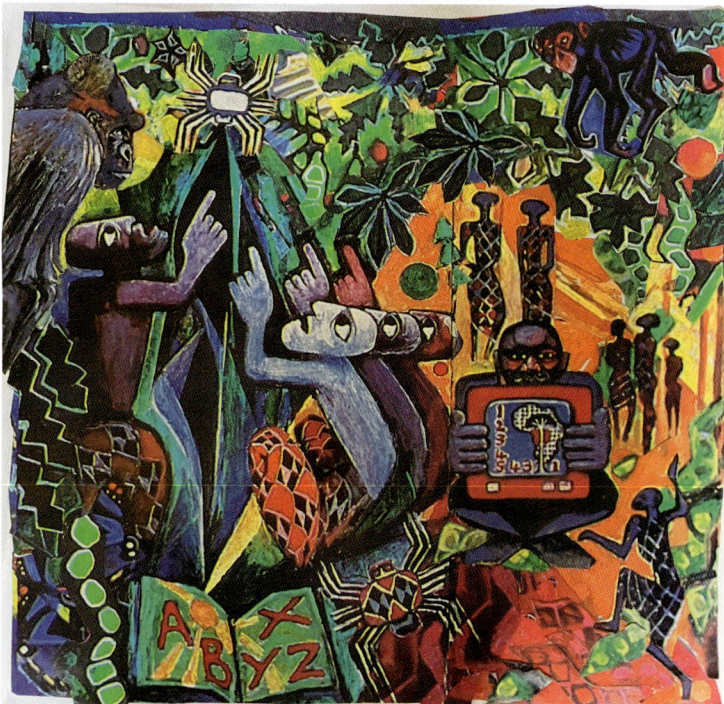

Education – In, for and about Nature.

Armageddon. *Terminal ceiling panel from Makerere non-denominational chapel.*

5 GLOBAL WOBBLES AND CLIMATE CHANGE

In which almost unimaginable fluctuations in climate test and shape the adaptability of living beings.

My father's balletic movements with our garden fruits, described in Chapter 3, tried to communicate the logic behind daily and annual astronomical cycles. Later we learned that every 100,000 years our orbit around the sun swings out from a near perfect circle into a just measurable ellipse. In the course of that widening circuit, the Earth changes its tilt. Perhaps wobbles of a toy top should have been Teddy's simile.

Modelled on the scale of a kitchen garden and a toy cupboard, all this seemed domestic and safe enough. However, when these departures from perfect circularity were first plotted, each 100,000-year wobble was found to correspond to a global Ice Age.

A mega-wobble occurred 30 million years ago when a mega Ice Age terminated the more equable Eocene as well as finishing off many animal lineages. The existence of vast piles of ice over both poles proves that we still live within an Ice Age, but well away from those icy wastes we can pretend to be tenants of

LEFT: *Moss forest.* RIGHT: *Waterfall.*
OPPOSITE: Below the Glacier – *Bujuku valley.*

an 'interglacial period'. Humanity has been supremely lucky during my lifetime to be alive during a period so much more benign than those most of our ancestors inhabited. World climates may appear delicately balanced for long periods but they have their own vulnerabilities and are easily disrupted.

Could the big-population, civilian societies we are born into have arisen during an Ice Age? Imagine trying to build New York City under an iceberg as big as the State, or Lagos in a hot, waterless oven.

Nonetheless, the alternation of glaciations and interglacials has left a rich legacy of natural selection in action, proving just how adaptable plants and animals can be.

Two etchings from Rwenzori.

Returning to that 100,000-year circuit with its great Ice Age dramas, my mountaineering and geologist friends, Henry Osmaston and Dan Livingston, first told me those wayward circuits were called the 'Milankovitch Cycle' after the guy who first figured it out and linked it with the comings and goings of glaciers and ice sheets.

It was expeditions in pursuit of an ancient afrothere, the endemic Rwenzori Otter Shrew, that first took me up into the fabled 'Mountains of the Moon'. I also wanted to see Rwenzori's high-altitude plants and landscapes, so totally unlike those that grow all around and below these mountains.

Before setting off on my first climb in 1960 (with Makerere student David Mwiraria and champion mountaineer John Maatay), Henry and Dan told me that the Rwenzori glaciers had come down to about 6,000 m at the peak of the last Ice Age, and that I should look out for signs of past glaciers.

A 'quicky' ink drawing of groundsels on glacier floor – Kitandara.

LEFT: *Rwenzori tree hyrax (*Dendrohyrax *sp.) 'singing'.* RIGHT: *Tree hyrax (etching).*

They never mentioned listening to the sounds of that retreat but it was my ears that first heard the death-rattle of our last Ice Age. I had carefully chosen my camp-site on the Fairfield Pass between Bujuku and Kitandara, where great colonies of furry tree hyraxes, fat on the rich alpine vegetation, took turns in shrieking insults at one another (or so it seemed to me at the time). Behind the hyrax yells, some kind of castanets kept up a tuneless background cackle. That morning I had walked past several kilometres of overhung cliffs, where the icy claws of glaciers had scratched deep oblique grooves along glacier-sculpted valley walls. All the distractions of steep day-time hiking had muted the sound of gravity nibbling away at every little overhang. That night a light frost must have helped dislodge tiny pebbles, cannon-ball rocks, or even more substantial rock-falls in bouncy tintinnabulations or booming percussion.

Pathway of glaciers.

TOP: *Moss forest.*

ABOVE: *A scramble of* Erica *and* Sphagnum.

The faint swish of water was a reminder that those blue and white blocks, way up in the highest valleys, certainly were melting, and melting fast – for the past several days I had subsisted on cold, glacial meltwater. Hints of mountain sickness kept me awake so there was plenty of time to ponder on the weighty power of altitude and ice to constrain puny animal existence. A mere 19,000 years ago, the Fairfield Pass was under many hundreds of metres of ice.

One of the distinctions of Africa is the restriction of its glaciers to a very few tall mountains, equatorial Rwenzori among them. Climatically, Africa's main distinction has been dryness, but that has been anything but consistent. Africa's plants and animals, including people, have been shaped by the perpetual tease of longed-for rains that might or might not fall.

Bigo bog (etching).

Here, I was direct witness to what a far-from-predictable, ever-changing and perpetually surprising planet I and my human family inhabited. I drifted toward wondering how my many ancestral lineages had lived through hundreds – no, thousands – of bitterly chilly years. Today we may include the Milankovitch cycles among the many predictabilities, but for the people who had to live through it all, both their existence and their survival have been one big crap-shoot.

Even modest little birds embody Earth history, as was illustrated by a bird's nest I once found near our kitchen in Mwanza, having had a dawn glimpse of the nest's architect when the bird sang from a discreet perch near the top of an ink-berry bush, a type of ebony. Called *mdaa* in Swahili, these bushes tend to grow around termitaries.

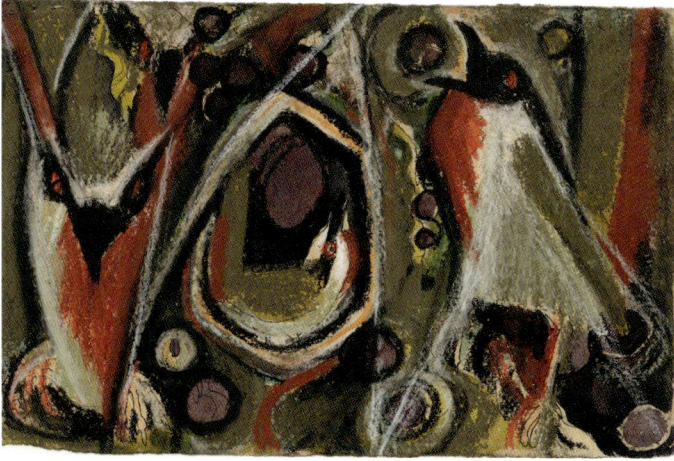

Mdomo mdaa, *Grey-capped Warbler* (Eminia lepida), *nest architect.*

For such a small bird, the volume and complexity of his song was astonishing. His aria began with an attention-catching, shrill whirring, followed by a prolonged stutter which turned into a sweet, sustained warble. Tail erect, the bird hurled his song to left and right while vivid brick-red flashes seemed to explode around his swollen prima-donna throat. This song emerged from a gaping, velvet black mouth – indeed, the birds are known locally as *mdomo mdaa* – 'ebony-mouth'. During his song his coal-black gape joined a broad black loop to effectively bisect the bird's head, seemingly dislocating the whitish-grey crown and upper mandible from equally pale cheeks and lower jaw below.

The young had recently fledged from this nest, and in spite of being superficially untidy, the carefully selected and logically layered materials were wonderful to find and tease apart. The basic globe was woven from coarse, resilient fibres, skilfully anchored to the liana-like sprawl of a morning-glory vine. The margins of the entrance were reinforced with a tighter weave. Inside, a few feathers shed by larger birds joined a layer of leaves that probably created a semblance of water-proofing. Inside that thin layer the nest was bulked out by moss, held together with stolen spiders' webs. Inside that was a well-consolidated insulating layer made from shredded thistledown mixed with strong, hair-like fibres, all tamped firmly down into a sort of dense felt. Finally, the entire globe was disguised by dangling fibrous trails of debris, all drawn in from the nest's immediate surroundings and tied with more spider silk.

Today, with geological and climatic histories to guide us, I can liken the building of *Eminia lepida*'s nests to the incremental elaboration of an ancient cathedral spanning several architectural eras. The elaboration of each nest-increment probably began with ancestral responses to new challenges. Thus, selection for water-proofing most likely occurred during a super-wet period, insulation in a super-cold period and the untidy camouflage during an era when the depredations of nest-predators were particularly severe. Living things, and the objects they craft, express a sequence of ancestral responses to all the chancy exigencies that every life, every lineage, must survive. Extinction is the alternative. Our sort, no less than the Grey-capped Warbler, has had to live through the same vicissitudes, but in our case the mark of these trials is not expressed in a choice of nest materials but in the convolutions of our genes, brains and manual skills – especially those that get passed on.

One theme in what follows concerns the distinctness of plant and animal life along Africa's eastern coasts. Here the warm waters of the Indian Ocean have sustained littoral forests and their biota along their entire length and over almost every fluctuation in climate. This has maintained an entire east-coast fauna and flora with numerous species that are found nowhere else (although many range inland to varying degrees). Among the scientists who have contributed to our knowledge of east-coast endemics is my contemporary, East African naturalist Leonard Mwasumbi, one of our unsung heroes. Beginning his career as a technician in Dar University's herbarium, Tanzania, he became one of Africa's most dedicated botanists, assembling field inventory collections for key 'biodiversity hot-spots', such as Pugu, Mbizi, Kimboza and Nyika. He has a magnolia-like tree named after him, *Mwasumbia alba*, itself one of the rarest and most ancient of plants endemic to the now near-extinct eastern forests. Leonard knew almost any plant he was presented with, having committed whole herbariums, specimens, collections, libraries and keys to his formidable memory, backed up by eyes that took their joy in botanical 'jizz', to borrow an ornithologists' term. Leonard took pleasure in the form and beauty of plants as well as in the order he found in their classification, but we both lived in a culture and an economy dedicated to their extermination. That is a painful contradiction to live with, like being bonded secretary to a cave-living Cyclopean monster.

The orderly and predictable was emphasised in my childhood lessons from Teddy, yet his own most formative years were dominated by the unpredictabilities of a disorderly and incomprehensible war, set off by a maverick grandson of Queen Victoria. There was no logic or inevitability about the predicament Teddy faced in 1918. He was barely two years off conscription when, as an 18-year-old boy, he could expect to become little more than target material for a new generation of war machines.

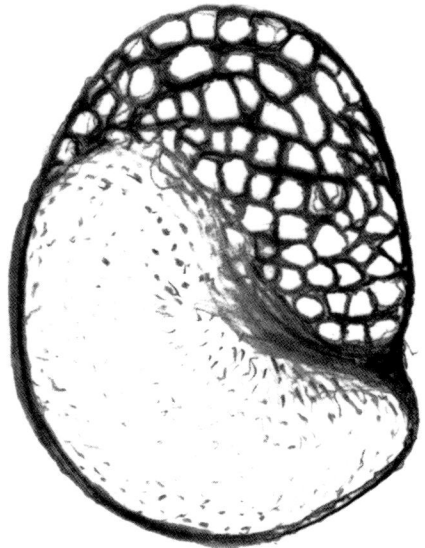

Mwasumbia alba *pollen grain.*

Small, independent families, such as his, got swept up in civil wars that were none of their making. When Lord Kitchener put Teddy's family and their ilk on a 'war footing', all the rules, regularities and morals that they had supposedly lived by were suspended.

Forty years later, when I put it to him in these terms he was indignant, especially when I introduced a biological metaphor he took to be insensitive.

I suggested an analogy in those superbly camouflaged grasshoppers that lead independent lives, grazing sparse grasses out in the deserts and arid pockets of Africa, Asia and, in between, in Arabia. When some semi-arbitrary El Niño-like climatic event heats up the western shallows of the Indian Ocean, the flip from the northeast monsoon to southeast trade winds can sweep rain deep into northeastern Africa and Arabia – even into the Sahara. Two months of rain are rare in these remote, arid regions.

Desert Locusts (for that is those grasshoppers' fearsome name) are superbly adapted to take advantage of such unpredictable rains. Every female drives her extensible abdomen deep into the sand, where she can lay between 80 to 150 eggs, all enclosed in a froth that dries out but keeps the eggs aerated until the flightless nymphs or hoppers hatch. During dry or near-rainless years the usual insectivorous predators soon reduce the hoppers to whatever scatter of solitaires predation and the sparse resources of the desert can sustain. If the rains continue, though, desert predators are soon swamped by the insects' fecundity, whereby each generation of fast-breeding locusts multiplies exponentially. It has been estimated that a single female can generate a 400-fold increase in two or three months of rain-fed desert grasses.

Hoppers hatching into a dense mass of fellows lack any trace of life-saving behaviour. Instead of brownish camouflage they are bright yellow (or orange) and black, and they hop away in a seething mass that travels more than 1.5 km a day. On the way, they eat every green thing they encounter. Once they are winged they can fly more than 100 km a day, but the direction they take seems to be determined by the height of the winds they select to waft upon. Swarms can cover immense distances without water, and tolerate exposure to great heat and continuous sunlight, but ground themselves at night. Typically, a swarm of 1,000 million can consume all vegetation over 20 km² in a matter of hours.

Teddy was particularly shocked by my suggestion that perhaps all animals, including humans, subject to the universal drive to procreate, get to behave in fundamentally different ways when normally small, family-like groups multiply into ravenous hordes.

Obviously, highly social and non-social species differ. Egyptian Fruit Bats packed cheek by jowl into huge roosts must behave very differently from more solitary collared fruit bats (*Myonycteris*) which, like snarling guard-dogs, are fiercely and loudly jealous of their individual space. What upset Teddy was my likening warring Europeans to locusts, where transformations in behaviour, even in sartorial uniforms, took place *within* a species, within a nationality, within a society.

We had ample opportunity to experience Saharan locusts during a series of swarmings that coincided with the years of World War II. Indeed, during two decades of catastrophe in Africa (1889–1909) uninterrupted locust swarms acquired the monickers *nzige ya zidihamu* ('the locusts of Armageddon'), or *ugonjwa ya panzi* ('grasshopper sicknesses'). To this day, plagues of locusts too often coincide with times of war, climate crisis and widespread misery.

For us in Mwanza, the swarm came in across the glassy waters of Lake Victoria. We had some warning and stood on the verandah watching a brownish band, well above the lake, that stretched right across the horizon. As it approached, the light dimmed and we were soon engulfed in a flickering storm of yellow, brown and black Desert Locusts. The whirr of their wings mixed with the crackle of their legs and the squeaking of their jaws as they descended indiscriminately on every plant from tree to shrub, weed to grass. Our mosquito-netted verandah gave some protection, but outside, scratchy legs crawled over every bit of you and got tangled in your hair. The horse and donkey did not like this at all and there was a lot of neighing, stamping and kicking in the stable. Women, wearing *kanga* sarongs, appeared with *kerais*, shallow steel basins in which they collected and cooked the locusts. I tried them, was unimpressed, and got the squits as a result. Several weeks of root-based meals followed.

Having settled on everything along a wide front of foreshore, the locusts ate every shred of green material by nightfall. The next morning they flew on south, leaving a monochrome, leafless landscape and a sludge of excrement and dead bodies. 'Oh, my poor flowers!' wailed Dorothy. But in truth her flowers were few and hardy survivors of an unending blitz from other insects, army-worms and vagrant goats which vanished them away without any help from locusts. So it was left to the wild morning glories to sprout and flower again, within days of the locusts' departure. Women, treating the locusts as a culinary bonanza, hung around for a while, catching laggards and late-comers, scorching and popping them in their own sizzling insect grease over twig fires.

In the locusts' wake were many birds, especially flocks of Wattled Starlings, most of them drab, smoky-grey with brown wings and tail, all of them energetic consumers of locusts. The females are very similar to other closely related starlings but what seems to have set them apart was their early discovery and ultimate near-dependence upon a diet of Desert Locusts and other insect swarms. They generally court, breed and raise their young in colonies that quickly assemble close to a burgeoning swarm of hoppers and, unlike other starlings, their sensitivity to locust signals seems to have influenced their vocabulary. Indeed, they have appropriated the sound of swarming hoppers – squeaks, screeches, rustles and bursts of hissing.

Among these starling flocks are much paler males, each with his own personal permutation of head ornament. Some scarcely differ from females but most sport various degrees of baldness and wattles of all sizes on crown, forehead and throat. The wattles are black and the birds' frenetic movements set

*Wattled Starlings (*Creatophora cinerea*) and their diet: locusts (Acrididae).*

them all ajiggle, like animated insect puppets. Black skin surrounds a helmet of bare, bright yellow skin that seems to match the colours of the locusts. The match is not just confined to colour – the sculpted forehead, thorax and kick-off legs of each locust are yellow, neatly edged with black margins or patches. Thus, some male starlings have pates and wattles that look like much-enlarged models of the foreparts of a locust or locust hopper.

Wattled Starling flocks are nomadic and have evolved to track and subsist on locust hoppers before they mature into flying swarms – once found, the super-abundance of their food sets off their own breeding behaviour. This includes wattle-growing and jaundiced baldness in males. Courting animals of many species, including people, offer delicacies as part of courtship, often in some form of 'food mimicry', but a male starling offering one more hopper to a female suggests the analogy of a couple on a pebbly beach where the would-be swain offers a stone to his inamorata. Perhaps the enlarged ornamentations of some male Wattled Starlings makes them resemble those fellows who can offer a whopping great diamond in a gold ring – one more stone, but what a gem!

Every available sense is involved in the search for whatever sustains any animal. Ancient hunters painted the silhouettes of longed-for prey on cave or shelter walls (the carvers of equally longed-for female humans were presumably male). European colonists and traders, suddenly rich from the spoils of the Orient, commissioned giant canvases depicting silks, carpets, exotic foods and spices, sometimes as courtship gifts and always as tokens of social standing. Our contemporaries tolerate giant billboards in which innuendo links cookies with sex – human artefacts always express the preoccupations of prevalent cultures, ours being dominated by industry, commerce and trade. I contend that Wattled Starlings, fired up by surging hormones, express and respond to signals deriving directly from their preferred, perhaps obligate, food – locusts.

How can any single male attract attention from within a mega-flock of hundreds, perhaps thousands, of starlings? Jazzing up the crudest visual features of the female's favourite food seems to have sparked a fetish dramatic enough to turn or draw her on – and submit to the indignities of avian copulation.

Three 'locust' panels from the ceiling in St Francis non-denominational chapel, Kampala, East Africa.

We had watched the locusts and the starlings descend on Mwanza with Hugh Elliot. Many years later his son, Clive, reported a large colony of breeding Wattled Starlings abandoning their nests and flying away after a pest-control team exterminated the swarm of hoppers that had triggered and sustained the starlings' effort to breed (probably poisoning insects, birds and soils in the process). The agencies spreading insecticide (originally dieldrin, DDT and other long-lasting chlorinated hydrocarbons) have yet to monitor the long-term effects of their poisoning tracts of Africa's remotest wildernesses.

Wattled Starlings resemble a great many other bird species by breeding in response to a single, ephemeral but immensely abundant food supply. For example, millions of migrant birds of many tropical species fly all the way to Siberia to subsist and to fuel their own reproduction on brief summer bonanzas of tundra mosquitoes, gnats and their larvae, or tundra berries and blossoms.

Populations of such birds have been shown to pulse in synchrony with the relative abundance of their prey, and seasonal fluctuations have probably governed the fortunes of countless 'summer visitors', birds migrating to Eurasia for the spoils of summer. It is very likely that populations of Wattled Starlings wax and wane in sync with the ups and downs of locusts and other swarming insects. Would-be controllers of locusts have shown scant interest in these astonishing birds.

In 1942 we felt less threatened by locusts than by Mr Hitler and his hordes, ostensibly seeking, in his own word, *Lebensraum* or 'living space'. Wire netting protected us from most insect invasions but a more persistent breach of our defences came via the radio. The house was small so I was never more than a room away from the BBC evening news bringing seemingly endless news of murderous invasions and counter-attacks way up north. There, Armageddon was man-made – down here locusts were but one of many agents of Zidihamu.

My induction into the predictabilities of the solar system, as illustrated by Teddy's selection of garden fruit, was designed to reassure and offset the many fears and insecurities of an infant listening to parents or the BBC and watching

Four ceiling panels depicting the Horsemen of the Apocalypse.

friends march off to lonely deaths far away. Many years later, I was asked to join two of my students, Ignatius Sserulyo and Peter Binaka, in decorating the ceiling of a Makerere University chapel. For a theme appropriate to our war-torn times, the Book of Revelation was suggested. The ever-present potentials of plague and pestilence offered one theme, the four horsemen of the apocalypse another. Our efforts on that panelled roof have survived their own various vicissitudes.

The task of initiating young people into adult, parental or religious society is generally allocated to a respected teacher. From my filial perspective, both parents filled that role but, as a teenager, I had the good fortune and privilege to be among the guests and participants at a ceremony designed to be as reassuring as Teddy's solar fruit salad. For our hosts the ceremony was also designed to make survival in a cruelly hazardous world more bearable. Similar domestications of an unruly existence have probably existed for tens of thousands of years all over Africa.

Panelled ceiling above the organ loft in St Francis non-denominational chapel, Makerere, Kampala, East Africa.

Rain-maker Gunda Anyampanda of Isanzu was a celebrated high priest for the Sun God and I witnessed him invoke rain at an important annual ceremony at Kirumi. Wearing long robes and a loosely wound turban, he was a dignified but approachable elder (he died a year or so after I met him). His powers and duties included being a judge, leading prayers and conducting ceremonies to ensure good harvests and fruitful marriages. He decided the times for planting, harvest and circumcision as well as being an authority on herbs and medicines. He was descended from a long line of famous rain-makers belonging to the Anyampanda clan. If you live within East Africa's drought corridor, where rain is fickle, any claim to be able to make rain is a very serious matter.

Teddy had taken me along on a visit which coincided with the single most important ceremony of the year. A crowd of hundreds had assembled around the 'rain shrine' at Kirumi and Gunda, wanted reassurances about the government's attitude towards Isanzu and towards their sacred rain-making ceremony. The missions and all previous governments had long been unremittingly hostile. Where did the District Commissioner stand?

In answer, Teddy seized a spear, took up a defensive position, shook the spear and said he would defend the Wasanzu and their customs. His answer was applauded by the crowd and I was quick to photograph this little display, so unexpected and so uncharacteristically spontaneous for Teddy.

Gunda had put out on rare display four ancient, black wooden figures, each gleaming with oil, guardians of the 'rain-stones', a collection of multicoloured beads of varied size and shape that were 'owners of rain' (actually tokens for the assertion that rain was owned by Gunda's clan). The largest sculpture served as both an actual seat or throne, and a statue of a regally seated figure on whose lap Gunda might sit. Gunda claimed these sacred sculptures, normally hidden

The Owners of Rain.

Teddy strikes a pose, Kirumi, 1953.

from view, had accompanied the Isanzu during their fabled 300 km trek from Ukerewe Island more than 150 years earlier, which, if true, made the images at least that old. They incorporated concepts of the Sun God and his Moon Wife.

To be literal about Gunda claiming to 'make' rain would be like dismissing Teddy's little lesson in astronomy as a fruit salad laced with mumbo jumbo. All of the crowd assembled inside and around the rain shrine compound understood that 'rain' had multiple meanings. For a start, there was hard male rain and soft female rain. The rain-maker connected male rain with the Sun God and female rain with his wife, the Moon Goddess. Sunrise was cosmic birth. Sunset terminated assorted conflicts as well as terminating life itself. If the rain-maker made rain then that was a comprehensible skill because male rain combined with female rain to make babies. In Isanzu the many unpredictabilities of life became subsumed in the mysterious fickleness of real rain. Somebody had to seek out its mysteries – Gunda's Anyampanda clan had appropriated that niche very early on in Isanzu history.

If there was an analogy between rain-language as used at Kirumi and my father's use of fruit to visualise the interplay of Earth, sun and moon, it was that rain-making tried to domesticate a universe that was just as awesome for the Wasanzu as for Teddy. They differed in that Gunda connected this with the mysteries of fertility and reproduction. Here the ceremony became concrete and down-to-earth.

On our arrival we had been shown a loosely thatched, log-wood lodge near one corner of the rain shrine compound. We were told that two Anyampanda rain-making apprentices known as *mutaata*, one male, one female, both stripped naked, would be incarcerated that night-fall within this lodge, called *mpilimo*. Inside that dark, encircling prison of planted staves, the man will scent spring soil and he will hear the voice of rain. His hands, conjured by hers, will come to help

The mpilimo lodge,
fronted by The Owners
of Rain.

hoe her field. His rain and her rain will flow together again and again. Then, having performed the central acts of 'making rain' on behalf of Isanzu's multitudes, they will sleep a season. They will wait for the Wasanzu to gather again to sing in praise of their rain-making. Their people will live on through their rain, their issue, their brief being among the rocks and corrals of Kirumi. A baby, born of this rain, will make manifest the community's only hope for continuity as a people.

Teddy respected the people of Isanzu, their ancient rituals, their metaphorical translations and what some might call their idiosyncratic use of language. He liked to quote a Nyakyusa proverb – 'ritual rids the soul of anger'. Words and rituals bind even the most rebellious and sceptical of contemporaries to some sort of community.

Rituals dominate the lives of pastoral people in the Nile Sudd, South Sudan. Once hosted among Dinka cattle camps, I offer below a pictorial memoir of their hospitality.

We may have a lot more choice than our parents and forebears but the real world is of such frightening barbarity that most, like Teddy and the Wasanzu, opt for institutions and mental models that emphasise domestic order. We seek it even when it flies in the face of grotesquely cruel and disorderly wars, or subordination to a witches' brew of dictatorial institutions (on both Left and Right), human follies, and natural or biological disasters.

The adaptability of animals and plants to extremes of climate is about to be tested again but this time it is an ex-African primate that has engineered a searing hot-house by burning oil and coal, reinvigorating the sun's most scorching rays by wrapping our planet in a gas – CO_2 (carbon dioxide). Industry and industrial agriculture are the contemporary faces of our new and very own era – the Anthropocene.

As Paul Crutzen, the creator of this term, was the first to point out, our actions are leaving their signatures on both terrestrial and sea floors. These deposits are already inescapable markers for a new era – they are comparable to dinosaur footprints marking out the Jurassic. The Anthropocene has already laid down its own unique traces just as other fossil lives demarcate past geological periods.

Our era is currently laying down a chemical record of the follies of our barbaric, so-called civilisation. Whether this proves to become another protracted 'age' or a blip, some brief 'event', remains unanswered.

Age or event, man-made climates are about to test humanity's capacity for co-operation and self-control. Either way, its traces, its Mpilimo lodges, will express our capacity and taste for language, for ritual and for domestication of our fears and follies.

Forest Undergrowth *(colour coded).*

HABITATS DEFINED BY PLANTS

In which no animal exists without plants. In which thieving chain-saws slice through centuries in minutes, without arrest or censure.

The history of life on our planet finds a miniature replay in the aftermath of volcanic eruptions when Earth's first *terra firma* cools as lava fields. Volcanoes and lava fields occur in many parts of eastern and central Africa and small-scale eruptions are common. The very first colonists of lava fields are usually ferns, primitive pioneering plants whose fossils date back to some 350 million years ago. Grim black slopes, often densely punctured by holes that resemble well-heads, betray where forest trunks resisted long enough for fast-cooling lava to bank up on their upper side before the wood turned to charcoal, then ash.

Winds over wet highlands, loaded with fern spores, soon clothe such bleak, black landscape in a thin film of brightest green.

It is a scene borrowed from the earliest days of life learning to live on dry land, and I have seen just such a replay on the slopes of Mount Nyiragongo. On those same slopes, clumps of ferns can look disarmingly like potted plants in an endless garden centre, because some species grow best in the pot-like spaces left by vaporised tree trunks. Other early colonists are fig trees that arrive by a different route. Pigeons, barbets, hornbills and fruit bats all relish figs and will fly long distances to find them. When these birds or bats fly over a new lava field, any falling fig seed-filled excreta that falls on lava is effectively free of competition from other trees. Growing fast on mineral-rich lava, doused by daily showers or dew, they reward their dispersers by creating new fig-filled forests.

The muscular roots and trunks of fig trees can even lift and carry rocks – a telling symbol of life's triumph over the mineral. You can retrace the tentacle roots of such colonists as they reach out in their eccentric harvest from nutrient-filled crevices of Mount Nyiragongo's lava.

African elephant skeletons entombed in lava (as remembered from a photo).

Quite the most melodramatic image I have ever seen of plants re-foresting the Earth was a photograph shown to me by my friend, Alan Root. He snapped it from the cockpit of the small aeroplane he was flying, solo, over Mount Nyiragongo some two years after a major eruption in January 1977.

A torrent of lava, fluid as hot tea, had swept down the steep mountainside, roaring through the forest at the speed of a race-horse. A family of elephants clustered close to one another as they fell over into the white-hot tide. Frosty nights helped cool the mountain as its lakes of lava subsided. Then vultures and ravens gathered to pick the elephants' skeletons clean, each one encased in its own velvet black, life-size, death-shaped coffin. Each elephant lay as close as it could get to its mother or brother. Rain washed the bones ivory-white. The only colour in this horrible cemetery grew out of the abdomens of this fallen family. Somehow seeds had survived being chewed and par-boiled to germinate into yet another momentary pause in the apocalyptic process whereby new continents are born. Opposite I offer my own rendering of Alan's photograph.

One day, ocean waters will flow over what were once the volcanoes of Virunga, but we can be sure that plants will still inhabit the new continent of Nubia and its offshore Victoria Microplate and Zanj Islands, future siblings for Madagascar and the Seychelles. We cannot be anything like as sure about the survival of our own kind.

Kondoa Irangi. Recolonisation of a Mineral Landscape.

In Kondoa Irangi, part of the province my father oversaw, cultivation of friable slopes and removal of all indigenous vegetation for firewood, followed by wash-away of all fertile soils, had left large areas infertile and empty of people. I remember being impressed by how quickly plants had recolonised this once inhabited, newly mineral landscape of eroded gullies and rock-strewn slopes. I painted a scatter of pale rocks over a surface in which scrubby emanations briefly green, but mostly grey, stood proxy for unseen pioneering root systems, while thin vertical stems measured out their own spaces and intervals in an otherwise desolate and chaotic scene.

Being aware of plants returning life to man-made wastelands offers some sort of an affirmation of continuity.

For an exhibition devoted to rainforest conservation at the London Museum of Natural History, I contributed two sculptures, one featuring a real chain-saw being recycled by its victim, an entangling vine, while the second modelled a fig tree sweeping a logging truck up into its stranglehold. Both pieces affirmed the certainty that plants must eventually triumph over the follies of humans.

FAR LEFT: Retribution: Liana Takes Its Revenge. LEFT: Fig Trashes a Logging Truck.

Searching for pictorial expression of plants as the foundation of life on Earth, as habitat, I put together a series of book jackets in which miniature outlines of my animal subjects were framed by colourful vignettes of flowering plants.

For a reminder of just how durable plants are in spite of all the vicissitudes they face from changing climates, predators, diseases and people, take *Dalbergia melanoxylon.*

Two dust-jacket margins: (left) Mammals of Africa; *(right)* Arabian Mammals.

Zachary Kingdon, pupil to Hendrick Thobias.

BELOW: *Outline drawings of Makonde sculptures.*

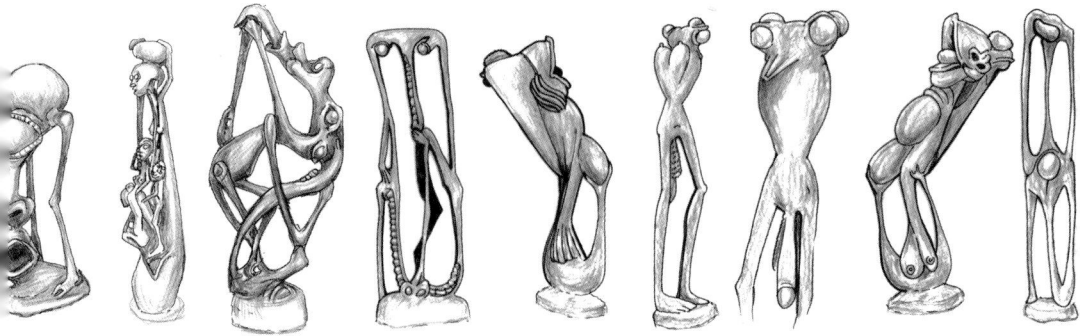

This plant's heartwood was the material my son, Zack, learned to carve as he sat with his tutor-maestros, Chanuo Maundu, Dastan Nyedi, Hendrick Thobias and Jacopo Sangwani, under mango trees in Pugu or on the Makonde Plateau in southern Tanzania. This jet-black wood is called *mpingo*, or African blackwood, and it is the raw material for some of the most original and innovative sculpture in the world.

Hunters and gatherers know that they are being sensed – seen, scented, felt or heard, by the countless beings that surround them. Makonde sculptors have brought this ancient awareness up to date. Their work has been described as 'haptic', while their wit and love of dance and ritual are legendary. For them every experience can be given formal shape in acknowledgement, in pleasure, astonishment, even gratitude for the senses that allow any one of us to bear witness to life as we know it and live it. Makonde have not allowed centuries of persecution and coercion by foreigners or neighbours to dim their independence of spirit and creative intelligence. These qualities helped Ben Mkapa, friend and one-time student at Makerere, to become President of Tanzania. His people are, in my estimation, one of the most admirable on Earth.

Returning to *mpingo*, its wood has long been valued for the production of woodwind instruments, piano keys and bagpipe pipes. At 25,000 dollars per m³, it is as valuable as ivory (before the trade in elephant incisors was banned). Today there are some 80 species of *Dalbergia* distributed throughout the tropics and it has begun to be cultivated as a commercial crop.

Some 60 million years ago, during a particularly hot period, this warmth-dependent tree grew in many regions, including Greenland! Such are the surprises of life on our far-from-predictable planet, under our equally unpredictable climate and collision-prone universe.

Dalbergia is interesting in a more general context because, like peas and beans, it is a legume. Legumes dominated the flora of Precambrian Africa because their crucial alliance with colonies of rhizobia within their roots enabled them to survive the lack of nitrogen in the leached, pushed-up soils. Rhizobia have developed mutually beneficial symbioses with the family Leguminosae (now called Fabaceae by botanists but still legumes for the rest of us).

Leguminous trees cover quite extensive regions of Africa. They have equally exclusive plant–animal dependencies that are of the greatest interest. Their southeastern communities are known as *miombo* (a mix of *Brachystegia* and *Julbernardia* trees, liberally laced with *Dalbergia*). Many species of animals are exclusive to *miombo*, most notably the magnificent Sable Antelope and long-faced Nkonzi Hartebeest, a woolly-tailed genet, and numerous birds. Up north, in a wide belt that runs from the Nile to the Atlantic, open woodlands are called *sau* or 'Soudanian *doka*' after their dominant legume, *Isoberlinia doka*. Here the most famous but now nearly extinct endemic is the Giant Eland, a long-legged, extravagantly dewlapped bovid, currently being exterminated by the livestock industry and motorised poachers.

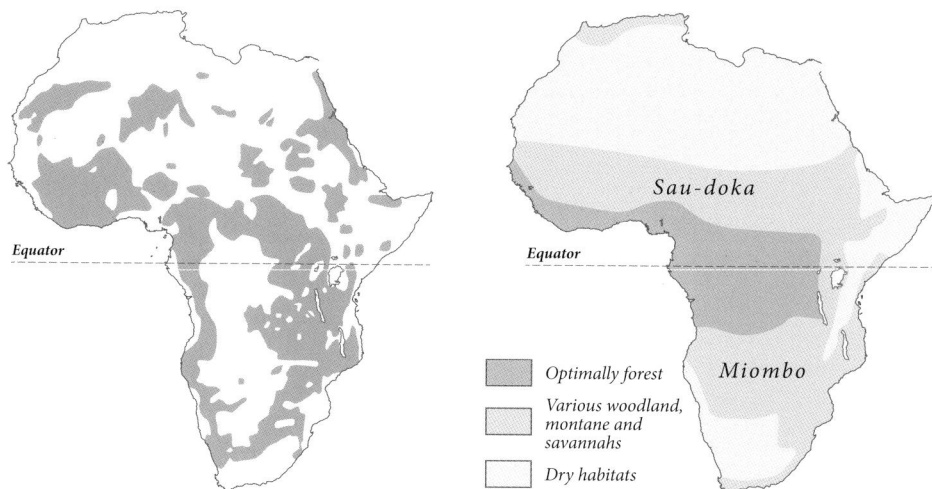

ABOVE LEFT: *Precambrian surfaces.* RIGHT: *Sau-doka and Miombo distributions.*

*Sable Antelopes (*Hippotragus niger*).*

Miombo surrounded me from my birth in 'The City of Camel Rock and the Springs of Kazeh' – Tabora. All around, fields of domestic beans and pulses were already replacing their wild leguminous cousins. *Miombo* grows on shallow, sandy soils over much of central Africa and, together with Myrrh, *Commiphora*, covers most dry Precambrian areas south of the equator. We savoured exquisite

Kora – Commiphora *country.*

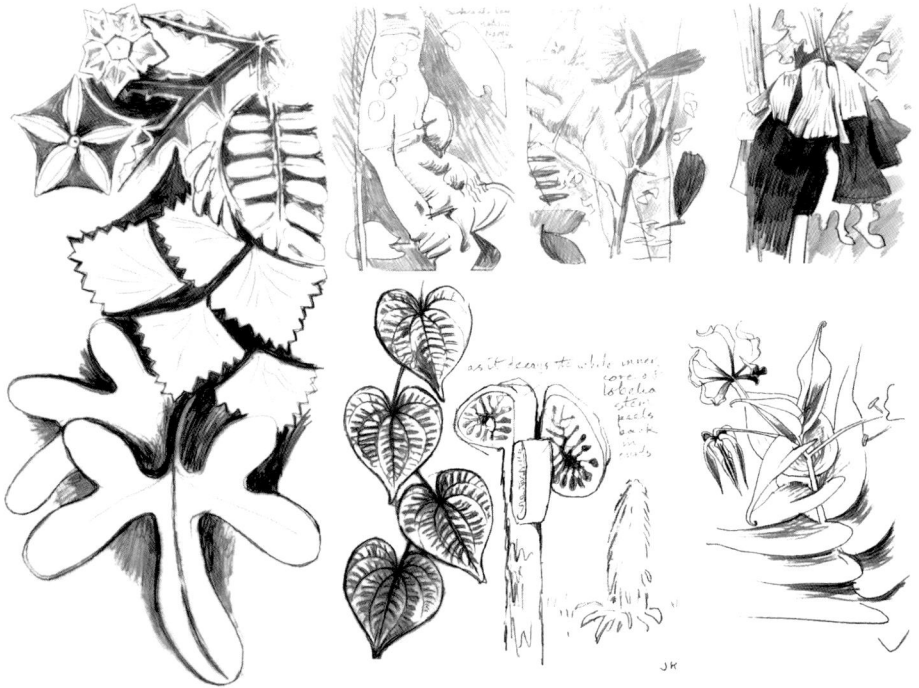

The geometry of plants.

miombo honey as if we were Italians in thrall to the diverse flavours of olive oils. With the onset of the rains in November or December, the naked *miombo* woodlands burst into leaf – yellow, pink, orange, scarlet and every nuance of green. In a *miombo* spring you walk through a jazzed-up Flemish tapestry of colour. Apparently the bright early colours are markers for chemical and antibiotic defences that protect young leaves from most of their potential consumers. It is a colour kaleidoscope that reverses the seasons of Eurasian or American autumn woods. The colours are also some measure of the real diversity of leguminous plants in *miombo* – more than 200 leguminous species can occur in quite a small region, including lots of acacias, albizzias, cassias, beans, clovers and others.

Dorothy's response to *miombo* was pictorial – she wrote, 'The long dry season is coming to an end. The woodland trees and bushes are already putting out new leaves. After months of drought the earth is parched and the brittle grasses are pale ochre all over. Suddenly the stark grey trees grow tired of waiting for the rains and array themselves in flame-coloured foliage. All the autumn shades shimmer, from pale yellow to richest gold, from russet to Indian red – among them a blur of emerald green. Each leaf is soft and burnished so that the sun shines through, giving all a translucent quality, a brilliance like stained glass. This is a yearly miracle – spring dressed up in all the glory of autumn, but fresh and sparkling to greet the first rains.' More than a hint of animism can be seen here, as if she could share vanity, an artist's palette and impatience with a tree. She had the eyes and the temperament to stand awhile and savour the glories of the present. For Dorothy what was past was gone, what might happen tomorrow could scarcely be guessed – it was what you did here and now, what you saw and celebrated in the moment, that mattered most.

When Dorothy exclaimed at seeing *miombo* spring colours, she was responding directly to coloured foliage such as she had never seen represented in any school of painting. She drew individual trees, she made abbreviated colour studies but never got around to painting a canvas of *miombo*. She did try but I suspect she needed some sort of cultural context or niche to work within. She shared, I think, something of the bafflement expressed by a metropolitan artist attached to an early expedition to the Rwenzori Mountains. Frustrated, the artist ended up reproaching the vertical crags for their insubordination to culture, exclaiming, 'They would not compose!'

As for me, I could never escape the naturalist's 'why and how?', even while responding to Dorothy's emphatic 'just *look* at that!'

In mitigation of my scientific bias, I did once see a flowering *Erythrina* red coral tree, being switched into rhythmic swirls during a raging thunderstorm. Without much more confidence than Dorothy, I painted a vermilion red and ultramarine coronet of resilient limbs, like talons squirting scarlet gore all over the canvas. It still survives on the wall of my office, above the desk where I now sit.

Erythrina, Coral Tree in a Storm.

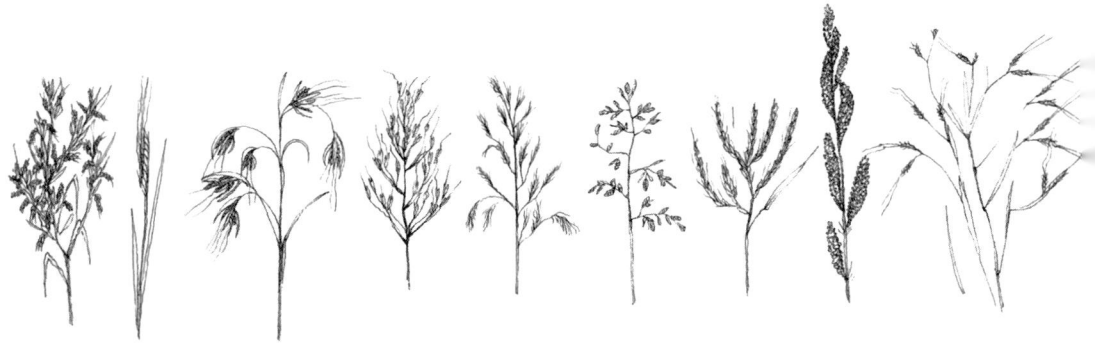

A selection of African grasses.

As a teenager, I camped in Tanganyika's central block of *miombo* woodland while the *miombo* were in flower. The buzz of bees was relentless all day but the scent was mellow and good and a light canopy made for pleasant hiking. Because the area was uninhabited by humans I had expected to find common large birds such as Ostriches, Kori Bustards and Crested Guineafowls, but there were none. There were no gazelles, no gnu and small antelopes like Steenbok or Suni, and dikdiks seemed to be absent. At night, no hyenas called and the drawn-out 'oo, oo-oo' of the common African Wood Owl was never heard. When I trapped for rodents and shrews, they too were rare and localised. Even the classic *miombo* antelopes, Sable and Nkonzi Hartebeest, were widely scattered, usually in small, very mobile groups, as were Warthogs and zebras. When the rains were over, it was easy to see why animals, most of them dependent on drinking water, were so few. The soils, being thin and sandy, quickly drain away the last May rains. Then strong June winds desiccate the foliage and fan the thin fires that sweep through, fed by sparse grass, dead twigs and leaves. These early burns leave the woods open and waterless, but the lightly scorched trees simply wait out the next four or five months. *Miombo* legumes are the true survivors, the life forms that have adapted to drought, fires and 100 million years of soil starvation.

A rare archaeological site from a *miombo* area showed that, given proximity to water, it was a habitat where prehistoric people could survive – but only just, and precariously at that. One significant detail, revealed by analysis of cereal debris embedded in a grinding stone, was that people must have spent long herbivore hours gathering tiny seeds from the wan, short-lived *miombo* grasses. At the time those people were living, any small finch was probably better at grass-seed gathering than this pathetic prehistoric precedent for combine harvesting, finger-pluck by finger-pluck.

There must have been many better places to live (as might be said of contemporaries in the Arctic or in the scorching cities of the Middle East). Once there are enough people for different communities to compete, some get pushed to the margins. Some may seek to leave overcrowded centres. Most find themselves stranded along the road to somewhere else.

It is not widely known that herbaceous shrubs and trees only began to give way to grasses less than 10 million years ago, yet grasses are today so ubiquitous that few realise how recently they have come to define African landscapes. Compared to trees and shrubs, grasses are relative newcomers, and extensive palaeo-grasslands only became widespread about 6 million years ago. In concert with this new dominance of grasses, mammal molars responded to the challenge of more silica in their diet by getting deeper, harder and more complex. An enormous range of plant-eating species – elephants, hyraxes, zebras, rhinos, pigs and, independently, a whole slew of bovids and even some rodents, all rebuilt their behaviour, their digestive tracts, their skulls, even their gross proportions, to accommodate to both a new diet and to new, often more exposed, grassy environments.

More abrasive grasses in the diet of elephants selected for a multiplication in the number and deepening of flattened enamel-coated dentine plates, all bedded in dental cement.

It is possible that an elephantine preference for leafy browse, and the associated bough-breaking and trampling by once-gigantic herds, created more tinder for fires, all of which opened up space and opportunity for grasses.

Other species might have benefited – take rhinoceroses, which have Eurasian origins. Fossils reveal that a common ancestor to the two African species lived around 10 million years ago, an animal that was broadly more like the browsing 'Black' Rhino than the grass-eating 'White' Rhino. By about 5 million years ago, grass-eating rhinos had become a distinct lineage that has kept improving its adaptation to a grass diet right up to the present – the Grass Rhino is no more a throwback to prehistoric times than a horse is. Often described as 'primitive', both African species of rhino were, until very recently, highly successful, abundant and wholly modern species. They have been pushed into near-extinction by human brigands from near-prehistoric, primitive and non-African, Eurasian cultures.

Now we turn from the evolution of graze-teeth in rhinos to their equivalent in bovids. Fossil teeth are the most durable and frequent remains, so the early classification of bovids was built upon two categories of molar teeth. 'Boodonts' had shallow, less complex molars while 'aegodont' molars were deeply folded – browser boodonts and grazer aegodonts were first treated as if they were independently evolved taxa.

Today it is clear that within any one major bovid lineage the more conservative species tend to be more boodont, while some of the more recently evolved species have become grazers with appropriately complex aegodont molars (not to mention rumens).

A relationship between grass diets and deeply laminated molars even applies to rats. Murine rodents came into Africa about 10 million years ago and seem to have split into a conservative, more omnivorous lineage and a more herbivorous one. The latter followed the now classic division into boreal centre–west

LEFT: *African elephant (*Loxodonta *sp.) molar.* RIGHT: *Vlei rat (*Otomys *sp.) molar.*

and austral southeastern lineages. Down south, little grass-loving vlei rats have evolved teeth that resemble tiny little elephant molars – deeply folded concertinas of hard enamel embedded in cement.

The late spread of grasses can be deduced from countless other forms of life, especially insects and birds where entire taxa have co-evolved with grass, from grass-hoppers to grass seed-eating finches and pipits.

Wherever grassland enclaves have displaced thicket or forest, a stay-at-home local pipit seems to have evolved, demonstrating that even sprawling grasslands can become ecological islands.

The relatively late emergence of grasslands has allowed a true 'grass cat' to evolve, as well. Watch a Serval catapult itself up out of tall grass in a soaring arc, as though, for less than a second, it is a swooping buzzard, propelled behind unsheathed, outstretched claws. Seconds later the cat will land in the long grass, and then emerge, a grass-rat dangling from her jaws.

Both the Serval and its prey have their separate histories written in their genes. Becoming dependent on grass-rats selected for the cat's soaring leap – the higher the grass, the higher the vault, for which those long, rangy limbs win over shorter legs such as those of a stocky Sand Cat. Meanwhile, only grass-rats with sharpened senses and the habit of still, silent freezing have escaped cats with satellite-dish ears ever more finely tuned to the bandwidths of grass-rat

*Serval (*Leptailurus serval*) snarling.*

existence. That particular rat was a bustling looser, but predator and prey have evolved in tandem.

I once raised two Serval kittens (named 'Spick' and 'Span') and kept them for more than 10 years. One day Span inflicted a painful lesson in the strength of her evolved routines when my fingers, fidgeting in long grass, were suddenly impaled by the smack of her unsheathed claws. Her obsequious response to my yelp of pain was even more surprising – as she slid her flanks against me her whining, salivating contortions resembled nothing less than a cat's *mea culpa*: 'Oh! I'm so sorry!'

To return to forest plants and their accommodation or symbiosis with animals, I once painted a panel in homage to my friend Bill Hamilton in which I depicted ant-plant hosts and their tenant insects in simulation of a medieval stained-glass window.

Trees on the sunny super-wet equator, in spite of growing on leached pre-Cambrian soils, can become rainforest giants. In equatorial Congolian forests, an intimate relationship has developed in which a small, relatively short-lived rodent could almost be described as husbanding the one species of giant, very long-lived tree – the Awoura or *Julbernardia pellegriniana* – on which it appears to depend.

This giant legume is among several that face very different challenges from their seasonally stressed *miombo* cousins. They must compete with several hundreds of other tree species for living space. The tiny rodent that assists them in this is an African gliding 'squirrel' (belonging to the anomalure family),

LEFT: Ant-plants, *a painting dedicated to 'Bill' W. D. Hamilton.* RIGHT: *Anomalure on Awoura* (Julbernardia pellegriniana) *trunk, framed by gnaw-pruned twigs.*

handsomely tailored in chinchilla mottle and equipped with enormous whiskers and big onyx eyes in a black furry mask.

A peculiar sap wells up from Awoura bark after it has been wounded and this sap offers a staple nutrient for anomalures. You could almost describe the glider as a sort of vampire that laps up this especially nutritious 'tree blood'. The anomalures use habitual glide-paths between their aggressively defended tree-hole dens and their regular sources of food.

However, tree branches have the inconvenient habit of obstructing glide-paths, so pruning is a major part of the anomalures' nocturnal activity. Indeed, during daylight, I found that pruned branches dropped onto the forest floor are the main indication of anomalure presence.

Seeing that a favourite 'blood-sucking' site was around the base of an Awoura trunk, I noticed that anomalures had expended much time in pruning away any vegetation close to the Awoura's trunk. I also discovered that one effect of this sustained cutting was the eventual death of young trees trying to grow close to the Awoura.

In effect, the anomalures were killing off any potential competitor that intruded into the Awoura's precious living space. Many other legumes protect themselves with toxic or semi-toxic bark, but not Awoura. Selection had brought together two separate needs in legume and anomalure, to their mutual benefit. In those forests where Awouras are still abundant, so are anomalures.

This little vignette can only hint at one minuscule detail of a complexity and forest natural history that lie way beyond human comprehension. Much of that history must remain unknown but surely the pursuit of knowledge must be preferable to private gain for the armoured few? Awoura logs represent money, and vast tree abattoirs have sprung up all over equatorial Africa, expressing the intrinsic values of a contemporary culture where Death Trumps Life.

The triumph of complex communities of plants over millions of years of drought and deprivation should earn our deepest respect, and a concerted international effort to study and learn from plants. Instead, Twa hunters stand by, watching the carcasses of their habitat roar away on giant trucks. A world hijacked by unholy alliances between rapacious businessmen and corrupt

Truck in a logging yard.

Vegetation offers countless patterns.

Our survival has long depended on vegetable foods, some indigenous, but mostly a rich variety of fruits and vegetables from South America and Asia, where systematic cultivation has a long history.

For my family, these foods were bought in the local market or grown in the garden. Being taken along to buy vegetables and fruit was a very frequent pleasure that began so early on I cannot remember any beginning to it. It is as though there were always markets, always stacks of multicoloured plants and, behind them, always-jolly market women who would tease, hug or kiss me while flirting with Saidi. In Tarime, there were daily visits to our own little fenced vegetable patch to watch the tomatoes and cabbages grow, only to find that caterpillars relished them too. Together with Dorothy and Saidi, we became *wadonoa wa dudu* – insect-pickers.

LEFT: *Fever Tree (*Vachellia xanthophloea*) and swamp palm, elephantised.*
RIGHT: Cordia *trees shading coffee.*

When my father bought the lease of a small farm near Mbeya, he took over a dozen or so hectares of coffee trees planted out by an adventurer with the nick-name of 'Zambezi White', who had inter-planted the rows of coffee with what became a veritable woodland of 'snot-berry trees', *Cordia africana*. Their func-tion was to shade the coffee (which is a common understorey shrub native to

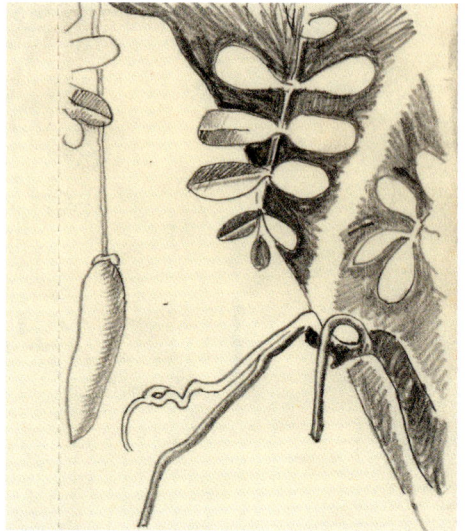

LEFT: Kibondo *Kigelia, a painting by Dorothy Kingdon, 1933.*
RIGHT: *Hanging sausage fruit (*Kigelia africana*).*

Africa). Snot-berries are tall, indigenous forest trees with leathery, plate-sized leaves. Their unappetising nickname is for the sliminess of their little yellow or orange grapes. Twenty years later, coffee and *Cordia* had sucked in an entire ecology all their own. *Cordia* foliage was sometimes turned into lace-work by the larvae of various insects. Dense clusters of sweetly scented white flowers attracted swarms of bees, flies and butterflies, which, in turn, brought in chameleons, praying mantises and small fly-catching birds. These tempted visits by snakes and lizards, snake-eagles, Lizard Buzzards and sparrowhawks. When the *Cordia* fruit ripened, the whole plantation swarmed with berry-eating species of pigeons, hornbills, parrots and barbets, and was subjected to furtive raids by Vervet Monkeys. At night Greater Galagos wailed plaintively, living up to their alternative name of 'bush-babies'. Fruit bats flew in from afar, and one, heavily pregnant, got tangled in netting over the peach trees and strangled itself. More than 20, now 70 years later I published my teenage drawing of this casualty (opposite).

Fruit bats are the principal pollinators of *mwegea*, or 'sausage trees', *Kigelia africana* (*kigeli-keia* in an eastern African dialect). There was a single sausage tree on the farm, close to the main road, a conspicuous marker on our northern boundary – this Cape to Cairo highway ensured that any acquaintance passing by used this tree as the landmark to take a break and stop for a cup of coffee, tea or a beer.

At first this tree set many 5 kg sausage fruits, most of which hung clear below the main canopy, for up to a year, attached by sinewy stalks several metres long. These cables began as dangling chandeliers of odd-scented, meat-coloured,

*Epomophorine fruit bats (*Epomophorus *sp.), 1951.*

nectar-filled flowers that opened at dusk and closed shortly after dawn. The very large flowers bloomed on the promise that there might be another *Kigelia* tree somewhere in the vicinity that would send its pollen via some nocturnal agent to fertilise the pistil of each meaty flower. Fruit bats were the most efficient night pollinators because they ranged sufficiently widely to visit several flowering trees during a single night. The numerous birds and insects that came for nectar at dusk and dawn were less prone to visit more than one tree, thus making cross-pollination less likely. Over the years our *Kigelia* produced fewer and fewer fruits, and eventually none. I think this must have been mainly due to the bustle of nearby houses and ever more road traffic deterring the bats. In addition, other *Kigelia* trees were declining, being felled, or had already become too few in our vicinity.

We owe much of our knowledge of plant biology to gardeners, estate owners and plant collectors that served their taste for the 'exotic'. Explorers who left scientific cultures in metropolitan countries to catch or collect in nature have always been subordinate to the tastes, the levels of education and the motives of patrons rich enough to fund their explorations.

I can speak with some authority on this because my father's first cousin, Frank Kingdon-Ward, a passionate botanist, spent his life negotiating this existential conundrum. Frank's father, Henry Marshall Ward, inherited the School of Botany and botanical gardens at Cambridge University from Darwin's great friend and ally, John Henslow. Frank graduated under his father's shadow and then, in the footsteps of Joseph Banks and Joseph Hooker, he became a botanical explorer. He ranged through the Himalayas, southwestern China and the mountains of Southeast Asia, his expeditions almost entirely funded by seed merchants such as 'Suttons of Reading' (a business, begun in 1806, that had grown out of selling corn and corn seed to farmers, a line of business that once supported my mother's family).

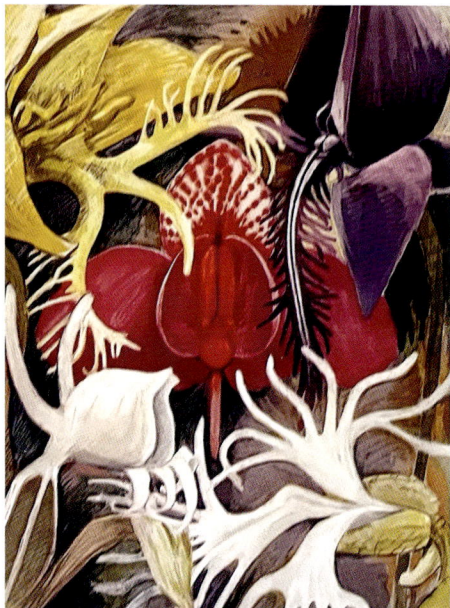

Kitulo orchids.

I met Frank just as he was planning yet another expedition to Burma. Of the flower of a lily, blooming way up in its Tibetan fastnesses, he wrote, 'A delicate shell pink outside, like dawn in June, with the sheen of watered silk; inside it was faintly flushed alabaster.' Not the language of a seed merchant, but 'needs must', as he might have said.

For botanist Frank, flowers were avatars for sex and this dimension came to mind when I tried to depict some of the rare, endemic but extraordinarily suggestive orchids of the Kitulo plateau, a favourite weekend jaunt of my childhood.

This chapter has been partly illustrated by sheets of sketches made during long hours in African forests, sketches that got incorporated into paintings that tried to hint at the complexity of life in our infinitely diverse forests.

In 1989 I joined a team investigating a little-known locality, the Matumbi Hills in coastal Tanzania. Years earlier I had flown over this small coastal block, which had once been an offshore island, like a miniature Mafia or Zanzibar. Our Matumbi survey was lucky enough to employ Mganga Mtalami Mbondei, the

Mganga Mtalami Mbondei on his home ground.

last in a long line of medicine men, *waganga*. He claimed that his lineage went back more than 30 generations – as a youth he had learned the names of each of those forefathers, and he told me that the earliest of his doctor ancestors had lived beside Lake Chilwa, in today's Malawi, and that they had come to Matumbi some 200 years ago. As the only speaker of both Swahili and English in the expedition, I acted as translator between Mbondei, the traditional botanist, and Douglas Sheil, our chief scientific botanist. Mbondei became my trusted friend and ally who, when understandably suspicious locals wanted to chase us out, or worse, helped me carefully explain, in Swahili, the nature of our work. Once he had taken our measure he approached me and announced, with considerable solemnity, 'Here am I, the last of a long line of doctors, medicine-men, of which I am very proud. Not long ago I would have refused to tell you anything, but not only do you know our animals, you are a teacher and your students from Dar es Salaam and abroad are making sure that future people will know about all the plants and animals that live here in Matumbi. You've seen my two boys – all they are interested in is girls and working in the Dar docks – they will never learn anything from me. So I will tell you our names for every plant, and I will even tell you how I and my forefathers have used each plant.'

He was true to his word – he did indeed have names and uses for a very high proportion of the plants we collected. Mbondei's herbalist ancestors, like the Hadza, the Ndorobo, the Sandawe and other non-agricultural people, compiled extensive and detailed pharmacopoeias and larders based on plants. Our oldest forebears had practical uses for hundreds of plants, even if they could not know how or why their chemistry evolved. Right up to the present, such gatherers have persisted next to large, settled populations of agricultural and urban peoples. It is important to remember that learning about plants is a process that never ends – there is so much more that is unknown than is known, even than ever *can* be known.

Over the centuries, wanderers and explorers have picked up scraps of knowledge from living herbalists such as Mbondei – primary sources for our often life-saving knowledge about plants and their innumerable uses. One of the tragedies of our age is the systematic extermination of the world's botanical heritage by industrial agriculture, loggers and chemical industries, backed up by purblind politicians.

Detailed copy of a rock painting from Zimbabwe, undated, but probably in the region of 8,000 years old.

7 EMIGRATION AND IMMIGRATION

In which an Ellis Island Register documents intercontinental comings and goings.

The honour and privilege of being invited to be a visiting professor at Kyoto University introduced me to two uniquely Japanese concepts. One was their ancient shamanistic recognition of *kami*, invisible entities expressive of the inseparability of living beings from the natural, material world in which they (including us) are embedded. My fellow ecologists told me that naming, deifying and placating *kami* had helped ancient Japanese people to tame and domesticate some of the terrifying unpredictabilities of their island home, prone as it was (and is) to tsunamis, typhoons, volcanic eruptions and occasional death-dealing plagues or killer winters.

OPPOSITE: Voices of Ancestors. ABOVE: *Bear confronted by the arrival, in America, of humans (bronze).*

The related concept of *kamamatsuri* describes a dimension occupied by spirits and wild deities. I learned that the Ainu people of Hokkaido Island, after ceremonially sacrificing a living bear, enlisted their champion archer to propel a single, beautifully ornamented arrow as high and far as was possible. That arrow symbolised the return of the deity-bear's spirit back to *kamamatsuri*, back into the hills whence it came.

Some years later I was asked to devise a memorial for Ruth Wainwright, the wife of a very great friend. Ruth grew up in the mountains of British Columbia and I intuited that she had kept, deep inside her, a wild spirit that had longed to return to her mountain birthplace. Remembering the Ainu bear ceremony, I made two confrontational bronze figures, one human and the other a bear. Death reuniting them would be her *kamamatsuri* arrow, the symbol of her return.

Bear confronts shaman (bronze).

In modelling a Rocky Mountain bear confronted with a female shaman, tambourine in hand, I imagined the very first human ever to tread American shores. I imagined some Ice Age moment when the bear, a fellow mammal whose kind had, up to that moment, occupied a large part of what was destined to become the human niche, over two continents. I imagined those continents empty of humans. I imagined a bear, faced, for the very first time, with the agent for a complete reordering of nature in the interests of a single species.

The bear was about to have its niche stolen.

That Old-World arrivist Abel soon gave way to more territorial, agricultural Cains. Their much-augmented descendants continue to vandalise two entire continents.

Now reverse the situation and imagine bears arriving on African shores. It actually happened, more than once, and almost certainly during one or more Ice Ages.

Fossils of bears in Ethiopia and South Africa betray these animals' predilection for cool uplands or temperate latitudes but, ultimately, bears could not hack it in Africa. Why? Because hominids (with back-up, perhaps, from hyenas, baboons and giant pigs) were already exploiting much of the bears' ecological niche.

Consider first some comings and goings by birds. In my biogeographic essay, *Island Africa*, I made an analogy between oceanic islands and isolation on mountain massifs within Africa. This was illustrated by the distribution of a particularly elegant array of gamebirds that East Africans call *quaali* (an onomatopoeic rendering of their reveille/evensong call 'quaaali-all-aaha-aaha-you-are'). They go by a confusion of non-specific labels, including francolin, spurfowl and partridge, all names imported from Eurasia, but let's stick with the African and less confusing quaali.

Recent research on the genetics of quaalis has confirmed Reg Moreau's contention that mountains often offer refuges for once-widespread, even once-dominant species. For those Ice Age populations that adapted best to cold, each epochal retreat tended to strand them in pockets of coolth on or around mountains, even close to the equator. Similar but more extensive ranges were often possible way down south. Less hardy populations which had developed other advantages could flourish over more extensive regions during warmer periods.

Serendipitously, quaalis turn out to be giant, intensely sedentary offspring of those champion little migrants – quails. Able to overfly the Saharan or Arabian deserts non-stop, yet reluctant to fly far once settled, an Asiatic quail ancestry helps explain *how* and *where* the quaalis' ancestors first arrived in Africa.

Quails are fast runners and also assiduous litter-scratchers, but with a dry-season reliance on the buried tubers of sedges and other tough corms, roots and

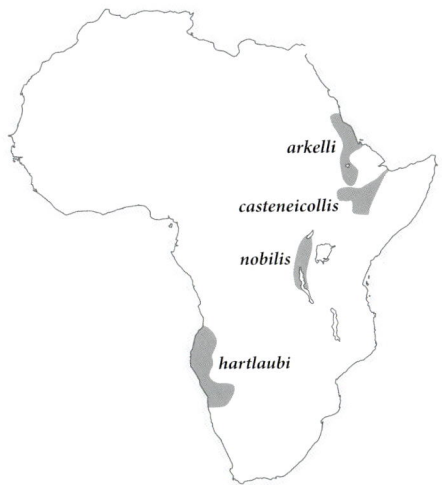

LEFT: *Spurfowl or quaali (Pternistis sp.) head, drawn in 1951.*
RIGHT: *Distribution map of supposed relictual, early members of the quaali radiation.*

seeds. Their descendant quaalis evolved enlarged, muscular bodies with strong, well-clawed feet and somewhat hooked beaks.

As it happens, quaalis have a strong, long-time association with elephants, sometimes following them, somewhat comically, in small retinues – why? Like scavengers on a garbage dump, quaalis compete to scratch out bonanzas of pre-masticated, pre-digested delicacies from within smoking dollops of dung. We can be sure that this scavenging habit must have been even more marked when elephants were ubiquitous across Africa.

Significantly, at least three localised species cluster over northeast Africa, the most likely entry point for incoming Asiatic quail ancestors. For a better under-standing of the quaalis' evolutionary progression, the 'Island Africa' map of tiny disparate enclaves can be broken down and augmented. A second map charts the ranges of but five 'intermediate' species.

LEFT: *Yellow-necked Spurfowl or quaali* (Pternistis leucoscepus).
ABOVE: *Quaalis in the wake of excreting African elephants.*

The quaali radiation does not stop there. A third burst of adaptation has re-sulted in birds that effectively inhabit all of sub-Saharan Africa. Sub-dividing, as ever, into boreal and austral populations, quaalis are actively switching their fealty to the humans that have usurped Africa from elephants. For rich sources of plant and animal fodder, soils macerated by hoes or ploughs are displacing vegetable matter macerated into giant turds. For the birds, it's a poor deal. Over an increasingly tilled continent, quaalis now feast upon sprouting Mexican maize and Chinese soybeans. In their persistence over some 20 million years you could almost call them 'super-quaalis', but for a great many Africans they are better known as easily snared garden pests that serve up as super-delicious quaali suppers.

Another group of bird voyagers between continents is, appropriately, called bee-eaters (but we always preferred their short genus name, *Merops*). Related to kingfishers and rollers, these birds enchanted me when I chanced upon a

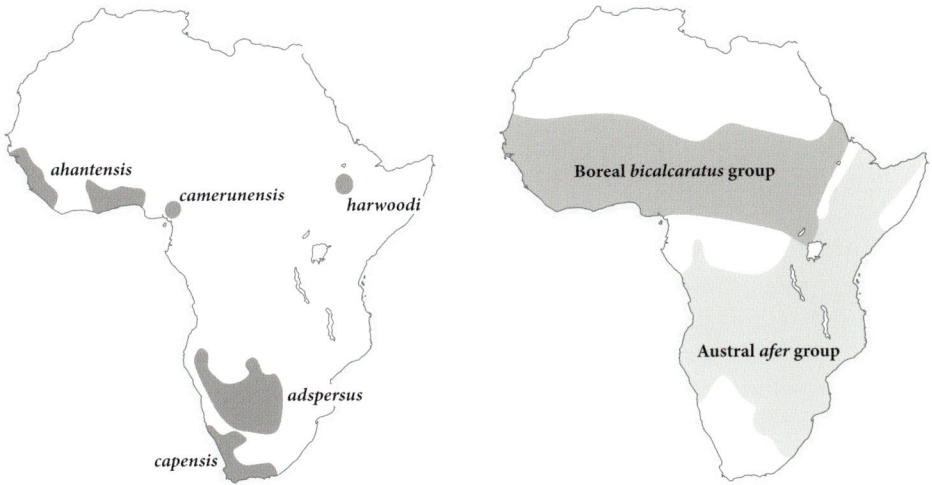

*Quaali biogeography. (left) Intermediate species of quaali with limited ranges;
(right) currently the most widespread, bare-necked species.*

colony of a hundred or more White-fronted Bee-eaters breeding in an exposed, cliff-like bank of the Kiwira River (close to the Malawi/Tanzania border.)

Recent research has shown that these are among the most conservative of *Merops* species, but all have refined their special capacity to feed off venomous arthropods avoided by almost all possible competitors. In South America, the motmots resemble *Merops* but also retain features that demonstrate their common ancestry with kingfishers and rollers. This hints at something important about differences between continents as theatres of evolution.

Throughout the sciences, there has been such heavy emphasis on revealing universal laws that important distinctions between continents have been neglected, even though it is now generally accepted that most of human evolution took place in Africa and could never have occurred in the Americas

*White-fronted Bee-eater (*Merops bullockoides*) on the Kiwira river.*

or Australia. Here in Africa, plants and animals have been forced into ever more specialist niches, ever more responsive to the ecological complexity of our continent. The evolution of humans in Africa is no less the outcome of that niche-multiplying process.

To return to *Merops*, these birds are 'flashers', showing off patches of brilliant colouring. I have never forgotten the sight of scores and scores of masks, all perfectly circular and perfectly symmetrical, poking out of perfectly circular nesting holes above the Kiwira river. Their very memorability suggests that evolution must have selected such unforgettable geometry to enhance their survival. The idea that visual geometry can enhance the ability of brains to remember such 'flashes' is explored in more detail in a later chapter (*see* pp. 386–393) but as schoolboy naturalist and picture-maker I just gloried in the flashing of green backs and wings, ultramarine rumps, bandit masks, and undersides that matched the ochres and apricots of that sandy cliff, perforated as it was by so many *Merops* burrows.

Some *Merops* species migrate back and forth to Asia, where there are also a few more local species. One African species has even learned to visit postglacial, seasonal-bonanza Europe to breed, while another has colonised Australia. However, it is clear that Africa, with some 20 species, has remained the most consistently welcoming continent and evolutionary hub for these exquisite birds. Their principal advantage has been their ever more finely tuned ability to consume venomous insects, which most other insectivorous animals avoid.

All rural sub-Saharan Africans are familiar with a much more distant relative of *Merops* – the hornbills, which range from giant, booming ground-hornbills to heraldic, talkative 'tock' hornbills in drier areas. The canopies of forested regions are alive with the nasal braying, social honking and whooshing wing-cranks of magnificent forest hornbills.

An assertive and fascinating presence in my childhood, hornbills have remained a source of endless wonder. In the general vicinity of Mongiro hot springs, in Bwamba-Semliki on Uganda's western border, hornbills joined other forest, savannah and montane fauna and flora in what is likely to represent the greatest local concentration of terrestrial biodiversity on Earth. There I eventually got to see and watch nearly a dozen hornbill species. Their black (or brown) and white patterns varied in the distribution and proportions of these contrasting colours.

A trip to Sumatra and views of the magnificent Rhinoceros Hornbill invited a translation of the birds' 'jizz' into a sort of origami game whereby cuts, folds and geometry invoked the process by which each species conjures distinctive signal patterns, from the raw material of patterns constructed out of contrasting colour patches.

Africa's radiation of these highly conspicuous (and often fearless) birds is impressive but it is almost matched by the diversity of hornbills in tropical

Hornbill heads examined: (left) 'Nkoba' head; (middle) ground hornbill;
(right) Rhinoceros Hornbill.

Asia. Where did they originate, and when did the ancestral exchange happen? Wood-hoopoes, restricted to Africa and Madagascar, show more than a passing resemblance to some of the smaller hornbills. Confirmed as their closest relatives by recent genetic analysis, the same studies also reveal that both have African origins. That hornbills formerly ranged more widely is revealed by their fossils in Morocco, Germany and Bulgaria, some from 47 million years ago (well within a prolonged period of global warmth that would have encouraged the spread of forests and their fauna). Hornbills maintained a strong preference for fruit on both continents (and some even exported their feather-lice). Tens of millions of years later, an African hornbill that typically associates with monkeys (hawking after the insects they disturb) may well have accompanied ancestral colobine monkeys in their colonisation of Asia (incidentally exporting insectivorous appetites not at all typical of other hornbills). There have been unlikely suggestions that invert this migration, proposing that these white-coiffed black birds represent a back-eddy from Asia to Africa.

Hornbill origami.

Hornbill casques probably evolved to serve more than one evolutionary imperative. For example, Asian Rhinoceros Hornbills seem to balance chitin copies of a large red and yellow fruit above their nostrils – their casques may also amplify their maniac cackles and nasal honks. For females, about to immure themselves inside prisons and become dependent upon mates that will keep feeding the prisoner for months on end, the evolution of such an artifice is as logical as the plastic food on display outside Asian eat-houses. My own take on the ground hornbills' brain-bonded casque is that it may well augment their ability to 'hear' the distant booms of fellows. In any event, these fabulous birds have fascinated me since early childhood and invited numerous exploratory drawings and sculptures.

I now want to explore those moments when the African continent regained overland connection with Eurasia because, excepting afrotheres, most of our surviving mammalian fauna originated after Chicxulub. Eurasian colonists came in across temporary bridges that were both physically and environmentally narrow, because it was only animals that could tolerate conditions on and around the bridge that could, in the first place, make it into our continent (unless they had wings or fins). Those bridges were our Ellis Island Control Points and their records, etched in fossils and genes, are becoming open to scrutiny as never before.

Returning to the intercontinental movements of mammals, quite apart from bears there were other lineages that failed to cross narrow bridges or could not compete in the face of well-established indigenous mammals.

Take tapirs. These often semi-aquatic, hippo-like horse cousins were, until recently, common to the tropical regions of Asia and the Americas. Why are they absent from Africa? They may well have had opportunities to make it into Africa but a close convergence with early elephants (tapirs even have short but scarcely serviceable trunks) and somewhat more distant resemblances with hippos suggest that, even if they crossed the bridges, they were soon or eventually out-competed by natives.

Then there are moles, which are unknown in Africa. A shortage of mole-friendly bridges to a notoriously dry continent might have deterred or inhibited their colonisation but it is also possible that the existence of native, afrothere golden moles and rodent blesmols helped tip the balance against Eurasian mole survival.

Turn now to the mammals that *succeeded* in colonising Africa, beginning with our own lineage – the primates. The very earliest primates evolved in tropical Asia, a proposition reinforced by the discovery there of fossil proto-primates as well as survival of living proto-primates, most notably the elegant, slender colugo or 'flying lemur'.

Given how thoroughly sea-girt Afro-Arabia then was, how primates got to our continent remains a mystery. Nonetheless, by about 50 million years ago primates were sufficiently well established and diverse to have entered the fossil

Surviving rodents		Rodent invader

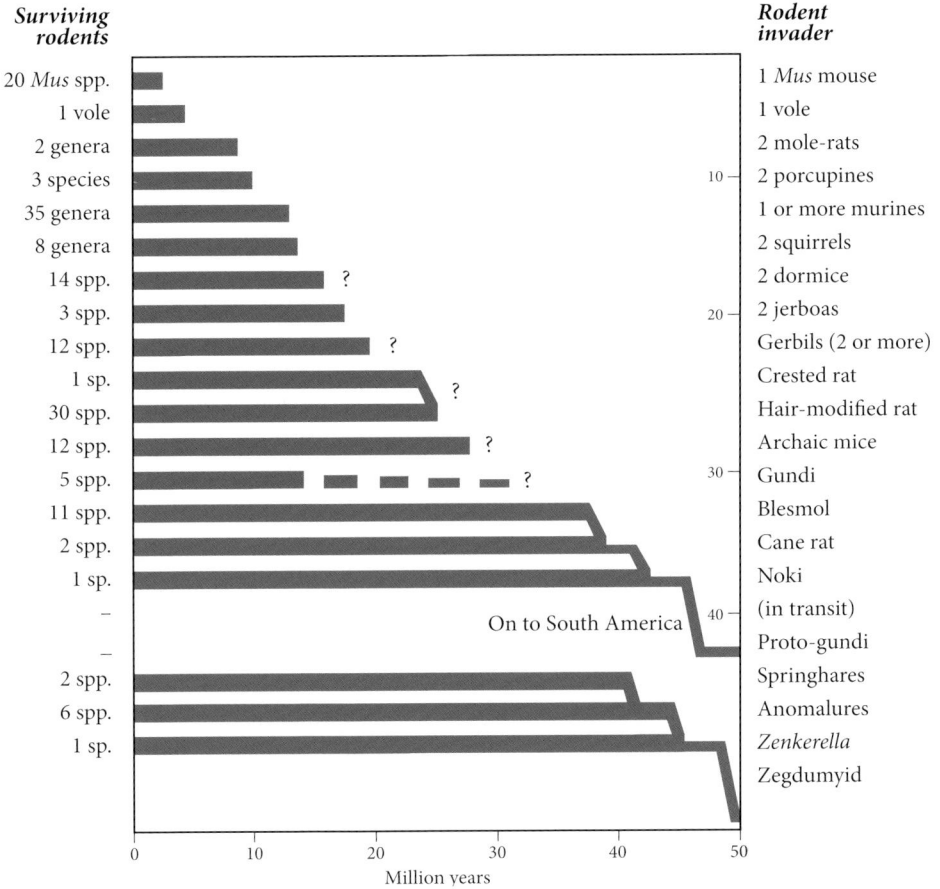

Surviving rodents		Rodent invader
20 *Mus* spp.		1 *Mus* mouse
1 vole		1 vole
2 genera		2 mole-rats
3 species		2 porcupines
35 genera		1 or more murines
8 genera		2 squirrels
14 spp.	?	2 dormice
3 spp.		2 jerboas
12 spp.	?	Gerbils (2 or more)
1 sp.	?	Crested rat
30 spp.		Hair-modified rat
12 spp.	?	Archaic mice
5 spp.	?	Gundi
11 spp.		Blesmol
2 spp.		Cane rat
1 sp.		Noki
–	On to South America	(in transit)
–		Proto-gundi
2 spp.		Springhares
6 spp.		Anomalures
1 sp.		*Zenkerella*
		Zegdumyid

0   10   20   30   40   50
Million years

Procession of rodent invasions, deriving from a single (rarely more) species of Eurasian invaders, from 50 million years ago to the present.

record; evidence for an ancestral arrival at some still earlier date. The careers of African primates await further exploration in later chapters.

Eurasian rodent immigrants into Africa have been particularly persistent – a history of wave after wave of arrivals. Each newcomer (likely, in most cases, a single species) either slowly replaced its predecessors as its own descendants proliferated and speciated or, more interestingly, squeezed precursors into ever narrower, ever more specialist niches and ever-shrinking ranges. This seems to have happened to Africa's most conservative of all rodents, *Zenkerella*, where the eons of time betrayed by its survival seem scarcely credible. Looking superficially like a big dormouse, it is, like them, both arboreal and nocturnal. Fewer than 20 have ever been collected by scientists, yet these specimens come from a vast block of forest, ranging from the Bight of Benin to the central Congo. Forest hunters who have eaten them say they taste awful.

The great rodent pioneer? Zenkerella.

I once visited Cameroon in the hope of finding living *Zenkerella* because I was interested in the evolutionary origins of gliding (and ultimately of flying, as exemplified by bats). My quest was *Zenkerella*-less. Now, some of the secrets of this truly ancient rodent survivor have begun to emerge from the research of my friend, Erik Sieffert, and a nicely inquisitive gang of colleagues. They have shown that this mysterious rodent is NOT the degenerate relative of the gliding anomalures that it most resembles, but that the two lineages diverged some 49 million years ago and *Zenkerella* has scarcely changed in all that time. Furthermore, *Zenkerella* has no traits that might anticipate a glider; instead it seems that these dormouse-like animals forage within very dense vine tangles at various levels, sometimes close to or within impenetrable swamp forests (which could be pertinent to their survival). Local trappers claim that *Zenkerella* sometimes comes down to ground level, but holes up for the day in tree hollows. Here it probably hangs vertically, with the weight of its body taken on a short tract of tough pointed scales on the underside of the bushy black tail. *Zenkerella* shares this scaly tail-prop and many features of its skull with the 'anomalous gliders' or Anomaluridae (mentioned in the previous chapter as champion forest pruners). Up to now taxonomists have always treated *Zenkerella* as an aberrant anomalure.

That was reasonable enough so long as comparative anatomy was our main guide, but now molecular clocks have changed everything. Suddenly, *Zenkerella* reveals itself as a little-changed survivor, deriving directly from the very earliest group of rodents to make it into Africa after the Chicxulub meteor hit, having first evolved in Eurasia from just one of four pre-KT ancestral rodent groups. The *Zenkerella* niche may well have been narrowed by later-arriving dormice, which are smaller, faster and more active competitors. Later-arriving arboreal predators such as genets or mongooses might have found them easy prey.

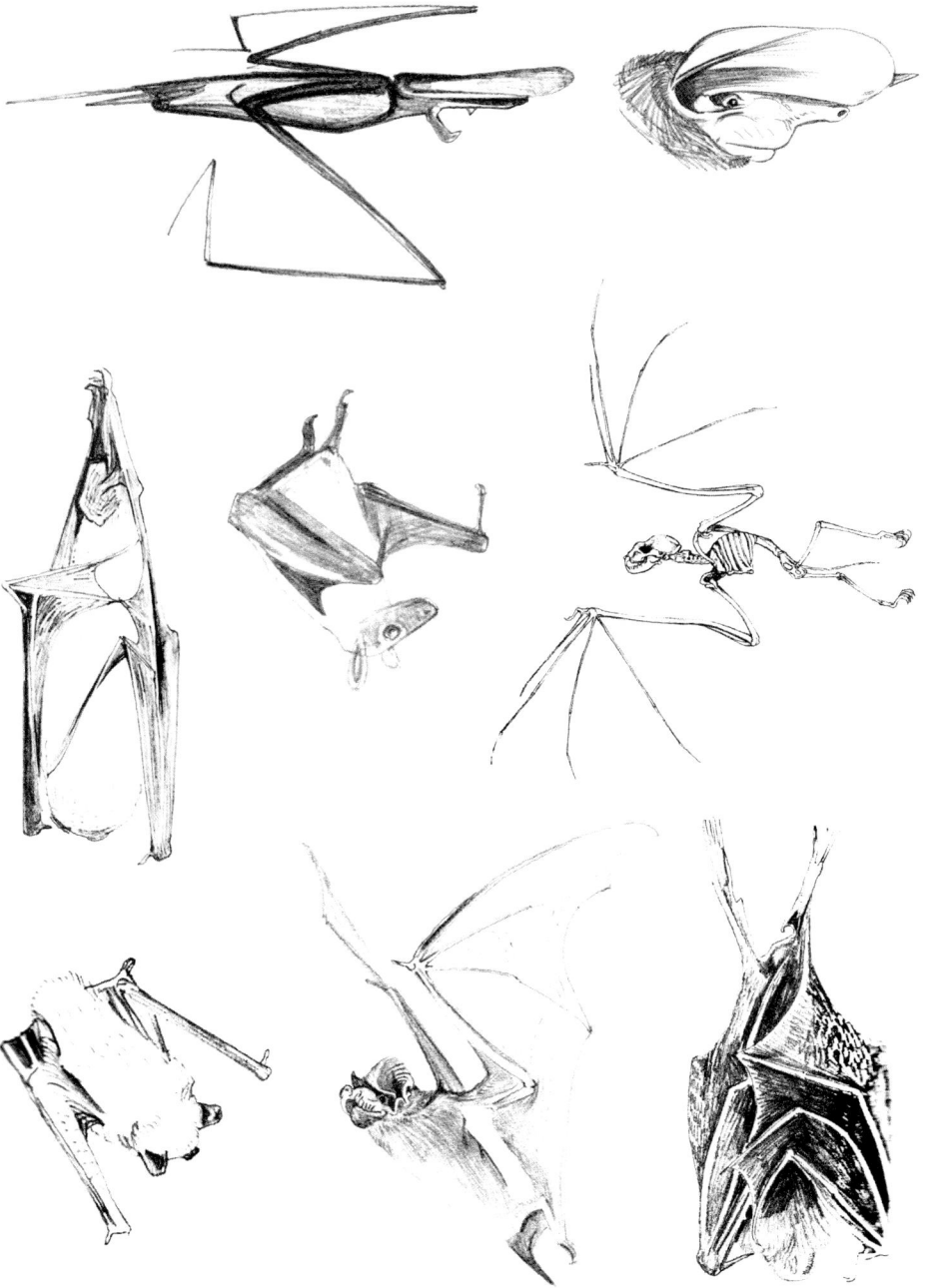

Exploring flight; a page of bat drawings.

Long after they diverged, the anomalures probably became gliders in wood-lands (as opposed to *Zenkerella*'s restriction to dense, closed-canopy rainfor-ests, typically habitats entangled by abundantly liaising lianas). Connections between spaced-out trees are not easily spanned so, if arboreal mammals are to travel between open woodland trees, they must descend to the ground to reach the next tree, inviting a host of hazards. It was this dilemma that faced the ancestor of anomalures, as well as that of the 'flying' lemur (in Asia) and, much earlier, the first gliding ancestors of bats. The dilemma reached its apogee in Australia, where open, liana-less woodlands have forced several independent lineages of arboreal possums to evolve gliding membranes. (An aside – as a child I named those butterflies that preferred gliding over flying, 'buttergliders', to my parents' amusement.)

From my desk in Bogor, Indonesia, I could watch gliding lizards launch their little splat-flattened bodies off one tree trunk to glide, fast and effortlessly, to some quite distant tree. There are gliding frogs and gliding snakes – all finding the same solution to a fundamental problem posed by spaced-out, liana-less trees.

The aces of all fliers are flies. When my friend and colleague, John Pringle, retired from our zoology department I was asked to devise a monument to his inspired leadership. As a leading researcher into the mechanics of insect flight, John showed how muscles attached to strategic points on flexible chitin walls provided the necessary purchases for flight. In our conversations he described clumsy equivalents in both row-boat rollocks and the mechanics of skiing. So my memorial took the form of a metal wall projection – a 3D construction

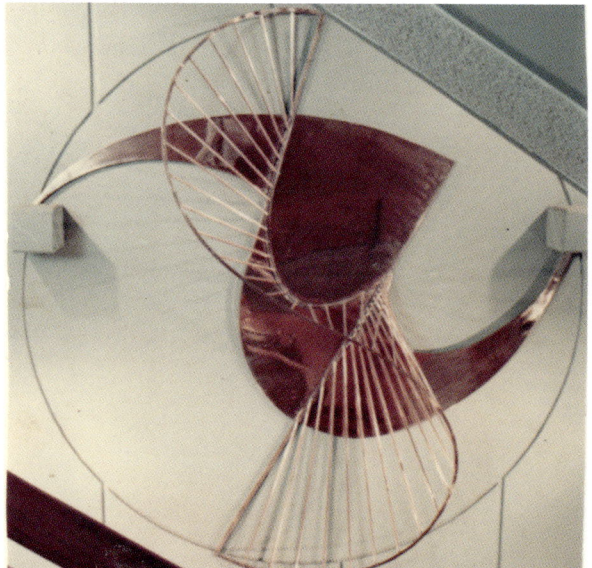

LEFT: Insect Flight, *a monument to John Pringle in Oxford Zoology Department.*
RIGHT: Aquatic Invertebrate Motion, *a bas relief for ADC headquarters building, Nairobi.*

modelled on Pringle's seminal research. It retained traces of an earlier relief where, in welded copper, I mimicked the undulating membrane of an aquatic invertebrate. My efforts ornament the previous page.

To return to our theme of immigrants and emigrants, antelopes offer a wide range of comings and goings. I grew up listening to nicely symphonic words because many Bantu languages use *paa*, *mpalla* and *swala* as root words for 'antelope', and in place-names for localities (such as Paraa, Pala, Palala and Kampala) where antelopes were once found in abundance. Probably an onomatopoeic sound-word for the antelopes' sudden exhalation that signifies alarm, this is the sort of word that agricultural Bantu most likely adopted from earlier hunter-gatherer lexicons.

The ancestral antelope was a single invasive species of small, heavily scented and scent-sensitive herbivore with a rather dainty short face, long, very slender legs and, among males, small, sharp horns and a willingness to use them in defence of one or more females or of a territory. Its main advantages were an ability to pick, choose and digest leaves, one by one (selecting the most nutritious from even the thorniest bushes with the sparsest foliage). It had very alert senses and, when needed, a fast get-away gait.

Its invasion of Africa and ability to cross a narrow, possibly very dry, bridge was also likely to have been helped by an exceptional tolerance of drought and heat. Arriving in Africa before 15 million years ago, it gave rise to today's 70 descendent species of African antelopes, a radiation that has culminated in a scarcely credible diversity of head-shapes. In every case, heads and skulls have been subordinate to the exigencies of male-on-male competition. Charging, wounding, throwing or wrestling have required differing battle techniques and a selection of head-wear that gun-men continue to steal and debase into boastful 'trophies'. I sometimes hear a mounted head whisper in my ear, 'Yup! I got clobbered by a cowboy.'

*Grysbok (*Raphicerus melanotis*), progeny of antelope pioneers.*

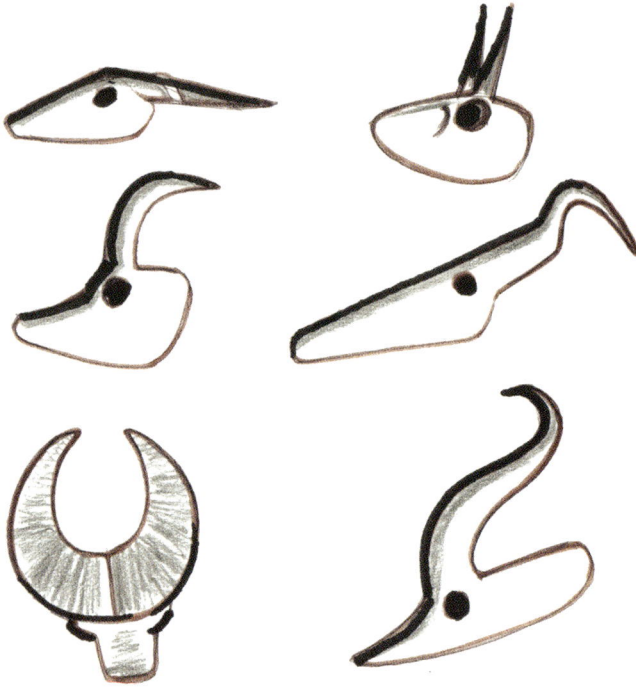

Outlines of antelope heads and horns.

One of the incoming antelopes' earliest divergences was to develop both a larger and a smaller form, in which females lacked horns while male foreheads developed robust, corrugated horns and reinforced skulls. The lightly modified descendants of those two early African antelopes survive to this day, as aromatic as ever. The dwarf one is called *paa* or *suni* in Swahili (Suni in English too). The Zanzibar Island provenance of the first specimen may account for its scientific name – *Nesotragus moschatus*. 'Smelly sea-nymph goat' could hardly be more insulting for such an elegant animal. Even so, having reared a Suni I can attest

Dik-dik (Madoqua sp.) head in bronze.

*1953 sketches of Ugogo Dikdik (*Madoqua (kirkii) thomasi*).*

to its pervasive musky smell (and to the ferocity of the male's defence of his female). Close Suni relatives, extravagantly crested dikdiks, with air-conditioner noses and facial glands bigger than their eyes, were other childhood pets, their emphatic, sculptured heads inviting many a sketch and sculpture.

The larger descendant of that primary antelope divergence is the Impala (*swala pala* in Swahili) and my quarry on many teenage hunts. It is another extraordinary survivor. Yet another antelope, the Gerenuk, has kept the tiny face and mouth parts of a dwarf while lengthening legs and neck to absurd yet elegant proportions.

Antelopes or *palas* were once distributed throughout every African habitat, from open desert to the densest forests. The livestock industry and their accomplices, intolerant of any competition, have made it their business to create the conditions that must exterminate antelopes – yet the fact remains that antelopes became the most efficient and most economic exploiters of almost every African vegetation type.

At least three habitats defeated antelopes. Hooves failed to become fins or spades, nor could they climb trees. Barring one (the Klipspringer), they shunned the great piles of granite boulders that erupt over so many of Africa's pre-Cambrian landscapes. These were habitats they were forced to leave to their afrothere predecessors – hyraxes, sea-cows, Aardvarks and giant elephants.

Before that modest little Eurasian antelope arrived, the mammal herbivore niche was shared between elephant and hyrax lineages and, briefly, some herbivorous proto-sengis. Some hyraxes resembled small rhinos (hyrax skulls and teeth have enough resemblance to those of horses and rhinos to have once suggested a direct relationship, but a closer look reveals a nice case of convergent evolution).

Our Ellis Island registry is crowded with entries from Eurasia. Rhinos came in about 20 million years ago, and equids (horses) also arrived, walking on ever-fewer toes in several staged invasions, of which zebras will feature later. Like-wise, pangolins, pigs and giraffes also came in tranches. Giraffes developed a variety of forms in both Africa and Eurasia. In their current determination to exterminate giraffes, you would think that humans have a special antipathy for the giraffe family. In fact the principal contemporary exterminators, directly or less so, are those involved in the livestock industry and charcoal burners.

Giraffes at Sunset.

Hedgehogs arrived more than 20 million years ago. Shrews have flourished, almost without measure, with a single genus (*Crocidura* – the white-toothed shrews) generating more than 100 African species. Hares are thought to have come in two waves.

The weasel family's arrival from northern Eurasia began with the invasion of an ancestral Ratel or Honey Badger, possibly as early as 11 million years ago. The fabled *nyegere* owes its onomatopoeic name to its rattling roar, from which even Lions have been seen to flee). A feral adventurer acquaintance of mine claimed that an angry *nyegere* once went for his privates. He thought a compassionate god had intervened when his shorts came off in the jaws of the *nyegere* who 'went off with them, roaring angrily as if to shake balls out of my pockets – what with him boasting the biggest balls in the animal kingdom I guess he thought I represented the competition'.

Hyenas, another group of carnivores of Eurasian origin, first entered Africa about 15 million years ago and they have made periodic crossings back and forth between continents ever since. Hyenas have, literally, iconic significance in my personal history and experience.

It was 1943 and my father's boss in Dodoma was District Commissioner Ralph Varian, a passionate hunter and trophy collector. Dodoma was close to patches of very dense bush and several large granite *kopjes* where there were communal Spotted Hyena dens. At night, hyenas visited the town and could sometimes be watched from the windows, trotting down the sandy street on disproportionately large but noiseless pads.

There was even a Leopard here that started to take people's dogs. Varian took on the task of trapping the Leopard with a tent-shaped tunnel made of stakes driven into the ground. Varian had two boys roughly my own age and the three of us watched, spellbound, as this crude trap was baited with meat and a treadle was carefully installed on the floor, connected by string to the trigger of a loaded rifle. This was mounted with its barrel pointing down and the trap was guarded by an *askari* until the streets were empty. This, we were told, was to avoid any repetition of an ancient but notorious incident in which a drunk had crawled into a similar trap and got a bullet through his head.

There was a loud bang during the night and I rushed round before breakfast to see what had been shot. The victim was a magnificent Striped Hyena, and Dorothy arrived with her sketchbook to do a drawing before it was skinned. She stared hard at this unfamiliar animal, trying to make sense of its shape, flamboyant crest and dramatic colouring. Then, with her pencils flickering over the paper she quickly sketched out a likeness of the dead animal just as it lay in the dust, then hurried away to avoid what was to follow.

Nothing could induce me to leave with her. I watched as other fingers, now holding a knife rather than a pencil, flickered over the prostrate form. As the

LEFT: *Striped Hyena* (Hyaena hyaena*); a quickie.* RIGHT: *Memories of a flaying – Striped Hyena.*

skin was coaxed and pulled away, a long-legged hound with huge muzzle and ears emerged as if from a hot bath, pink but very elegant in its lithe, spare physique. Its skin was to join the many other trophies collected by Ralph Varian.

A few nights after the death of the Striped Hyena, there was another bang and this time it was a Spotted Hyena that was shot. Again, I was among the fascinated crowd that watched the skinning and I well remember registering the very different proportions of the two hyena species. 'Stripey' had long, slender, neatly striped legs, a fox-like tail and a big head with a muzzle and neck as hefty as that of a mastiff. 'Spotty', instead, wore rounded, stain-like blotches, had a shorter black-blobbed tail and huge pads on the end of furrier, more muscular legs. His head was deeper and broader, and his face shorter.

The trap was then moved and some time later the original Leopard target was killed. Once more, I joined the Varian boys in watching the careful skinning and preparation of yet another pelt to join the Varian collection. The revelation of the Leopard's skull as ears, eye-lids, lips and nose were carefully peeled away with the rest of the skin was an object lesson in the layering of reality, cloaked in a subtle meld of camouflage and black and white contrasts.

The very peculiar experience of skinning trophies instilled in me a larger curiosity about what lies beneath all appearances. It has been a curiosity that can never be fully satisfied but one of its beginnings lay in an odd equivalence between my mother's pencil sketching and the skinner's knife.

Each of those peeled death masks stared fixedly into the sunlight, each wearing its own toothy carnivore grimace. Like countless other animals, their deaths were an inevitable price for their persistence while scruffy pioneering settlements took over their territories.

Memories of a flaying
– Spotted Hyena
*(*Crocuta crocuta*).*

Painted maw – Leopard (Panthera pardus) head.

For all the fascination that a corpse, a scientific specimen or a filmed biography offers, it is only physical, self-reproducing lives that have any real, sustained meaning.

It is one of the great wonders of our age that we now, for the first time in all of history, have the tools and the aptitudes to begin to understand the processes that have generated so many diverse living beings, including ourselves.

Animals and plants are a lot more mobile than we think. Many have come and gone over immense spans of time but the import and export of species is no left-over from ancient history. Countless birds leave Africa every year to feed and breed on the huge superfluities of northern summers. There, cuckoos announce that the Africans have arrived, with spring.

Now winged and faster than any swift, swallow or cuckoo, the ideas, techniques, actions and diseases of humans, for good or ill, will continue to flow in and out of Africa. African voices will be and are being heard far from African shores – and vice versa.

The voice of pan-Africa, South African activist and singer-songwriter, Miriam Zenzile Makeba Qgwasha Nguvamo.

Mural in North Carolina Zoological Gardens.

DIVERSITY – HIGHS AND LOWS

In which the fortunes of plants and animals are sketched, and the hubris of agricultural monocultures is fingered.

Biodiversity is a word invented for a conference. Ed Wilson wanted to wake people up. The sound of biodiversity is the song of every bird, the sonic boom of every whale, the vibration of every insect, feather or vocal cord that ever evolved, the sound of every coming, of every going.

The smell of biodiversity wafts from every flower, every gland and, yes, every female mammal as she comes into season, and the scent of her newborn baby. It is also the miasma of every corpse, potential punishment stored in every skunk's or Zorilla's anus.

The shape of biodiversity is manifest in the architecture of a giraffe or empires of underground tree roots, in the geometry of diatoms and the symmetry of spiders, deep-sea fish and peacocks' tails.

The colour of biodiversity flashes from the sides of every courting bird or flirting female, from the petals of every plant and the changing tints of chameleons as they ponder their next moves. Antlers on the noses of these combative lizards come in more inventive shapes than do medieval military helmets or ninja mutant turtles.

Chameleon geometry.

Bronze-winged Courser masquerade.

Living beings have untangled and reassembled light itself. Sunbirds, frenetic at midday, throw back restructured reflections of sunlight in all manner of ostentatious feathering. Moonlight off the metallic tips of Bronze-winged Courser wings divert Gnu and Topi traffic around the birds' precious clutch of eggs as surely as a cop's illuminated batons at a midnight road-block. The same feather-tips can fake-flutter in the nose of a foraging jackal, to distract from newly hatched chicks as they crouch flat in the burnt stubble.

That iridescent reflection off the courser's wing must also catch the naturalist's eye, because such split-second flashes reveal senses far more acutely tuned than our own, slow, lazy eyes are used to. We miss the sexy eye-caress reflecting off the coronet of a Scarlet-chested Sunbird's forehead, we fail to startle at the momentary fanning of a Namaqua Dove's heraldic tail, nor can we begin to register the extraordinary permutations of geometry splashed across many a bird's flashing wings.

The flash of a Namaqua Dove's tail.

Fruit bats (Pteropodidae), champion forest pollinators and dispersers.

The carmine coruscation of a turaco's wing features elsewhere in these essays, but it too sends multiple messages. One of these is the authority of the moment in which the viewer's eyes are dazzled by that blast of refracted light. Turaco subsistence routines, their endlessly repeated, bounding ascents up the spiral staircases of tropical trees, create the moments in which the flasher must judge his flash exactly in relation to both his female target and the sun behind him. All three, sun, actor and target, must converge to bring about that moment during which she becomes susceptible, open to that moment of ecstasy or paralysis in which their separate gametes unite.

Very different signals govern animal existence at night, but most are just as fast and even more hidden from us. Because we are such very diurnal animals, we sleep while nearly 200 African bat species and billions of individuals are awake, pursuing their starlit lives. Many millions of fruit bats pollinate, fertilise and disperse unknown numbers of the trees that loggers are chainsawing into oblivion.

The naturalist's attention must be alert to such split-second signals, because they reveal the role of senses much more highly tuned than our own. Every hound remonstrates at our insensitivity to the fragrant subtleties of urine, but we are less willing to acknowledge our shortcomings in vision, hearing and touch.

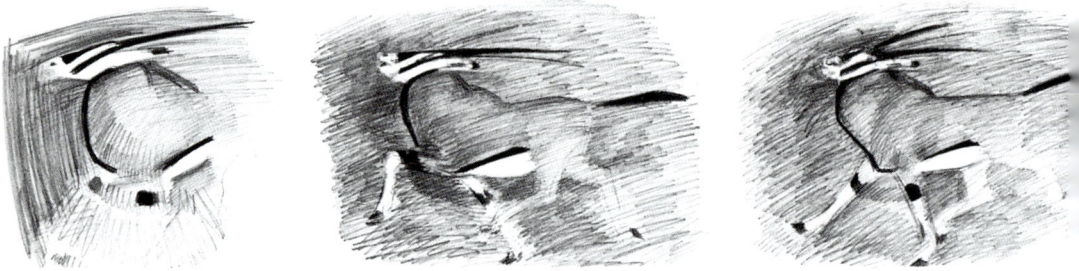

Beisa Oryx in fast 'tournament' gait.

I remember watching young oryx as they romped around the perimeters of their herd, breaking into a fast hackney gait in which heads, above bunched necks, swung from side to side in synchrony with their fast, circling gait. I think it possible that such hurtling antelopes are 'taking a fix' of their surroundings. We know that individual oryx migrate between highly localised pastures in circuits that are maintained over decades. How else do yearlings learn to map their vast, arid environment into such geographically sophisticated brains?

For lazy human eyes, the speed and complexity of natural life is baffling, but missing the action has no consequences. For wild actors, he (or she) who seizes the moment wins. Both his love-life and the survival of his kind hangs on the acuity of his senses as well as the lust aroused by the flaunting of his fitness.

Mandrill sketches.

Life on Earth, even life within a single hectare of African forest, can assume shapes way beyond the wildest of human imaginings. No science fiction writer could out-swagger the rainbow of a mandrill's swanking arse, nor outdo the outrageous, sculpted, painted snarl that passes for a mandrill face. A flock of Piapiacs or a covey of Crested Guineafowl have vocabularies, complexities of syncopation, burdens of meaning to match a human composer's range of composition, a sensitivity to pitch that would please Art Tatum.

Frogs and toads have the mutability of plasticine or rubber – evolution has modelled their instruments of sound into countless experiments, every one a different conjugation of vocal cords, amplified by playgrounds of ballooning throats, necks, cheeks, chins and flanks, all primed to pop.

There are hot-houses bursting with diversity, but also ice-houses and dark cellars where life is hard to find. It is lost with the pressure of a digit on a trigger, in the minutes of a meeting in a bank or, as 'collateral damage' in a war among bipedal primates.

Mara estuary (coloured sketch by Dorothy Kingdon, 1938).

The vignette of diversity that follows is taken from a 1938 letter Dorothy sent to my nana, Bibi Minnie, at the end of August, following the first rains after two very dry years. 'Teddy drove us to Mwanza across the Serengeti plains. The first part of our journey was very beautiful in the early sunlight, all along the Mara estuary, long shadows, a white sheet of water deepening to an indescribable colour and the blue-grey hills of Tarime mounting into a mother-of-pearl sky. We reached a painted notice announcing the "Serengeti Game Reserve", which was surrounded by baboons. Jonathan had just woken up, all fresh and lively: you can imagine his delight when we stopped the car and watched the "muckies" as he calls them, and his imitations were very comic.

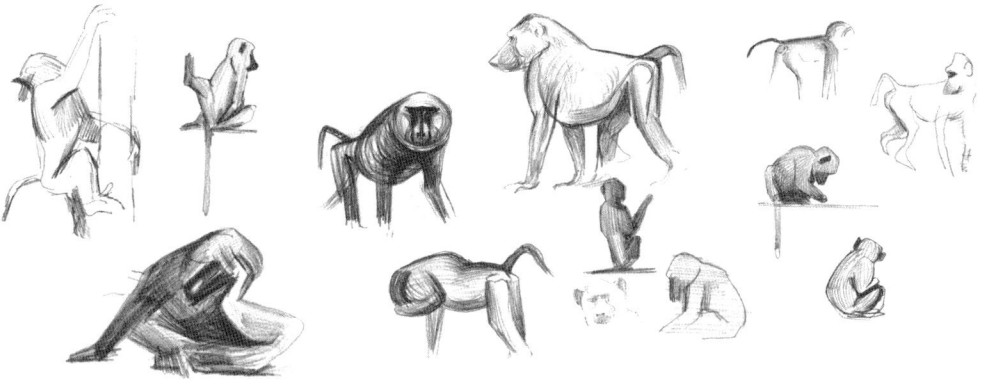

Baboon sketches.

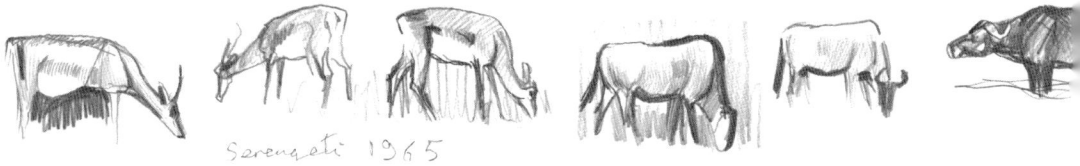

Serengeti 1965

Antelope sketches.

The plain was teeming with game of all sorts. It is a great open plain dotted with thorn trees, the ground absolutely flat, so we left the road and charged about all over the place. Great herds of zebra would stand and watch until you were within 30 yards, then they would trot off to a safer distance, turn around and watch again. Everywhere were the little golden Thomson's Gazelles with their delicate horns, eternally flicking short black tails as they scampered away or stood and watched for our next movement. Topi there were, with their heavy forequarters and rather nude-looking behinds, and enormous herds of gnus. Further on we saw hundreds of Grant's gazelles which are bigger than the Tommies but much the same colour and have very beautiful horns. A herd of impala leaping over low bush is pure poetry brought to life. Once seen, you could never forget. Sometimes we would see a black blob in the distance. This would prove to be an ostrich. [Dorothy always thought of these birds as generically female.] She kept pace easily with her colossal strides while the old car panted along flat out – or as flat out as the rather rough ground would allow. Jonathan was in a tremendous state of excitement, imitating all the animals, galloping, bouncing and nearly cracking his head, shouting and making a noise like a horse every time we saw a zebra [I guess I wanted them all to reply!] It was all the most joyful adventure. Shadows under the thorn trees were animals resting in the shade. Beautiful creatures grazing at their leisure. Everything was peaceful in the midday sun.'

ABOVE: *Jonathan and friend Hazel on Serengeti wagon, 1939.* RIGHT: *Ostriches.*

Our journey took place just after two years of drought, and very large numbers of animals had died or were finally succumbing to prolonged starvation, providing easy pickings for abundant meat-eaters of all sorts. Dorothy, writing to her mother, then turned to clouds of scavengers swirling around on outstretched wings, drawn in by fresh carcasses strewn over the plain. Here her words echoed Darwin's 'clumsy, wasteful, blundering, low and horridly cruel works of nature'. But Dorothy had been raised to see the struggle for existence in moral terms, the 'lovely ones' being preyed upon by the 'cruel ones'. Her letter had described a scene outside anything either mother or daughter had previously experienced, in spite of Minnie being a farmer's daughter, accustomed to raising animals for slaughter; calves with doe eyes, ewes that bleated for lost lambs.

Minnie's farmer forebears had also been domesticated, by the clergy, beholden to County grandees, in liege to kings and the countries they had conquered or inherited. For Minnie, animals, minds and concepts lived in pens. Whenever undomesticated words tried to jump into a conversation, single, seemingly innocuous words like 'Darwin', she would shut the gate with a pursed mouth, change the subject and ask, with a touch of asperity in her voice: 'Would you like another cup of tea?'

In her letter to Minnie, Dorothy claimed to have lost sleep after our drive over the plains. In Europe, death was commonly consigned to paintings, funerals, distant trenches, war memorials and slaughterhouses.

In Tanganyika living, dying and copulating were undisguised, undomesticated, in your face. During Dorothy's upbringing, the major intellectual framework equipped to comprehend all aspects of Nature and its fecund 'biodiversity' had been doused with a cup of tea. So the daughter, to her mother, wrote, 'I hated the Serengeti for its savagery.'

To almost everyone else she described our trip as a great, wonderful thrill – she never did get to confront her own mercurial inconsistencies.

It was just such experiences, seen through the amoral eyes of near-infancy, that led me, blink by blink, towards what might, today, be called a Darwinian vision of being and meaning, of existence bedded in deep time. Assent at being a single evolved member of a small planet's 'biodiversity'.

Well before our Serengeti trip, I had repeatedly hopped around and gazed, enthralled, over columns of army ants as they disbanded and spread out over vestigial flower-beds and shrubberies, even the floors, walls and thatch of a house, spreading terror amongst every small living thing they encountered. I watched as platoons of pincer-armed, eyeless, deaf soldiers converged on prey, cutting up a squirming caterpillar, slicing up a living cricket into portable provisions, wounding, ever more lethally, a newly hatched lizard that thrashed its life away before my eyes. Those blind *siafu* were much worse than anything out on the Serengeti plains. Back then, in pre-war East Africa, none of us could imagine that *siafu*, wild dogs, crocodiles, the fiercest forces of Nature, were about to be overtaken by the knowing ferocity of Nationalism in all its cultivated malevolence. Way up north, in the cold, ploughshares were being melted into cannons and, later, so we were told, human fat made into soap.

After 'Hitler's War' was over I was sent, for schooling, to a Europe ruined by civil war. It was not long before I felt that it was some of my kind relatives, my school-fellows, and the demobbed, ex-civil-war Europeans that most needed schooling. In a spirit of teenage self-assertion I put together a sort of lavishly illustrated, hardback 'Birder's Diary' of my 1951 trip home. It is only with hindsight that this schoolboy initiative could find any relevance for my documentation, many years later, of the mammalian biodiversity of East Africa (quite a sizeable chunk of our continent) – its subtitle was *An Atlas of Evolution in Africa*.

A literally central component of Africa's biodiversity surges out of what, in another of my books (*Island Africa*), I called an 'evolutionary whirlpool' – the Congo basin.

One intensely biodiverse sample of Congo biota overflows the margins of that whirlpool along the eastern banks of the river Semliki (the waters of which end up in the Nile). I spent a decade making frequent visits there, most often in the company of a rambunctious cantor Mbuti (pygmy) called Tumbo. Knowing that I was a teacher and that I respected Mbuti knowledge (and especially their wildly creative, heart-warmingly beautiful music), he trusted me to share his own hunter's wisdom with my students and readers. When sober, Tumbo took my Atlas enterprise very seriously and he helped me capture one of Semliki's most elusive of inhabitants, the Water Chevrotain. It was the first authenticated record of the animal from eastern Africa and I remember how blithely Tumbo waded along beside me down a shallow tributary of the great Semliki, while several fellow Mbuti and their dogs tried to drive a chevrotain over the bank where they reckoned I might get a quick pot-shot before the animal submerged. Tumbo reassured me that he would give me plenty of warning if a crocodile came after us!

LEFT: *Chevrotain.* RIGHT: *African Forest Buffalo.*

He was with me once when we almost collided with an apparently pregnant, richly russet African Forest Buffalo. Tumbo was disgusted that I refused to shoot her.

Having made numerous safaris across the Semliki to meet up with relatives in the Congo, Tumbo was well acquainted with the Okapi, an exquisitely beautiful, shorter forest relative of the giraffe. Tumbo and the elders of his village remembered when Okapis, as well as the ox-like Bongo, Bay Duikers, red colobus monkeys and Giant Forest Hogs ('all delicious'), lived on the Uganda bank of Semliki. Even then, the elimination of biodiversity had begun in populous Uganda, while the ancient constraints on human numbers – disease and internecine warfare – still operated in the Congo.

Semliki was a thrilling locality for any naturalist. Below a fruiting giant fig you could be sure of seeing all the forest hornbills, barbets, turacos, chimpanzees, several species of monkey and, perhaps, the drama of a Crowned Eagle swooping in to snatch an unwary young mangabey. In Bwamba-Semliki my entomologist colleagues at Makerere found themselves in an arthropod heaven but, yes, the horseflies were painful and the mosquitoes a serious hazard.

Part of this extraordinary concentration of equatorial species was due to a small zone of overlap between mountain and lowland biota, around some hot springs at Mongiro. Here, within a radius of 10 km or so, was the greatest concentration of primates in the world. I listed sixteen species. Enlarge that radius to embrace both banks of Semliki and the northern pinnacles of Rwenzori and you would get close to the greatest measure of terrestrial biodiversity on planet Earth.

Biodiversity may actually be greatest not on land but on reefs, along the long shores of Earth's continents and around islands. I am not at ease in water, having been born and bred very far from the ocean, but scuba and diving technology

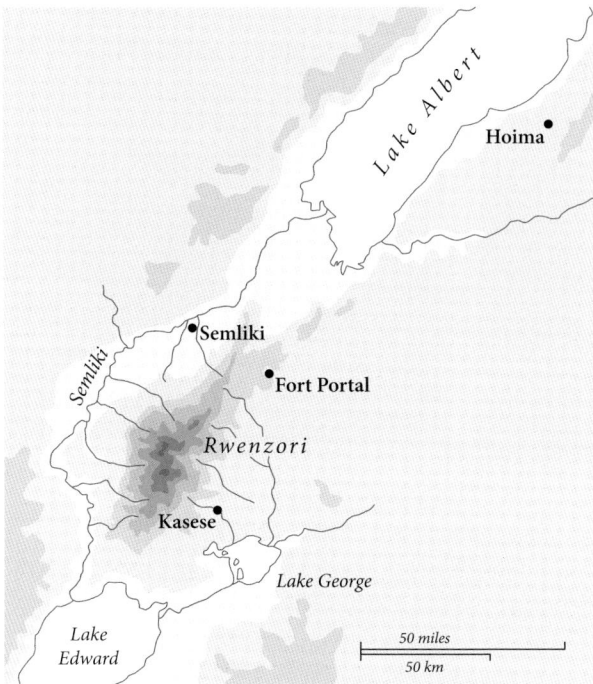

The Semliki River and its drainage lands, a cauldron of biodiversity.

are now allowing us to explore the deeps of our planet. As a keen surface goggler I have relished the visual, scientific and creative challenge presented by reef fish. My amateur enthusiasm for sea life has developed just as our most piscivorous and voracetaceous (my word for whale-whacking) nations deploy ever more sophisticated technology to hoover up the seas' more edible bounty, smashing up the sea-floor and exterminating albatrosses as they go, giving more than a bad name to all commercial, large-scale versions of hunter-gathering.

I remember finding a broken-backed dolphin, dead on Kunduchi beach, and, gazing into its open eye, seeing there my own miniature reflection. That reflected image came alive some years later, when an aquatic mammal made a very memorable impression on me.

Whales make annual migrations up both sides of Africa and that blustery day off the Kwazulu-Natal coast a nursing Humpback made sure she kept herself between our boat and her calf. Another whale-watching trip allowed me to make momentary eye contact with an animal many times larger than any elephant.

Maybe some 15 m long, certainly longer than the little boat we were in, what I was told was a 'youngish' Humpback Whale broke away from its pod to investigate us. We felt the boat lift as the whale plunged cross-wise under our keel, then it turned, so broad-side that boat and whale were almost (but not quite) scraping each other's flanks. I was hanging over the gunwales as the Humpback's head drifted by. Both our eyeballs swivelled in their sockets as whale and human tried to maintain eye-lock for the few seconds of our encounter. Then the whale

Breaching Whale *(linocut by Afra Kingdon).*

turned and allowed its giant snout to point straight at the three humans in their tiny boat. The captain leaned out to try to touch that enormous upper lip, but I had the distinct impression that the whale backed off enough centimetres to avoid actual physical contact. However, there was no doubt about its curiosity, nor about its awareness of space all around it – down to the level of centimetres.

I knew that sound maps a whale's ocean spaces and co-ordinates its actions, both social and strategic, but here two sets of very different eyes had clearly retained their ambition to read meaning into that passing stare. I sometimes wonder whether that whale still remembers our momentary meeting, but why should I be any more interesting to her than a barnacle?

Of course, she was a mere midget compared with a Blue Whale (at 30 m long and up to 200 tons, the largest animal ever to have inhabited our planet). Plants may also grow as huge but no other animals, not even dinosaurs, have ever been bigger than today's Blue Whales.

In the early 1970s I was often in Nairobi, witness to the battling and bargaining that went on just before the launch of the 'International Convention on Biological Diversity'. Making my own views known to one of the drafters of the convention, I soon learned that agriculture's lobbies were calling the shots. To put their position into perspective, there had been very real fears of widespread starvation in the post-war years, and the green revolution had put farmers into driving seats that they were unwilling to share (let alone exchange or relinquish). In short, the Convention looked like trespassing on their newly global territory – they

had even commissioned maps in which the entire world was parcelled up into areas suited to one or other, sometimes several, crops. Agricultural organisations insisted upon many changes to the original document of intent, the most fundamental being presentation of every natural habitat as potential farmland.

The mirage of 'Reclamation' presents a world being returned to its rightful owners – farmers. The document that emerged was not permitted to escape this Neolithic perspective but it was the first international acknowledgement that ecology and natural communities had any kind of global value beyond the economic.

In spite of its very watered-down nature, all but the least enlightened or captive of nations signed up for the convention. The motives of those that refused were significant. Religious bodies cited some Neolithic god (who made man after his own likeness, or was it vice versa?) and dispensed dominion over every living thing. Only if their man-like God's authority was invoked would the Vatican's Holy See sign.

In Somalia, the only African country to decline signing, self-appointed generals give orders, never take them.

As for the United States of America, business lobbies, like primitive Wild West settlers dedicated to deterring *any* constraints on *any* of their activities, pressured their government to refuse even token support for biological diversity. So exceptional was the USA that this cauldron of industry and innovation, this once slave-dependent libertine, was the only major country not to sign the convention.

Those were Cold War days and there were ample examples of agricultural 'experts' creating niches for themselves among competing 'donors of overseas development aid' programmes. Most had their eyes on raw African products that could be exchanged for industrial surpluses, generated in their own faraway lands.

OPPOSITE: The Last Quagga. ABOVE: *Giraffe head as knobkerrie – wooden club (bronze).*

This was not my first encounter with the enemies of biodiversity, nor with their opponents, and I found myself modelling an epitaph for the death of biodiversity as 'The Last Quagga' and 'Memoir for a Giant Hog'. I tried to celebrate lost presences in paintings entitled 'The Elephant's Shadow' or 'Chyulu, Rhino Cemetery' (see pp. 197 and 399).

Less vincible, Woodland Kingfishers announce early morning matins with explosions of sound shrilled from scarlet beaks, beneath flashing wing spreads of blue, black and white. I enlisted this familiar expression of diversity into a stained-glass window I designed for a small chapel in Tanzania (see p. 175).

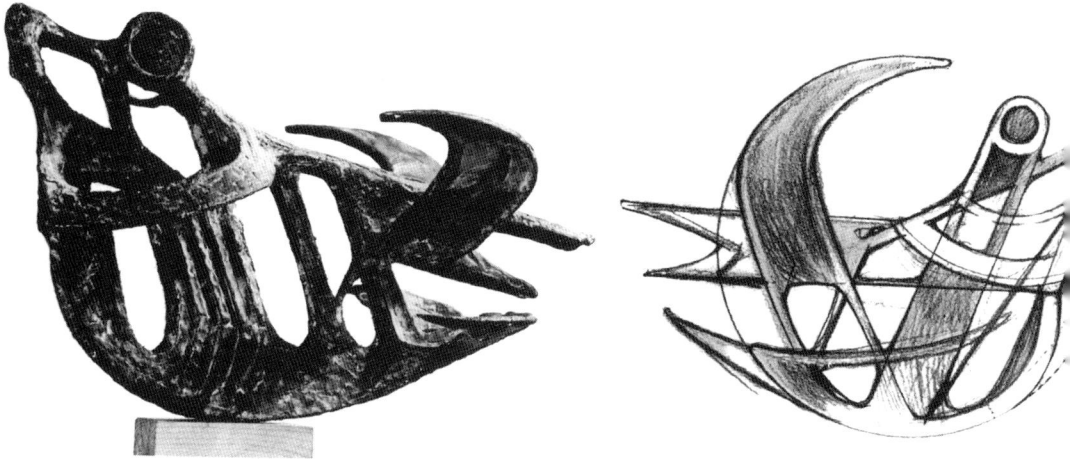

Memoir for a Giant Hog *(bronze and drawing)*.

Returning to biodiversity's need for warriors against the ravages of monocultures and money-men, one singular eccentric was Game Ranger 'Iodine' Ionides. Taller than any stork, Ionides invited his Swahili title, *Mzee Karongo*, 'Old man Stork', from his ground-searching gaze and long, deliberate pace. He had round, raptor eyes behind a nose that was more Imperial Eagle than Marabou Stork. Whether his famous predilection for capturing snakes and other reptiles was driven by raptor-like predatory instincts or the needs of herpetological science was a distinction that meant nothing to him, nor to a great many fellow naturalists and collectors of his generation.

When 'Iodine' stalked into Teddy's office in Utete in early 1947, the reputations of both men had preceded them and both were predisposed to like and respect one another. Some years later Ionides gave me copies of his field notes, all laboriously transcribed by his own neat hand. His notes and records included detailed observations on mammals, great and small, from all over the then largely unexplored but ecologically important southern provinces of Tanganyika. When we met for the first time he had greeted me with 'Oh! Yes, you're Teddy's boy, aren't you?'.

At the time they met in Utete, Ionides had put himself at the centre of a fierce dispute that went to the roots of both Imperial and Neolithic hubris. Local leadership of dependent territories had always been decided in Whitehall. Grey men umbrella-ed in the Colonial Office had (perhaps foolishly, more likely strategically) promoted a sulky career official in the Dar es Salaam Secretariat, Sir William Battershill, to become Governor of Tanganyika. Known locally as 'Battered Bill' and more politely described as 'liverish', Sir William owed his rise to a metropolitan patronage system that was abhorred by the new British Labour government. A serious shortage of vegetable fats during Hitler's blockade of the British Isles led careerist chameleon John Strachey, Minister of Food, to launch a 'groundnut scheme'. He endowed the British Government's Overseas Food Corporation (OFC), a new state conglomerate with a multi-million budget, to grow peanuts in Tanganyika.

At this point, allow a short digression. The groundnut (or peanut) *Arachis hypogaea* is native to South America and its domestication in the Gran Chaco area probably began more than 4,000 years ago. These nuts came to symbolise farmers' birth from and return to the soil (providing a tasty staple food in between). Groundnut facsimiles, meticulously observed, sometimes greatly enlarged, were cast in solid gold, nutty glorifications of the kings of ancient Peru.

Groundnuts were elevated in 1947 to become the solid gold symbol of monoculture's rescue of humanity from famine. In spite of its spectacular failure, the groundnut scheme did bring ephemeral wealth on a scale we East Africans could scarcely imagine. It also brought a helpful transition to a new sort of civilian life for many thousands of recently demobbed African soldiers (some soon to become executives of national independence) and about 6,000 Anglo equivalents, soon to be known as 'Nutters'.

For three post-war years, Tanganyika was dominated by a forerunner of what is now called the 'development industry'. Such programmes are still funded and justified by the appetites of big, non-African countries for our resources.

With the itchy trigger fingers of recent soldiers, those OFC Nutters were soon shooting some of the several thousand elephants that inhabited the Mbemkuru valley headwaters. This was part of *Karongo* Ionides's range and one of southern Tanganyika's principal ecosystem reservoirs for indigenous fauna and flora. Ionides, outraged at this invasive and illegal slaughter, wrote a letter of protest to the *Times* of London. His letter precipitated a determined protest from many conservation-minded people, including vocal members of the House of Lords. This enraged London's then mainly Marxist ministers, and the head of OFC demanded Ionides's immediate dismissal.

Nothing could have mobilised local opinion better than an incompetent Londoner versus an experienced local. Ionides was a popular figure in spite of (or perhaps because of) his many eccentricities. A few short years later Strachey and his minions were in disgrace while tough old Ionides survived in his chosen job as Game Ranger of Tanganyika's Southern Province.

Ionides and the Nutters were in contention over a stretch of little-known habitat that supported several thousand elephants and a broad range of *miombo* woodland fauna and flora. Ionides had covered much of the area on foot, gathering original information on the ecology of animals both large and small. For the Nutters, this area was Block B, an unvisited patch on a wall map in a London office. All of the elephants in Block B, many thousands of them, were to be shot and the woodlands felled to make way for a giant Soviet-style mechanised state peanut farm, projected to be a model for many more to follow.

Stalin's model for feeding a hungry world only began to wither away with his death in 1953, but since then 'Food Inc.' has taken over a much more pervasive Stalinist industrialisation of agriculture.

At this time Ionides was a major contributor to the first ever checklist and gazetteer of mammals in Tanganyika, published by the East African Natural History Society, a non-government, non-business association of mostly amateur naturalists. A few years later still, Ionides provided much vital information for my *Atlas of Evolution in Africa*.

Every effort to reveal Africa's wealth in terms of 'wild' (that is, pre-adapted) animals or plants has been fiercely opposed by any and every individual or organisation with a stake in foreign domesticated plants or animals, and the gilded techniques that have proliferated around their husbandry.

Arrogant, ungovernable young warriors are often called 'cowboys'. In Africa, 'cowboy cultures' have not only proliferated, they dominate the politics and armies of many countries and hold sway over vast areas. Consciously or unknowingly, cowboy cultures have effectively declared war upon the natural communities of Africa.

A livestock official once condescended to put it this way: 'Proposals that favour indigenous biota over livestock muddy the water for us. Just get used to the fact that most African animals must go extinct, like anywhere else – that's progress!' Significantly, that elderly official was raised on a South American beef lot with churned mud underfoot and hoofed-up dust for air, his ambitions measured in gold peanuts and blutwurst – his career seemed to be dedicated to converting our continent into a poor copy of Uruguay.

In any event, livestock everywhere is beginning to assume a much-diminished role as beef, milk and leather get replaced with less damaging alternatives. Replacing nearly 100 species of hardy large herbivores and entire ecosystems for four or five convalescent exotics (that are treated as anonymous bags of walking cash) is not just a bad bargain, it's a pathway to misery, it's stupid and it is a meteor hit on the biodiversity that our better-informed descendants will value way beyond obsolescent meat, milk and leather.

I trust that old men, their vistas of bovine buttocks clouded with dust, their ankles in mud, will become no match for our young, the stewards of the richest and most diverse continent in the world, looking out onto horizons of unmatched diversity.

Stained-glass window, Rondo Seminary, south Tanzania.

In the late 1980s I found some satisfaction in persuading two young expeditioners of the importance of East Africa's coasts and 'eastern arc' mountains as cauldrons of evolution and super-centres for ecological diversity, stressing how urgent was the need for systematic research. Endorsed by Professors J. Kapuya, A. Semesi, K. Murira, K. Howell and Scandinavian Aid agencies, their Society for Environmental Exploration founded a research programme called 'Frontier Tanzania', which has gone on to become one of Africa's most significant multi-partner scientific research enterprises, drawing in the participation of many of our young people. It could and should be a model for research, education and collaboration in other regions of Africa.

All over Africa, primary, secondary and tertiary education should introduce our youth to the fact that our continent is the richest in the world. Common or rare, ancient or new, birds, antelopes, insects, spiders or plants – all combine to confirm that Africa has been a durable, continuous and patchily fertile habitat for living things, a place where every kind of organism, including human ancestors and their cousins, has existed and evolved within a mosaic of ecological islands. But for how long?

If thoughts raised in this essay have any virtue, it will be in further exposing the barbarism and magnifying the offence of stripping Africa and the world of its diversity – its biodiversity.

Tropical forest, the most dynamic and ever more complex of terrestrial habitats.

SUCCESSION

In which one elephant displaces another and a scavenging ape gets to inherit the world. Mysteries in other minds. Can joie-de-vivre be shared? Lies and lessons from rinderpest. 'Our own most ruthless cullers.'

The water dribbled out from cracks and crannies in a sandstone cliff that had been deeply carved. The carvers had left unambiguous signatures, in spite of their technique being little more than patient abrasion by some relatively soft agent, like the fingers of some obsessive stonemason rubbing away with over-used sandpaper. Here, at a rare spring close to Indian Ocean shores, rubbings by itchy flanks over unknown millennia had etched elephantine outlines onto the travertine cliff-face. More extraordinary, the exploratory probing of trunk tips, apparently into a minor crevice, had slowly abraded a long, neatly tapering, perfectly trunk-shaped tube from which water still issued.

ABOVE: *Elephants below Rabongo forest, Uganda, (1960).*

LEFT: *Mbondei beside elephant watering hole, Matumbi – rocks abraded over millennia.*

On the slopes of Mount Elgon there are a dozen or more long, deep caves like this, largely excavated by many, many generations of elephants in their quest for salt and other minerals. Here the miners' tools were mainly tusks, but imagine how many generations of deep-mining excavators it took to gouge soil away from what probably began as salt-licks dug into a salt-laden bank.

Elephants shape landscapes and history in many more ways than mining for minerals, hosing down water or scratching an elephantine itch on a path-side tree-trunk. Nor do you have to go that deep into the past to appreciate the impact of numbers (in this case of elephants) over time.

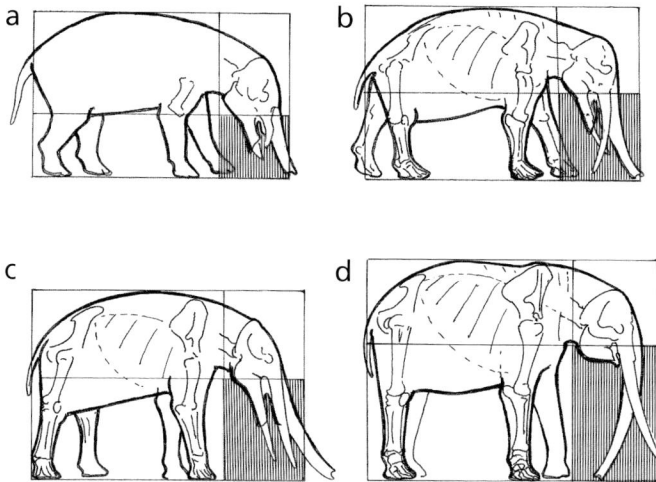

Evolutionary changes
in the proportions of
elephants (Proboscidea)
in which the level of the
mouth rises and the
trunk lengthens:
a) reconstruction of an
ancestral proboscid
Phiomia;
b) Gomphotherium;
c) Stegotetrabelodon;
d) African elephant
(Loxodonta).

I am among the few who could once look out over grasslands studded to the horizon with thousands of elephants, fragmented into herds, some of which held hundreds of individuals. Throughout my youth, both people and elephants were increasing at comparable rates. Both needed food, water and space, and often competed for the same food, water and space. As an adjunct to his duties, my father, Teddy, like my great-uncle Hal, had to shoot elephants in defence of peasants' fields and livelihoods. As a youth, I was persuaded to do likewise, acts that are now, in one mood, unconscionable, but in another are a reflection of the imperatives, even the indifferent inevitabilities of my time, our time.

Beginning, perhaps 70 to 80 million years ago, certainly by 65 million years ago, African elephant ancestors, the first proboscids, had a long headstart as hulky and ultimately the hugest of extant mammalian herbivores. In the interim giant giraffes, buffaloes, hippos, pigs and rhinoceroses have come and gone, while the elephant lineage has gone through multiple transformations, most revealing a steady increase in size.

This trend culminated in giant *Palaeoloxodon recki* (formerly *Elephas recki*; let me call this the 'African Grass-elephant'). In southern Asia *Elephas maximus* (today's much smaller Indian Elephant) was the survivor while mammoths, *Mammutus*, dominated the far north of the Americas and, in Asia, survived into historic time. Like several mammal groups that have had to contend with frequent periods of drought or cold in a relatively dry continent, more than one type of elephant species made a prolonged, steady adaptive shift away from hot, wet, marshy environments toward drier, more open and grassy habitats.

In Eurasia it could be said that the elephant niche first fell to 'Perissos', the rhinoceros-tapir-horse lineage. Among them were the largest land mammals ever, the indrics, *Paraceratherium*, but they never made it here.

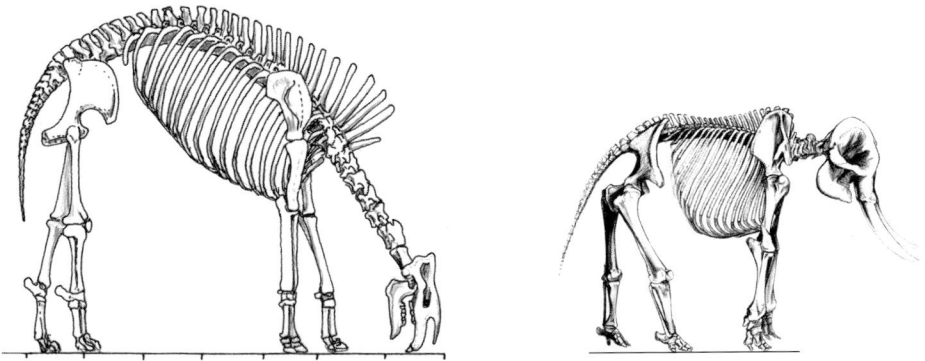

Paraceratherium skeleton (left) compared with that of a modern elephant (right).

The very idea that modern elephants were once dwarfed by a mammal that was taller than a giraffe and two or three times the weight of a modern elephant may seem preposterous. Well, some *Paraceratherium* species stood, at the shoulder, about 2 m taller than a modern elephant, and members of this Eurasian lineage survived up to some 11 million years ago.

One highly aberrant offshoot of horses entered Africa about 20 million years ago. Chalicotheres are not common as fossils but their remains are very widespread and they seem to have been associated with warm, wet forests, often close to volcanic areas. Their bones crowd shelves in the research labs of Nairobi's National Museum. In Eurasia, chalicotheres radiated into more than a dozen genera and survived until about a million years ago. Their astonishing anatomy suggests that they exploited a niche that no living animal exploits today. Their teeth betray a herbivorous diet, while powerful claws and muscular forelimbs prove that they were habitual diggers. This combination suggests exploitation of a gigantic resource that is still exploited on a massive scale, but by insects and relatively small rodents – roots.

I like to imagine these huge animals crouched over their excavations in soft, wet soil, grubbing up and swallowing down the tubers, bulbs, corms and storage roots of all sorts of plants, as if they were huge, animated potato- or carrot-harvesting machines.

Their skulls were unambiguously horsey but, like the earliest elephants, they clearly evolved from tapir-like ancestors, remarkably similar to the earliest elephant ancestors. As both lineages evolved, they developed somewhat similar enlarged skeletons, but their skulls diverged hugely.

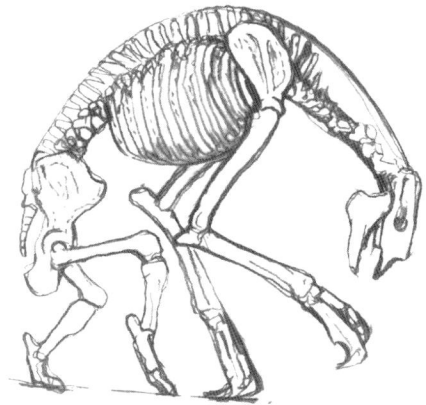

Skeleton of a chalicothere in digging posture.

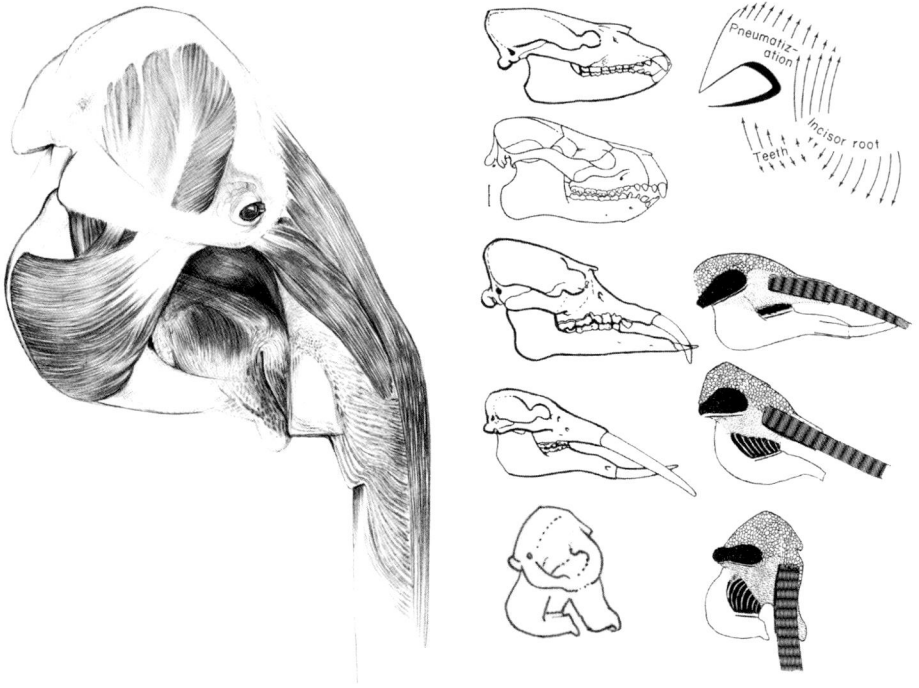

The transformation of elephantine heads through evolution.

Among Perissos, ancient tapir-like skulls could elongate into undulating waves of bone, yet similar proboscid skulls could compress themselves, over some 30 million years, into the extraordinary structure that is a modern elephant's skull, its tusks pasted vertically, and improbably, down the front of its face. The sketches above summarise some of my efforts at visualising that succession of evolving elephant skulls.

I once visited the sculptor, Henry Moore, in his Hogland studio where he housed an elephant skull that I must have first encountered, among many, at Park HQ in Tsavo. This skull was donated to Julian Huxley, and I next met it in Julian's Highgate garden. It was given in turn to Moore, who called it 'the most impressive item in my library of natural forms'. As a fellow lover of bones and fossils I could sympathise, but my own pursuits after the processes of evolution over-ride Moore's exclusively aesthetic, haptic approach. Interpreting the ecology and life-histories of an evolving lineage of animals is difficult but exciting. Each epoch embraces almost inconceivable swings in climate, and in opportunities and constraints for evolutionary adaptation.

My friend Alan Root once flew me high over Loitokitok and Amboseli, so high we could see Kilimanjaro's crater. Through clouds, through a propeller, through engine noise, glimpses of elephants came and went, their scale reduced to that of ticks on a buffalo's back (*see* p. 436).

One trigger to my interest in giant bones began with a large object that lay in the corner of a store-room opposite Teddy's office in Mbeya. Could it be the sculpture of a pair of gigantic bone bananas? Or two fat curved bean-pods with an odd little lip at their point of junction? The object's identity was disguised by the absence of any trace of teeth but I was familiar enough with elephant skulls to know it was the lower jaw of a large and exceptionally ancient, totally tooth-less animal. Teeth rooted in the horseshoe-shaped mandibles of ageing humans more or less stay where they begin, just wearing away until the last roots and the surrounding bone erode away, leaving a pathetic little stirrup of jaw.

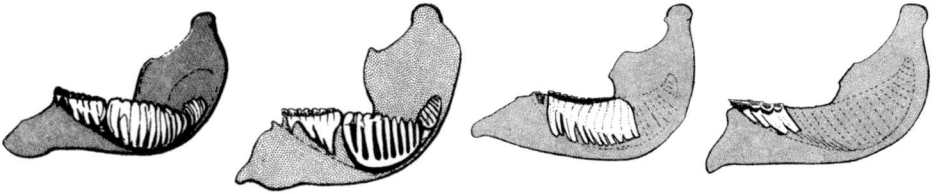

Elephant lower jaws at two years, eight years, 25 years and 60 years of age.

By contrast, each shallow arc of a baby elephant's bulbous mandible encloses the buds of six molar teeth, stacked like beans in a swollen, banana-shaped pod. Complex sculptures in their own right, growing molars push the vegetable anal-ogy still further. Each molar is an amalgam that begins with 10 or more separate enamel buds, each like an empty bag of corrugated material or a flattened sleeve with a serrated, somewhat conical profile. All steadily enlarging and hardening, these ten buds of enamel are stacked, like compressed pips in fruit-pulp, within the membrane of a compound tooth encased in cement. Within its banana, each growing tooth awaits its turn to shunt forward, all the time accreting more cement. Pitched at an angle, each set of teeth slowly wears away, front first, until the last root of the last lamination is gone and the animal is toothless.

That giant mandible was no fossil. It presented several possible deductions about a living species. The most obvious concerns age. Our Mbeya jaw came from an animal that was somewhere between 60 and 70 years old – not so far off the same age as similarly toothless old people.

Which end of that age range depended upon how abrasive that elephant's diet had been. How much dry, silicaceous grass had it been forced to chew?

That jaw also reveals that elephants of the 1930s and 40s were well able to evade ivory hunters and live to a ripe old age. To help understand her longevity, what sorts of social setting permitted her to survive after she had been reduced to masticating food on her gums?

Every elephant depends upon its mother who, in turn, looks to her offspring's grandmother, even great-grandmother, for a lifelong security that few other land animals can match. The females of any group that has survived or evaded ivory hunters are so close-knit that most members can expect to live as long as their teeth. With a little help from their friends, perhaps a little longer.

My own first intimate encounter with an elephant followed upon the death of a female shot in defence of a Gogo farmer's garden. Such reprisals have been routine over most of Africa for centuries and my great-uncle Hal, as a District Commissioner, was often asked to shoot crop-raiding elephants. Hal had begun his career farming Ostriches in the South African Karoo before buying a few hectares beside Lake Naivasha in Kenya. World War I transformed this tall, athletic big-bird farmer into a bush-savvy soldier. He enlisted as an officer under the command of General Jan Smuts, who had been his friend and fellow hiker in South Africa. He was among the victors that marched on Dodoma and Dar es Salaam to win what they called 'The German East Africa Campaign'.

His adventures were retold in letters to his only, much-beloved niece: my mother, Dorothy. For this intelligent, adventurous and gifted girl his photo-filled letters created a longing for East Africa where, perhaps perversely, she thought she belonged, rather than in the bland English suburb wherein she found herself; a cocoon that had only thinly shielded this 10- to 14-year-old child from a merciless, medieval European Civil War.

The images he sent were pasted into a precious album. There were photographs of Hal among his Ostriches, others of famished villagers being fed in Dodoma market square, of an eland caught in a pit-trap, pet dik-diks and Cheetahs, and Hal standing amongst a great crowd of Gogo tribesmen after helping them survive a war-induced famine (above). The photo to which Dorothy returned, over and over again, was of an orphan elephant, shuffling along beside the walls of Dodoma Fort (wherein Hal lived and worked). In the background stood the gardener who supervised and bottle-fed this tethered captive on

cow's milk. Hal reported the elephant's death some months later, to Dorothy's great distress. The photograph had somehow induced intense affection in the emotions of this sentimental 15-year-old. Elephant orphans were brought to Hal, to us and to our neighbours with some frequency – all foundling victims from a confrontation between their needs and ours that goes back deep into history.

My own first encounter with a very young, pink-toed elephant has become a peculiarly durable memory. I suspect Dorothy was remembering her own teen-age attachment to images of just such an orphan when she allowed her eight-year-old boy to spend time with a clumsy little round-headed, knock-kneed, big-eared and very hairy companion. He was kept in the Dodoma horse stables and maintained for a while on cows' milk and I was firmly convinced that he liked me as much as I did him – he was certainly greedy for constant compan-ionship. Furthermore, he gave every sign of having an innate love of play, with any animate being (and sometimes inanimate objects too).

It was no time before I became a responsive playmate and he would come and wrap his trunk around my head and, with its fingered tip, probe with gentle yet insistent movements into my ears, nose and mouth, even, very gently, my eyes. We watched one another continuously because our eyes seemed to be our greatest commonality. I had no trunk and my ears were pathetic compared to his. He was quite content to walk on four legs and did not seem to mind his lack of fingers. Still, he seemed to appreciate it when I reciprocated his longing for physical contact with hugs and scratchy rubbings over his face, behind his ears and around his neck and 'armpits'.

Less easy to handle was his propensity for leaning on me – difficult because he was a good deal heavier and a lot more firmly anchored on his four legs than I was on two. Nonetheless he soon seemed to realise that it took all my strength just to push back against the friendliest little jostle. Perhaps my el-ephantine wimpishness took the fun out of jostling because he soon desisted

Sketches of infant elephants.

and tried to persuade less delicate, inanimate objects to play with him. He sprawled, spreadeagled over a bale, only to let himself topple over before rising on very human bent knees. Suddenly he would catapult himself into a burst of activity that would end just as abruptly as it had begun.

He seemed to expect his audience to mimic his change of mood. The clue lay in his eyes, which watched mine attentively for any reciprocal activity change, from action to inertia or vice versa. When I matched his mood I was rewarded by rhythmic body rockings or gentle trunk touches. He soon learnt that I was ticklish – I would laugh and squirm to remove his trunk from inside my shirt, in response to which he gave every sign of delight. He also enjoyed wrestling my wrists and hands with his infinitely manoeuvrable little trunk. It was as if he was determined to demonstrate that there was nothing I could do with my hands that he could not do just as well with his trunk. Our fellowship was all about touching and being touched.

My most vivid memory is of an incident during the rains, when the enclosed stable yard was wet from recent rain and muddy puddles had formed. Distracted by something (perhaps it was some specimen of wet-season insect life), I momentarily ignored my significant friend in favour of some contemptible rival creature. Suddenly, my attention was reclaimed in no uncertain manner. The elephant had sucked at one of the puddles and, creeping up softly behind me, squirted a trunkful of muddy water all over the back of my neck, head and shoulders! My shriek was immediately answered by a squeal that I took to be a cry of triumph as the little elephant waddled, at top speed, to the other side of the yard, where he swung round, ears akimbo, trunk twirling around like a propeller, to face me off, little eyes rolling in what I interpreted to be a look of pure elephantine glee. When I charged across to smack him he set off, round and round the yard making baby trumpetings that had the stable boy running out to see what was going on.

What was going on can only be expressed by that very short word – joy. *Joie-de-vivre* was shared in that Dodoma stable yard between us both. Dorothy exemplified *joie-de-vivre* and was eloquent in its defence as a philosophical and practical principle. The little elephant was expressing joy in its purest form. I know he knew joy, because my presence in that soggy yard, my willingness to play, my lapse of attention, provided the certainty that I could be yanked back into the game.

Am I attributing my private emotions, and Indo-European words for them, to an elephant? Years later, when I retrieved this memory, Teddy reminded me that on any single day an elephant ate as much food as the whole population of a small village. Cauterised by the carnage of a World War I childhood, he thought Dorothy and I were both being 'mawkishly sentimental and anthropomorphic'.

I prefer to put it the other way round. For one moment I was being 'proboscomorphic', and was sharing pure emotion with an elephant. It was a friendship that reinforced my lasting sense of affinity with fellow mammals and a feeling that the joys of existence, of life, were both shareable and were actually shared. When he died I remember my grief and the inescapable comprehension that a personality as vital as myself was gone. The flies around his emaciated corpse signalled that whatever it was that linked us no longer existed.

If there was a lesson in that nameless elephant's passing, it was to value life as a precious fire inside oneself and to learn that its opposite, life

Elephants sketched.

extinguished, is but a return to the raw materials of life, suddenly breathless. Was there any purpose to his short life? If it was expressed in those moments of joy that we shared it had to be a joy that really belonged elsewhere – deep within his lost family.

For just a few weeks two infants had met. One was born from the largest animal still allowed to walk the soils of this planet. The other was born of the most intelligent and self-conscious animal on Earth, a species that named itself *Homo sapiens*. We both had social instincts so imperative that our momentary fellowship was as inevitable as our meeting was but a product of chance.

That elephantlet had been orphaned by his mother's enlarged appetite for vegetables – her last meal had been an entire plot of millet, the precious reward for a human family's labour following a fierce famine. Her long white incisors were auctioned and the proceeds entered the treasury that issued famine food for the farmer's family and a portion of my father's salary. Those elephant deaths were but details of a drama that had been played out across an entire continent, over as many millennia as human trade, weapons and agriculture have existed.

I have seen ivory and elephant-bone gate-posts in a Dinka cattle stockade and, in museums, seen mammoth tusks serve as roof rafters in Siberian huts. In museums there are animal carvings in ivory that are tens of thousands of years old, and on intricately carved ivory thrones you can imagine the popes and other kings that once sat there, smothered in splendid regalia. Carved cosmetic chargers, coffers, mummy masks, hair combs and ivory-inlaid furniture crowd museum exhibits of Ancient Egyptian tombs. Piano keys, billiard balls, picture frames, human false teeth and touristic nick-nacks are everywhere.

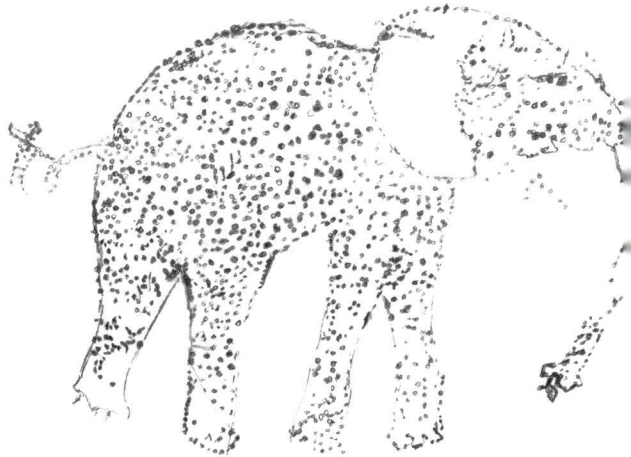

LEFT: *An elephant in chains.* RIGHT: *Pecked Saharan image of a man cutting or pulling an elephant's tail – age unknown but likely to be several thousand years old.*

Traditionally ivory was just the costliest, whitest, hardest, most durable and most beautiful of all materials – an animal analogue for ebony wood or oak, just as a living tree and a living elephant are ecological analogues in their very gigantism. Over the centuries, millions of tons of ivory have left Africa's shores. For centuries ivory, like gold and silver, has been Africa's ultimate currency. Today plastic, paper and electronic money have proved more than adequate substitutes.

If Dorothy had wept for the loss of our momentary companion in our Dodoma stables it was because she knew that milk, fellowship, love, flows from every mother – she saw no exceptions to that mammalian legacy. It was, perhaps, as much within her womb as in her brain that Dorothy lamented the theft of an infant's life as well as that of its mother. A woman and a she-elephant have similar life-spans. Equally able to generate a dozen or more descendants, they are the only guarantors of any future for their separate kinds. People and elephants have evolved to multiply, and multiply they will, whatever the odds. Past multiplications have been cut back, over and over again, by all manner of catastrophes – predatory diseases and epidemics (elephants suffer from tuberculosis, filariasis, bilharzia, all manner of worms, malaria and hepatitis, even heatstroke), climate-induced starvation, geological upheavals as well as continuous competition, expropriation and predation by other forms of life. For humans, add slavery, civil wars and all manner of other internal social stresses – recent history tells us that we, ourselves and our microbes are our own most ruthless cullers.

Ignorant of the ways, values and motives of shipborne pirates, villagers all across our huge continent yielded up those things that other, better-equipped and less scrupulous people sought. In markets, most of them way down on one or other of those cruel coasts, the slavers sold human bodies together with the heavy tusks that yoked heads had carried down all those hundreds of kilometres.

It has been estimated that some 10 million slaves might have survived transportation across the Atlantic. Unknown numbers reached the Mediterranean, the Middle East and south Asia. Over some 400 years, just as many more had died, within Africa or out at sea. The scale of obscenity only becomes imaginable when we take its measure beside the endlessly repetitive mobilisation of entire nations in what are always civil wars, always tribal wars.

Born into a land deep within its shore-mapped, elephant-shaped outline, I grew up surrounded by forts, market buildings, prisons, even trees, that retained scars from eons of slavery. I inhabited landscapes that had been elephant estates for millions of years before my lot, modern humans, evolved. Our two fast-evolving kinds have been accommodating to each other right up to the emergence of modern iterations in each of our respective lineages. Today, human appetites and developments very far from Africa have intensified a brutal confrontation between elephants and us, which is not palliated by occasional lapses into fellow-feeling.

ABOVE: *Marching slaves to the coast (Africa Story panel, 1959).*

LEFT: *Slave porter and tusk.*

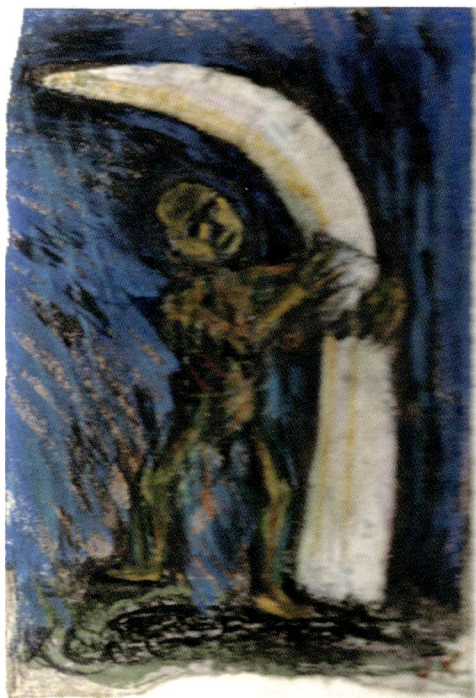

It is no accident that ivory and slaves became Africa's prime exports. Both species descend from ancestors that evolved within Africa. Over most of prehistory, numerous plant and animal species have had to accommodate, adapt and evolve in contention with great herds of elephants (sometimes of several species) and, increasingly, with even greater herds of humans.

Once centred on bones and fossils, my almost universal interest in ancestry and the vicissitudes of ancestors has now broadened immeasurably with the advent of molecular science. I have been gifted with a preliminary sketch of my personal genome that confirms African forebears but also tells me that I am a fertile mule. Some of my ancestors belonged to another species – the 'Neanderthals'. My Papuan friends can claim yet another long-extinct, prehistoric species, the 'Denisovans', among their kin. A shared ancestry among these peoples lumps them as 'Neandersovans'. Modern genetics serve to mock any person who puts too much store on recent pedigree.

Equally mocking of brittle contours around hard categories, we learn that elephants are no less fertile mules than I am. Contemporary genetics have revealed that African Bush Elephant genomes boast a complicated history of separation and admixture, not only with African Forest Elephants but also with huge, straight-tusked elephants of a separate genus (previously allocated to *Elephas*, the Indian Elephant genus, now to *Palaeoloxodon*). Modern African elephants took up their role as a continental super-giant relatively recently, and it may well have been modern humans that helped tip their still taller predecessors into extinction.

Human artefacts sometimes resemble shadows, in as much as their contours mimic the forms or contours of living, moving beings. With shadows as metaphors for the transience of life, and as witness to the steady elimination of elephants, I once made a painting in which mere hints of past elephant presence drifted past the white arc of Kilimanjaro's summit. This chapter ends with a print of that painting.

Up to about 15,000 years ago, and for perhaps a million years before that, Africa's dry woodlands, savannahs and steppes were dominated by *Palaeoloxodon recki*. I have drawn and catalogued this species' bones, now crowding the back-rooms and metal shelves of Nairobi's National Museum. I have examined their tall, deep-jawed, top-heavy fossil skulls, measured the dimensions of their teeth and counted their dentine cones. Whatever caused the extinction of *Palaeoloxodon recki*, there was nothing inevitable about it even if predatory *Homo sapiens* accelerated it. Its demise opened up a vast, then part-empty elephantine niche. Descendants of an earlier folivorous proboscid, *Loxodonta adaurora*, which was perhaps more elusive, more intelligent or more aggressive than *Elephas* or *Palaeoloxodon*, seem to have been sucked out from the well-watered equatorial forest zone they had occupied for a million years or more. Elephant progeny have now been shown to resemble modern humans in being mules, fertile mules, mules adapted to 'go forth and multiply'.

A strong preference for browse tempts them into over-browsing any environment they can access. In widely separate areas I have seen trees and shrubs battered out of existence by elephants, and colleagues have documented how African Bush Elephants from over-populous herds in grasslands live shorter, sicker lives than their deeper-toothed Asian cousins. The contrast between *Loxodonta* and *Elephas* teeth has led at least one biologist to call bush elephants 'maladapted improvisors', but that's how many a new lineage begins, how many a survivor wins. Elephants have often risen to become the pre-eminent, sometimes most influential species within any given ecosystem. It seems to depend upon the situation at whatever moment of opportunity occurs.

Nearly 100 years ago, my one-time colleague in Oxford, Charles Elton, demonstrated how a specific predator and/or disease can drive any herbivore's numbers to effective collapse. With access to detailed accounts of the Hudson Bay Fur Trading Company, Charles found a record of 152,000 Snowshoe Hare pelts taken in 1864. By 1868, hare numbers had fallen to near zero whereas pelts of the hares' principal predator, Canadian Lynxes, had peaked shortly after that of the hares, only to mirror their decline and collapse a year or two later. The Hudson Bay records revealed a 10-year pulse in both hare and lynx numbers that was maintained over the course of a century, revealing that the numbers of herbivores and their predators (or diseases) can rise and fall in a regular, predictable fashion. The fur industry was revealed as essentially parasitic on a cycle that long predated the arrival of humans in America.

Of all biological insights reconstructed from (even older) commercial records, a most challenging and thought-provoking set of statistics has been extracted from the log-books kept by huge and prosperous American whaling companies as they hunted for Sperm Whales in the early 19th century.

Once a spotter up in the 'crow's nest' of a factory ship had sighted the spouts of a Sperm Whale pod, the ship (or fleet of ships) converged on its prey and each launched 16 or so long, narrow whaleboats, each rowed by crewmen led by a muscle-man harpooneer who stood up in the bows, waiting for the whales to surface. Each pod might consist of about 12 highly voluble families, led by matriarchs with head-heavy bodies up to 16 m in length. Sperm Whales have the largest brains of any animal on Earth (8,000 cm^3 to our 1,300) while their life trajectories, maternal solicitude and versatile, adaptable societies are a match for those of humans. When it comes to subtleties of language or song, it is likely that Sperm Whale voices plumb depths, both literally and in content and complexity, beyond even our own voyaging minds.

In response to several million years of group defence against Killer Whales, their only major predator, Sperm Whales converge into tight, inward-facing star

*Sperm Whales (*Physeter macrocephalus*) hunted by whalers in the 19th century.*

formations up at the sea's surface. Any approach by Killers is met by a barrage of powerful tail flukes, each capable of delivering fast karate-chops.

Naïve whales responded in the same way to the attacks of whaling fleets and, although many an early whaleboat was smashed or capsized by such bladed flukes, the harpooneers took full advantage of this surface-clumping to make multiple kills. Profits from Pacific Ocean whaling soared and whaling fleets became so enlarged that the whaler/writer Herman Melville wondered how long whales could endure 'so wide a chase and so remorseless a havoc'.

The log-books (kept by on-board Captains or First Mates) certainly confirm what huge initial profits were made by the industrialisation and expansion of whaling. Then, within a few years, their listing of successful harpoon hits dropped by nearly 60 per cent, confirming what the crewmen had already noticed. The Sperm Whales abandoned their long-tested clumping defence and it would seem that underwater clues, revealing how dependent the big sailing ships were on wind, led the whales to flee, well submerged, upwind. Most extraordinary of all, it seems that escapees from 'remorseless havoc' were able to teach innocents to override any impulse to clump and get harpooned. Instead, those that escaped this unprecedented attack from the ceiling of their universe devised and taught a new getaway technique. All of this implies a very high level of analytical thinking and deduction on the part of Sperm Whales, as well as the ability to share and exchange knowledge.

Herve Glotin of CNRS has been a leading pioneer in the long-term study of Sperm Whales. In Francophone Africa I salute CNRS and leaders such as Theodore Monod, François Bourlière and André Brosset, admirable pioneers in establishing systematic scientific research in western and central Africa, sometimes making good use of old records kept by tidy predecessors.

In East Africa, Desmond Vesey-Fitzgerald, 'Vesey', a biologist of my father's generation, was a man of great insight, experience and influence on ecological thinking. This penetrating observer gifted young 1960s researchers on the Serengeti plains with concepts, even slogans, drawn from his decades of

Gnus (Connochaetes *sp.) sketched.*

experience. Pointing out that animal successions were closely integrated with those of plants, he wrote, 'The component species are always competing with each other for supremacy. Plant succession is a tremendous force that cannot be stopped, although it may be halted or diverted in a variety of ways. Grasses as a group are ideally suited to the struggle for supremacy ... We are living in a grass-age, which has been largely brought about by ourselves. Trees compete with grasses by cutting out the light, and grasses compete with trees by their effective root systems. The growth of *shoots towards the light* and of *roots towards the dark*, and all the intricacies of the production of nutrients by leaves bathed in sunlight, are basic functions of an ecosystem. An ecosystem is a section of nature functioning as a whole. It comprises the whole environment and every animal that lives in it.'

Momella lakes reflect Kilimanjaro.

Vesey and I spent many happy hours by the Momella lakes and on the rim of Ngurdoto Crater in the Arusha National Park, watching warthogs, Waterbuck, buffaloes and elephants, a community of plants and animals inhabiting a volcanic basin that was close to being one of the most densely populated and diverse patches of land anywhere on Earth. There, Vesey liked to recall Charles Darwin's brief visit to the Cape (his only African land-fall) and his astonishment at the number of animals able to inhabit an area that, to all appearances, generated so little food. Supposing that extensive annual movement by the animals might be involved, Darwin also stayed long enough to note that grasses were no sooner consumed than fresh growth took their place.

Vesey also impressed his young protégées on the Serengeti plains by pointing out how the digestive systems of grazers (especially the leaf-like absorbent lining of rumens) responded structurally, somewhat like deciduous foliage, springing into active growth for the duration of the rains. Then, come the long dry seasons when grasses dried out (only burning if left ungrazed), leaf-lined rumens wilted like autumnal foliage but their parent ruminant could still get by on grass that was now expendable rather than renewable. Until very recently, all over Africa's grasslands, there were successions among beautifully refined grazers in which one species followed after another, minimising competition because the grazers specialised in particular growth stages, each antelope species preferring a particular spectrum of locally evolved grass or herbaceous species. The Topi/Tiang/Tsessebe complex dominated the scene on flood-plains, while elsewhere gnus, zebras, buffaloes or elephants became 'Keystone Species'.

The great ecologist Bob Paine borrowed this descriptor from architecture to demonstrate that entire ecological edifices could collapse with the removal (as distinct from, say, the Snowshoe Hares' recoverable decline) of their keystone species. Ever more frequently, humans are the agents for such collapses but other predators or disease can also topple previously stable (even cyclical) ecosystems.

Disease can take the form of nightmare pandemics such as the late 1800s rinderpest, the Pesadilla flu pandemic of 1918, or more recently Covid-19. For Sperm Whales and elephants, piratic humans are the plague, the disease.

The year 1900 marks the date when a particularly pernicious and damaging mythology was propagated. It has shaped attitudes towards development, disease and 'wildlife' to this day, and it has empowered apprentice technicians within the livestock industry, out of all proportion with their real importance.

Indian cattle infected with a particularly virulent strain of cattle-measles (hereafter called rinderpest, its common name) are now known to have triggered the epidemic. It was imported into Eritrea in 1887, and trade sped up the spread of this Asian strain of cattle disease in Africa. In subordinating themselves to another species (cattle or any other domesticate) entire human communities, especially pastoral people, place themselves at risk of disaster, starvation and death from such virulent epidemics in their commensals. This is what befell all livestock-dependent people, and all cloven-hoofed wildlife, of eastern and southern Africa more than a century ago. This terrible epidemic swept throughout the eastern half of Africa, reaching the Cape in 1897.

With no history of exposure to this lethal imported strain, indigenous cloven-hoofed bovines, giraffes and wild pigs died in their millions throughout eastern and southern Africa. During my childhood there were numerous tales about how rare 'game' had been during the first decade of the 20th century. Fecund species such as the reedbuck and gazelle families rebounded quickly, but bovines such as Bongo, Kudu and even buffalo were slower to recover.

The very traders in cattle who precipitated this cloven-foot holocaust inverted its origins by persuading themselves and nearly everyone else that indigenous African animals were the culprits, livestock the victims. Most significantly, this fabricated myth was not only swallowed and propagated by South African cattle men, but by the commercial and mining interests that they supplied with beef.

The rinderpest epidemic brought Africa's topography into sharp focus. Long, deep lakes, like moats, repel invaders, define biogeographic borders and govern much else, including speciation. Such moats can also channel, even delimit, epidemics. The eastern side of Africa has some eight substantial rift valley lakes, most running north–south, and the longest and deepest, Lake Tanganyika, is more than 560 km long (about the same as the Adriatic Sea). Lake Malawi, almost as long, would become a southern extension of Lake Tanganyika, were it not for the space between them being the volcanic, mountainous point of junction between the western and eastern rift valleys. A smaller, shallower Lake Rukwa floors a minor cross-rift, the Rukwa valley. Otherwise there is a wide band of upland between the two great lakes. Rinderpest's tidal wave reached these uplands by the early 1890s then swept on to the Cape.

Getting together over a map-covered table in Johannesburg, a convergence of globally influential parties devised what they were persuaded would be an Ecological Great Wall of China – it would stop diseases from the north ever again reaching the south. A *cordon sanitaire* was proposed that would permanently bar any infection ever again coming down from the north. The only place where such a barrier was remotely practicable was in the corridor between the southern end of Lake Tanganyika and the northern end of Lake Malawi (then called Lake Nyassa).

This land was well known for its abundance of large wild animals and their traditional Nyamwanga hunters. It had only just come under German control but was still a theatre for local conflicts between the several peoples who inhabited the region.

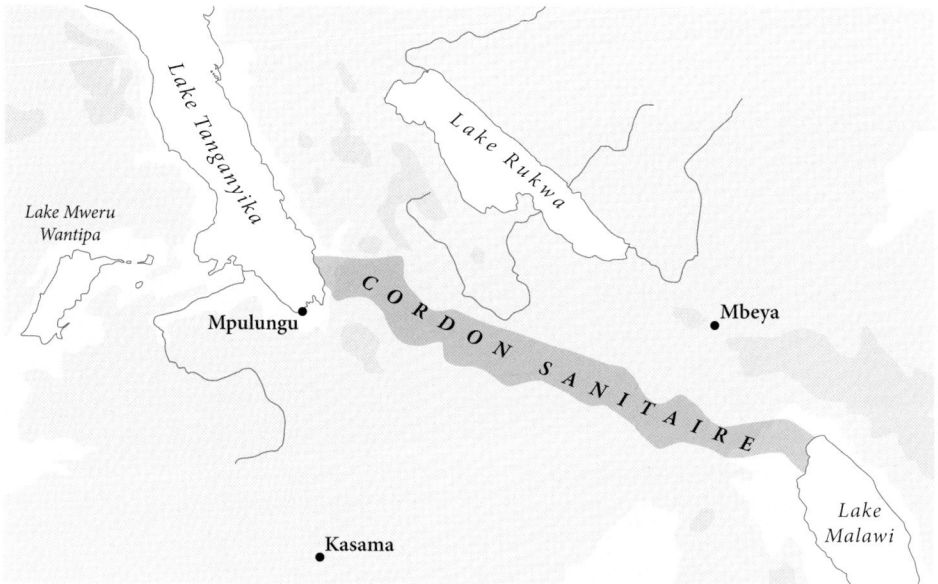

The cordon sanitaire *between Lakes Tanganyika and Malawi.*

With South African mining money paying for it and with the added bonus of a *cordon sanitaire* helping define the borders between German East Africa, Northern Rhodesia (today's Zambia) and Nyasaland (today's Malawi), German colonial authorities set up this cordon. The only structure consisted of stretches of wire-on-post fences, past which any movement of cattle could be controlled.

Teams of shooters were deployed on both sides of the fence, charged with killing every cloven-hoofed animal they could find along a strip 16 km wide and more than 320 km long. Some clearing of woodland also took place. This shooting was sustained for some 40 years and a veritable hailstorm of bullets succeeded in exterminating all the major indigenous animals that had originally inhabited the area. Just such exterminations, official and otherwise, have recurred over and over again, all across Africa, all ultimately at the behest of or to the perceived benefit of the livestock industry.

This corridor region had once sustained a variety of antelope species, zebras, warthogs and giraffes, and I remember finding bones of Roan Antelope and rhino eroding out of our Luanda River's banks, once one of their perennial sources of water. Lions and Leopards preyed upon them while their other major predator, human hunters, had long subsisted on some 16 species of large wild game, including wandering elephants.

Indeed, it was these traditional subsistence hunters, mainly Nyamwanga locals, that were recruited to become marksmen under one or more European hunter-supervisors. Given a gun, ammunition and the brief to totally eliminate his peoples' original sources of meat, it was one of these shooters, Hamisi Sikana, the grandson of a traditional hunter, from whom I learned much of this sorry tale. Hamisi moved on to become a senior sergeant in the Tanganyika Game Department, in which role we spent time together on several survey safaris.

Nonetheless, it was symptomatic of the times that no serious study was ever undertaken of wild animals, the animals best able to tell us about epidemiology and its dynamics. Cattle represented money and influence while 'game' was expendable trash, unless it yielded ivory.

Politicians who seek to enlist young warriors to their cause often invoke slogans. With Stalin and Mao Zedong setting the precedents, 'War against Nature' has been a most persistent political rallying cry but it is one that is exterminating or near-exterminating too many natural communities. Even more damaging, it is closing off much more intelligent potentials for our future relationship with nature in all its plenitude.

I encountered a stark example of this war in Uganda where cowboys,

Hamisi Sikana.

a.k.a. Texan cattle men, funded and oversaw an Ankole Ranching Scheme that involved the total destruction of all indigenous woody vegetation and the elimination, mainly by shooting, of an estimated 40,000 head of indigenous large mammals, including all carnivores, Giant Hogs, Roan and other antelopes.

The context for such brazen vandalism in the name of 'Aid Overseas'? Independence had coincided with the Cold War, pitching us into the ideological conflict between East and West, as separate powers vied for influence.

In Ankole, a locality once famous for Serengeti-beating biodiversity, some of the land became almost as sterile as a Texas beef-lot. In addition to this scorched-earth approach, the 'Schemers' (as they were widely known) organised land-grabs to enclose large tracts of the neighbouring Lake Mburo Game Reserve, specifically to appropriate essential water-points. I declined an invitation to join a couple of weekend slaughter-fests in which hundreds of antelopes, zebras, indeed any and every wild animal were up for target practice. My close friends in the Game Department saw lifetimes of dedication to the conservation of Uganda's incomparable wildlife heritage destroyed in hours as super-fast teams with posts, holers and barbed wire rushed in to ensure that the Ankole Ranching Scheme's theft of game reserve land and water was a *fait accompli*.

The externally funded Ankole Ranching Scheme exposed how any attempt to reveal, let alone conserve, Africa's wealth in terms of 'wild' (i.e. pre-adapted) animals or plants is fiercely opposed by any individual or organisation with a stake in foreign domesticated plants or animals, and the gilded techniques that have proliferated around their husbandry.

In crude mimicry of Wild West Texas, ranches have been created in Africa at much cost to become the property of a few influential individuals or their clans. Local beneficiaries of cowboy culture sometimes bear arms, wear Stetson hats and TexMex fancy boots in homage to the world's premier cult of the cowboy.

Ranching is a crude monoculture and large herds of cattle degrade land fast by their unstructured consumption of vegetation and by destructive trampling. Too many cows, concentrated in huge herds, especially around water sources, trample up dust that dry-season winds blow up into pink clouds. With the arrival of the rains, floods bear away thick soups of diluted mud, to be dumped in the oceans. It has been calculated that in a single year wind and rain can remove up to 200 tons *per hectare* of Africa's thin, precious topsoils.

My colleague and head of the Uganda Game Department, Adam Ruweza, was a well-informed critic of the Ankole Ranching Scheme. He saw it as typifying the warped mind-sets that are so often brought in with foreign 'aid' projects in Africa.

To help outdate such primitive mind-sets has been one purpose of this work. In the biological senses I have just hinted at, succession can mean that just one of nearly 100 surviving species of African primates can and has insinuated itself into virtually every habitat, while every other primate species is in retreat. An abundance of hominin fossils even tells us that *Homo sapiens* now occupies all the habitats once parcelled out among various, once co-existent hominin species.

Like any other species we are least tolerant of those most similar to ourselves, because their needs are closest to our own. Even over millennia, one lot succeeds while their contestants recede. That one group lives while others die is but one manifestation of 'succession'.

Preoccupied with our own day-to-day existence, we can have little inkling of the 'value' others attach to incidentals we scarcely notice, such as 'Rare Earths' that find their way into supply lines for semi-conductors and other expensive electronic goods.

It is said that children in Kimberley (South Africa) once played with and then discarded diamonds as pretty playthings lying around in the dust of their village backyards. They could not know what fortunes lesser cultures attached to such baubles. Nor could they foresee the miseries imposed upon their descendants by others' greed for mere minerals, buried beneath their humble village.

When the coveted resource has bulk, like tusks, timber or ore, the prizes go to those in possession of fleets of ships, planes or trains. As recently as 2019, my friend Paul Theroux cut straight to the quick: 'I saw how with backhanders or huge loans China bought dictatorships in Zimbabwe, Kenya, Sudan and Angola, in order to obtain ivory, gold, bauxite, oil, and much else, leaving the countries in deep and sometimes unpayable debt – indeed, debt slavery. But the United States still does the same in many countries, taking advantage of a government's indifference to human rights abuses.'

In the 1950s, as country after country in Africa became independent, we students expected much of our leaders, and we saw African Unity offering escape from servitude and confirmation of our dignity as contemporary citizens. The politicians let us down then, and they persist in doing so to this day.

The Elephant's Shadow.

DISEASES AND PLAGUES

In which the ultimate predators are exposed.

My own first inklings of life's chanciness and fragility were so early I cannot even remember them. An anxious mother confirmed that I was but a very small baby when malaria first struck, then again and again. These recurrent bouts were eventually assumed to have conferred small measures of immunity.

Since then I have found some comfort, even a sense of joyful fellowship, with other living beings, in the knowledge that chance, lucky escapes, good fortune, medicines and some innate *savoir-faire* has conditioned every triumphant survivor's existence – from myself to a House Mouse. Infectious diseases have been the nearly invisible selectors, the final arbiters as to whether any lineage survives.

For those talking the language of metaphor, here are the 'Hands of Gods'. For inventors of microscopes and anatomists of DNA, here are newly visible, newly nameable microbes, determining the course of evolution – who lives, who dies and where, when and how.

More than Lions or crocodiles, diseases are predators that attack us on scales of gazillions – trillions of microbes and parasites versus many millions of us. As with any other mammal, dangerous infections have plagued us and our ancestors from the start. That is evolution at its most fundamental. It is as primate mammals of tropical origin that we are peculiarly susceptible to diseases hosted by primates or other tropical animals. That the scientific study of tropical ecology still lags behind was shown up by months of helplessness in the face of the Covid-19 pandemic and before that in the 1918 Pesadilla flu pandemic (which was restricted to *Homo sapiens* but probably began with avian hosts, then swapped species somewhere along its devious history). It is at our peril that we ignore the fact that diseases are a major force in evolution. They have shaped our modern world far more extensively than is generally acknowledged.

In war time, lake-side Tanganyika we followed evening rituals that were daily reminders of our well-founded fear of disease, fear of mosquitoes and other insects, fear of too many unknowns buzzing, whining and hissing away out there in the night.

OPPOSITE: *Seven-headed monster of the Apocalypse symbolizing plagues and disease. From St Francis non-denominational chapel, Makerere University, Kampala, East Africa.*

The half-hour just before nightfall became a time of sneezes as Saidi walked from room to room in his white robe and fez, pumping a fine spray of 'Flit' (a kerosene-based insecticide) from his 'Flit-Gun'.

Even so, there were so many mosquitoes, so hungry for blood, so adroit at getting past every barrier. Sooner or later an infected mosquito got us and the words we used were plainly descriptive – we 'went down' with malaria. We were felled. Felled, no less than the tallest tree or greatest tusker. A trophy gunned down by a midget mosquitoess, her bullets the malaria parasites, *Plasmodium falciparum*. They invaded blood cells to circulate, in ever more millions, infiltrating every limb, messing up every faculty, compromising every function, even able to render you as senseless as a felled mahogany.

But my 'going down' was never the instant crash of a tree. My brain insisted on translating those replays of hitting the soil, those surges of over-heated blood, those reverberating pulses of my heart. Infected physiological functions translated into the drags of currents or waterfalls, the thunder of falling rocks, the churn of a whirlpool, the swelling and deflating of bladders, the tramp of a rampaging rhinoceros.

Very early on I got to know and dread the warning. It started with what is commonly called 'light-headedness'. Yet this does not describe the nauseous first phase in a process that feels like a dissolution of wholeness, of the integrity of body. Light-headedness, I suppose, tries to convey that sensation in which the head takes leave of gravity and floats above or away from the neck, playing a nasty game of hide-and-seek with an increasingly independent, unstable and disconcertingly fragile body.

I can understand why our ancestors invented souls, separate from bodies and supposedly separate from the senses. In the process of developing malaria (and, I guess, many of the other fevers that they experienced) an alternately sweating, aching furnace of limbs, torso and head turns to frozen jelly, shaking uncontrollably to become the inert and breakable shell around a loose and frightened spirit.

I know that my mind, gloriously inseparable from every other part of me in health, suddenly seemed, in a malarial fever, to become bewitched, threatening at any moment to become severed from my body. It was as if the mind, spirit, soul, or whatever you wish to call it, that living self that is all feckless 'behaviour', suddenly shrivels like an infected kernel in a nut, rattling loose within its outer shell.

With the progress of the fever, the sensation of disintegration continues until the mind, now as shredded as the body, becomes indistinguishable from mere physiological sensation. This thought/sensation flows like sticky, blood-laden water, swirling in slow motion over endless fields of jumbled, unstable, threatening, black rocks. Falling, always falling – mostly downwards but sometimes inexplicably lurching sideways or swinging about as if in a nauseating, shapeless roller-coaster. Sometimes the swirling is silent, sometimes it thunders through the ears and skull with a relentless pounding. Random senses collide and grind against each other behind a fevered forehead.

SISTERS IN DELIRIUM, DDUBI FOREST.

'*... it was the darkness of empuku, the hole in which the blind grub blunders. I heard a soft throb as from the throat of a threatening animal. Stinging fingers held me all over, hair, spine, shoulders, everything. Panic.*'

(Linocut and words by Betty Manyolo.)

One time I was in deep water, heaving boulders to dam a river. As I completed my task the boulders swelled in size, becoming a huge lakeside dyke. In a half-awake state there was nothing supernatural about gods, ancestors, or me making mountains or excavating river trenches.

Front and back of Janus sculpture, by Agustini Mugalula Mukibi.

One such frightening fever felled me exactly 10 days after an abortive crocodile hunt – I can almost place the moment when I was bitten. In August 1951 Teddy took me along while he conducted a *baraza* (public meeting) in the courthouse at Ipiana, a village on the very malarial northern floodplain of Lake Malawi. Meanwhile I needed no prodding to go and hunt crocodiles up-river with my small-bore rifle. A kilometre or two upstream I was preoccupied with scanning the river below when a powerful-looking, well-dressed man came round a corner of the pathway. On seeing me he stopped, struck his forehead with his fist and exclaimed, '*Mwungu saidiye! Nimeokoka hawa blurri wazungu kule kusini. Sasa nina kuta moja hapa nyumbani!*' ('God help me! I thought I had escaped the bloody whites down south, now I find one here at home!')

He was even more surprised when I replied, '*Pole rafiki, baba yangu anaongoza baraza kule Ipiana, tumesikia mamba wamekula watoto hapa, sasa nina winda hawa mamba.*' ('Condolences, friend, my father is holding a *baraza* in Ipiana, we heard that children have been eaten by crocodiles, so I'm hunting crocs.')

Submerged crocodiles (Crocodylidae).

Sceptical that a mere boy could really hunt child-eating crocodiles, but some-what mollified by our conversation being in Swahili, the man revealed that he had just returned from working in the mines near Johannesburg. Why, he asked, are there so many whites down there and so few up here? We agreed it must be something to do with 'History' and 'Big Boats', but the question remained effec-tively without answer.

Ten days later I went down with malaria, quite possibly from a mosquito that bit both of us while we talked. Unwittingly, my fever implied an answer to his question. His people had lived for many generations in an equatorial land. Countless numbers of his immediate lineage had been culled by insects, mi-crobes and viruses. That fit, strong young miner represented a survivor, now immune, or nearly so, to many of the tropical diseases that had assailed so many of his forebears.

By contrast, most of my more recent ancestors inhabited lands where animals and plants alike had been culled, just as ruthlessly, by winters. Culled by cold, by wood-smoke in the lungs, by a paucity of sunlight, by winter epidemics, by famines and in more recent history by feudal and then colonial servitude. In the near-absence of malaria in northern Europe, one tranche of my ancestors had bequeathed me the lack of any inborn resistance. All I could do was drink lots of quinine and go yellow.

Natural selection, a.k.a. malaria, zapped *wazungu* in tropical lowlands so, without quinine, they could only settle temperate, malaria-free regions way down south. Hence *wazungu* survived down there, but up here, pottering along beside the Kiwira river, was a single specimen, and a soon-to-be-sick one at that.

Back in Tukuyu Nurse Fastness was asking, 'Headache? Hot? Cold? Are you in pain? You're dehydrated, dry. No? You're feverish, you're soaking wet! Are you thirsty? How do you feel?' At such times my mother's fear for my life was undisguised and I remember my own alarm, some years earlier, when she passed out, crumpling up into a heap on the floor when a doctor, palpating my swollen abdomen, pronounced that my spleen, or my liver, felt as if they were on the point of rupture.

But swooning terrors were rare: my parents' awareness of the pervasive presence of disease and sickness in the air found many a mundane expression. Fear of malaria and other, more mysterious, maladies conditioned some of the standard rituals that punctuated our existence every day.

Most obvious was the mosquito netting around the verandah and over the windows, taking the force out of every breeze, upping the temperature and imparting a sense of shady, stuffy gloom to the interiors of iron-roofed, mud brick-walled and lime-washed bungalows.

Then there was the evening ritual of my parents pulling on their knee-high 'mosquito boots' of shiny black leather (which had distinct pretensions to style in spite of their professed functionality). These were paired with evening attire of long-sleeved, floor-length dresses for Dorothy and white shirts with cuff links for Teddy. Even 'flit-guns' came in economy and deluxe models. The former resembled something made from spare soup cans but the latter (which I only encountered at senior Government official houses) were all shiny, with turned brass fittings.

Much later, talking to my parents as an adult, I learnt that, for them, a primary source of such fears was the admonitions and health briefings that had been given to both of them before they left England. Although these inductions into the customs, duties and terrors of the tropics were part of a responsible Oxford-based programme, orchestrated by the Colonial Office in London, one can be certain that they had much earlier, more primitive, roots.

One line of descent must have begun with the gossip of 'old hands' hazing apprehensive country boys as they set off in wooden sailing boats to 'Company Lands' in Africa, the Americas or the Indies, East or West. For new recruits to the Royal Adventurers of England trading with Africa, 'the White man's Grave' was no scare-story. Between the 17th and 19th centuries, likely death was a stark and predictable reality for young Europeans entering the tropics. Even in Tanganyika, 'Ex-German East', the lonely graves of youngsters with European names in scattered cemeteries are reminders that death by disease has been an ever-present force.

Illustrations for school history booklets in Luganda.

A 'tradition of fear', especially fear of a lonely death in strange lands, was part of an imperial imagination that my parents had not entirely escaped. They were touched by it long before either of them ever left England but it was not enough to stop them going off to join Uncle Hal in Tanganyika.

Sitting around a campfire one evening, my father remembered how he had once been much preoccupied by being buried among his family in his ancestral land, but that this had faded away until he had come to take an almost opposite position. Why? Because Tanzania had become his home. For us locally born children there was no choice – wherever death finds us, we die Africans, no less than George Washington died American.

After living for many years in the healthy 'Southern Highlands' of Tanganyika and with the development of new drugs, my parents' subliminal fears abated and many of their elaborate precautions fell into disuse. Nonetheless, I have to acknowledge that those early years left their legacy in me. It is difficult to shed such intense conditioning and I still retain a sense of caution, even today, about gratuitous exposure to mosquitoes and other sources of potential illness or accident – all vestiges of my be-netted birthright.

Illustrations for school history booklets in Luganda.

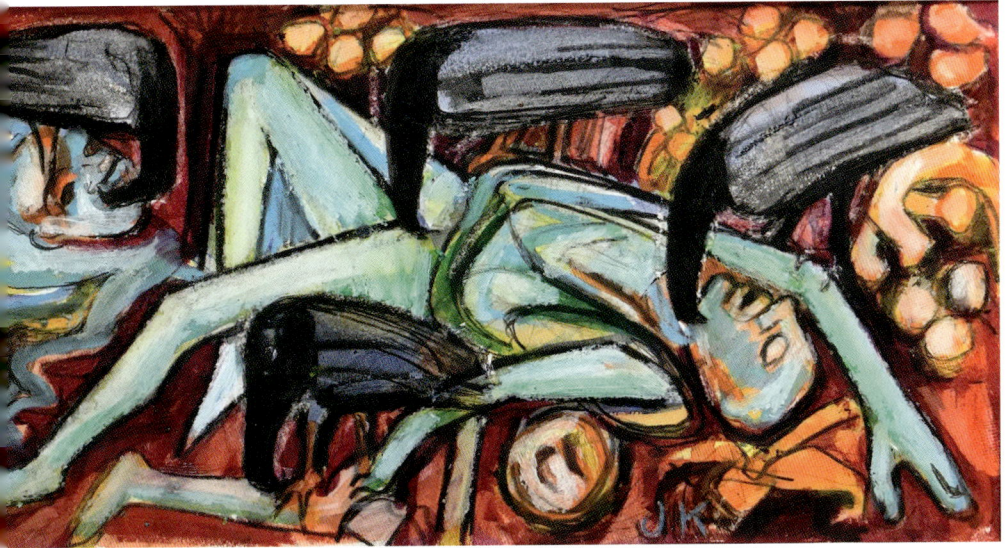

Crows on corpses – a riff on medieval horrors.

That preoccupation with health and the fear of death around the corner should not be seen as aberrant. Without acknowledging it, early colonials shared some of the almost universal perceptions of indigenous people in the tropics: a deep-seated and well-founded fear of – or at least caution about evading – a sudden death. It flew in from the darkness outside, unannounced, inexplicable and mysterious.

For pre-scientific victims of 'Mal Aria', literally 'bad air', some reassurance came from magical spells and incantations. Yet I remember *mganga* Mtalami Mbondei inviting me to sample a bright yellow bark that he scraped from the roots of a large *mtumbi* tree. It tasted just like quinine. Mbondei was adamant that he and his countless *mganga* forebears had long prescribed this root to ameliorate high fevers.

It is appropriate that Swahili uses the same word '*mganga*' for both traditional and western-qualified doctors. Faced by the most terrible agues, whom to turn to for help? Whoever makes the most convincing promises, and demonstrates the most effective cures, earns allegiance.

I actually witnessed something of just such a cultural shift during my childhood in Tanganyika. Within my lifetime, governments, in the shape of schools and dispensaries (reinforced by religious mission schools, madrassas and hospitals) very largely displaced the power of *mgangas*, traditional medicine men for whom

Mganga Mtalami Mbondei.

magic may have been a part of healing but whose herbal knowledge was often immense. Ultimately the greater part of that knowledge came from intensely observant hunting forebears, who could only survive and propagate themselves by being analytical observers and strategic gatherers of foods and medicines. Less observant farmers and pastoralists, late arrivals on the scene, adopted the hunters' names for most plants and animals. During my lifetime, demonstrated before my eyes, farmer folk sometimes depended directly upon more knowledgeable Dorobo or Twa foragers for their medicines.

The efforts of both missionaries and governments to discredit traditional *mgangas* has had limited success. I once heard an elderly village man claim that missions had simply arrived with drugs, weapons, cash-crops and written languages, each demonstrably empowering. It was those sources of power, certainly not biblical hellfire and damnation, that had persuaded so many people to submit to outsiders and abandon the legacies of ancestors often so much wiser than themselves.

That same elder reckoned the diseases that had most influence were those of the mind, mysterious mind-sets to which foreigners often seemed the most prone. In practice, some of the hunters' ancestral knowledge has been not so much abandoned as renamed, after being selectively analysed, reformulated and appropriated by science, religion or some such formal authority.

Given what a disproportionate preoccupation of our thoughts and energies were devoted to mosquitoes and malaria, it was scarcely surprising that among the questions I plagued my mother with at bedtime were, 'Did God make mosquitoes?' and 'Why?'

By claiming that God had made everything, Dorothy boxed herself in, so, unable to improvise an answer, she actually said she did not know. Once more, I was left to try and think the riddle through on my own. By seven or eight years of age I was suspecting, more and more, that if God existed, there was nothing to show that he, she or it was endowed with human (or even a nice monkey's) feelings and must be particularly malevolent to invent anything as demonic, yet tiny and innumerable, as mosquitoes.

Like the giant swarms of lakeflies, I had learned that mosquitoes were not born flying; I could watch their jerky, shrimp-like larvae in water which, Fairfax explained, had hatched from blood-fed mosquito eggs. It was as adult insects, and only female ones too, that they whined their way, in billions and trillions, into a brief airborne quest for blood. What sort of vampire-god thought that one up? Like the birds that also hatched from eggs only to fly from meal to meal and from mate to mate, the mosquitoes also fell back to earth once their appetites for food and sex were done and their bodies were worn out or eaten by a greater predator – a dragonfly, bat or bird.

Besides, mosquitoes were clearly oblivious to where they got their blood. I could see that dogs, even chickens (and, once, a sengi) were endlessly harassed by them. My father's horse, Jaggery, or my donkey Judy, or the neighbour's dogs provided just as acceptable a meal of blood as I did. So here was yet another reminder of my commonality with the other warm-blooded animals that surrounded me.

Biting insects joined crocodiles and Lions as familiar ogres of my childhood world, all indifferent as to where their next meal might come from – I, we, all of us were prey. In the mosquito-netted cocoon that I inhabited I learnt that my survival lay in denying searching predators their opportunity, or blitzing them with kerosene 'Flit', or bullets, before they got me. Teddy, whose work exposed him much more frequently to infection, regularly 'went down' with malaria and other tropical maladies, and on at least two occasions he nearly died.

At about the time that our terrors of malaria were abating I came across a Japanese wood-block print by Hokusai, which illustrated a story in which the ghost of a man murdered by his wife's lover creeps up to their bed and steals their mosquito net, leaving malarial insects to punish his killers. The ghost's gimlet-eyed skull, conjured by a few deft lines, grins as his skeletal fingers rip away the flimsy barrier between life, death and nightmare.

There is no knowing where a mind in fever can travel, and mosquitoes are not the only villains. Cases of 'sleeping sickness', originating from just a few kilometres away from Mwanza, on some of Lake Victoria's many peninsulas, fuelled yet further fears, this time of the dreaded Tsetse Fly.

Hokusai print of ghost stealing mosquito net.

One evening I eavesdropped on one of those conversations where the adults scarcely disguised their anxieties about both fly and disease (for which, at that time, treatments or antidotes were of uncertain efficacy). That night I stood, paralysed by fear, as the only child in an inward-facing circle of grown-ups. They too were immobile except for their eyes, which swivelled about in their sockets trying to follow the erratic flight of a loudly buzzing fly. Eventually the fly alighted on someone, whereupon the man died. His death was melodramatic: standing at attention, like a soldier on parade, he fell backwards to lie, stiff as a length of firewood, finished.

One by one the adults fell, like skittles, each deathly crash increasing my own paralytic terror until my turn came too. As the fly needled me I felt myself falling, only to wake, in a soaking sweat, but thrilled to be alive, even a bit malarial.

The sensation of falling, however frighteningly drawn-out in a dream, must belong to those real-life fractions of seconds when the body must somehow respond to save itself. Likewise, the opposite moment, of take-off into a jump, may belong to the illusion of levitating.

From a very early age I have had just such dreams. By invoking an intense dream-concentration I could fly, standing erect and without wings, in any direction. During flight I could view a panorama of diverse things from a more detached viewpoint. That detachment from the constraints of being an Earthbound animal, together with evocation of an internal need to focus my mind and body on freeing myself from gravity, somehow made these dream episodes better able to survive the sleep/awake boundary. My levitation became consciously remembered, often with a sense of disappointment that it was all in my mind and not a verifiable part of my Earthbound existence. The reappearance of this dream in adult life confirms, for me, that some parts of every person's mentality never grow up. Its persistence also suggests that other children, perhaps going back into the remotest prehistoric times, must have experienced similar dreams.

If that is so, it is not difficult to imagine that such illusions could be labelled as 'magic' and the dreamer (if able to articulate such experiences to a susceptible audience) could acquire the label 'magician', thereby feeding people's susceptibility to quacks and charlatans, as well as less culpable seers.

Regardless of how cultures have interpreted such dreams, I suspect that some measure of magic omnipotence remains somewhere at the timeless core of every individual's sense of self. What child does not long to be free of all the constraints that are imposed by physics and physiology, by the biology of having two flat feet and the burden of all the social boundaries that parents, peers and others enforce?

Today, cameras inside mini-drones can soar like larks, filming scenes way beyond anything that we kids once thought we could imagine. Allied to eyes that seek pattern and process in nature, my descendants can now absorb scientific and intellectual concepts in an aesthetic state close to rapture.

*Exercises in pattern formation – Vulturine Guineafowls (*Acryllium vulturinum*).*

I remember making quick notes beside sketches of Vulturine Guineafowls as they paired up just before the breeding season. Later, while elaborating my observations into paintings, I found those ostensibly naturalistic and analytical notes taking me into a sensory state that was closer to dreaming than noting.

Returning to the fevers (from which too many of my dreams have emerged), it is not widely appreciated how many (perhaps most) human diseases originate or have been caught from other species of animals, nor how deeply the world's history and geography have been shaped by disease, cholera, yellow fever, tuberculosis and smallpox. Smallpox alone is thought to have killed about 30 million people in the 20th century.

HIV/AIDS, Lassa fever, Ebola, Zika, SARS and Covid-19 (and more to come) are all thought to have been caught from other animals. In some cases, the very reason certain habitats were unoccupied by our kind may have been due to them being ancient reservoirs of disease. The adoption of *Homo sapiens* as a host for HIV, even human leukaemia viruses, can be presented as a direct consequence of the outside world's appetite for Africa's 'wild' timbers, which has brought people into ever-more-intimate contact with countless other forms of unknown wildness. We have Marie Curie and other dedicated scientists to thank for the

Illustrations for Luganda school history booklets.

blunting of disease's impact on us through the development of vaccines and antibiotics, which are but two of the most influential of science's many benefits.

Chance, and living at too close quarters with rats, gave rise to one of the greatest killer diseases ever to afflict humanity. The bubonic plague is caused by a bacterium called *Yersinia pestis*, injected in the saliva of a rodent-loving flea. It is enough to be bitten by an infected flea immediately after it hops off a rodent to suffer terrible consequences as the *Yersinia* invades and proliferates through the lymphatic system. Teddy, who had studied Classics at school, could expound on the Plague of Justinian in 542 CE, when half the classical world perished. Eight hundred years later, he went on, Europe's population was halved again when 75–100 million people died in the 'Black Death'. When this epidemic reached Paris in 1349, 'King Philip the Fortunate' ordered the University of Paris to seek out the cause. Their conclusion that cats, fat on rats, were to blame led to a war on cats, which probably permitted a lot more Parisians to be bitten by infected rat fleas.

*Black Rat (*Rattus rattus*).*

In East Africa we were well aware of plague because at the outbreak of 'Hitler's War' more than 60,000 deaths had been recorded in less than 30 years from Uganda alone, and throughout the war reports of plague deaths kept coming in from all around the lake, so any proximity with rats was avoided 'like the plague'.

In 1941, my parents' anxieties were obvious when, in common with most Mwanza children, I caught measles, a nasty infection thought to have first jumped from cattle to humans in the 12th century. Measles is also thought to have been a recurrent disease, normally restricted to bovines, that long preceded their domestication.

Is body size the measure of a plague? I have even heard the word 'plague' applied to people (as when wild, uncivilised men landed from boats onto African shores). For people trying to raise grain crops, a handsome, sparrow-sized finch can arrive in flocks that can even out-rival locusts in the rapidity with which they strip every precious bed of every head of millet, every sorghum seed. Here the word plague falls far short of the conundrum the birds pose. Of, supposedly, Earth's most numerous bird, I could search in vain for words to convey the marvel of 2 million birds scanning the wet Sudd flats far below them for today's most concentrated, most nutritious belt of seeds among the dozens of catenary zones within the Nile's Sudd. Ever re-morphing, flocks of Red-billed Queleas wheel, whirl, swirl, writhe and roll, filling all three dimensions of the sky with shapes that last as long as one note of a cantata. It is among the grandest displays, the most dazzling of operettas that nature can put on. It is a show that developed over the 10 million or so years that grasses have been probing their own super-diverse ways of surviving in a super-competitive world. The queleas evolved to harvest the ever-shifting seasons of Africa's abundance of grass-seeds; their foraging fields were growing wild rice and scores of other cereals before the genus *Homo* had begun to evolve. Now humans are in direct competition with queleas. We shall swarm with a lot less grace.

Visual Geometry in African Monkeys.

11

LOUDER, SOFTER, BIGGER, SMALLER, FASTER, SLOWER, DULLER, BRIGHTER

In which small changes in size, speed or colour have unexpected and consequences, all determined by senses. Picasso's 'Dinky baboon'. In which animal lineages 'shift channels' from voice to vision, from hiding to swanking, night to day, camouflage to flag-waving.

I was not the only East African kid to convert a baboon skull into a sports car. Mine was painted with 'Ricket's Blue' from the kitchen laundry, and I wedged shards of broken glass into its empty eye sockets to become windscreens. It joined a sort of safari rally, with screeching corner turns and loud revving as it raced up escarpment roads that I had carved into the banks of a drainage ditch beside the Mbeya School sportsground.

Baboon racing car, 1944 model.

Without our knowledge or permission, Pablo Picasso inverted our school model-making when he allowed a Dinky toy Dodge Saloon car to stand in for a baboon matron's head. Like Pablo, we too registered the expressive logic built into the architecture of aeroplanes and motor vehicles, but for him the toy became a skull while for us it was the skull that became a toy.

East African children of my generation had ample opportunity to examine road-kill so I acquired an early interest in likening animal skulls to the functional

*Pablo's Buick baboon
and baby.*

styling of Dinky or Hornby pressed-metal models of Dodge trucks, Rolls-Royce saloons, Bedford lorries or Jaguar sports cars. The distribution and shapes of spaces allocated to seeing (behind transparent windscreens), thinking or driving, the passenger seating and engine – they all differed from brand to brand and from species to species. All elicited satisfying noises, and I remember one kid exclaiming, 'Ahh! Lovely smell of engine.' Later I watched a youthful owner caressing the wheel guards of his MG sports car as if it was the frontage of a girlfriend.

Of all the skulls I examined, rodents offered the most eloquent and obvious expression of diverging specialisation. Earlier on I mentioned *Zenkerella*, which

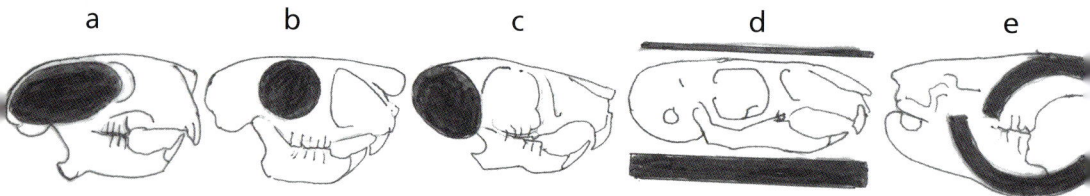

*Five rodent skulls showing capsules given over to: a) brain; b) sight; c) hearing;
d) flattening; e) non-stop biting.*

is relatively unspecialised and is possibly the most primitive of all living rodents in Africa. Compare its neat little skull with those of real specialists. Note the hugely enlarged ear capsules of desert jerboas, rounded sockets to encase the torch-reflecting orb-eyes of springhares, or near-blind blesmols that use enlarged incisors powered by big chewing muscles to tunnel through the earth like some sci-fi mining machine. Both pectinators (Somali equivalents of crevice-dwelling guinea-pigs) and Dassie Rats have flat, squashed skulls, all the better to squeeze into the refuge of narrow rock crevices. Every sort of adaptive specialisation has shaped, enlarged or shrunk whatever capsule its organ occupied. If the relative size of crania is a guide to intelligence, *Zenkerella* has the feeblest of molars but quite a decent brain-box. Could a marginally superior intelligence have contributed to their survival over tens of millions of years? Even helped them evade genets and nosy biologists?

If sensory and other functions leave their mark in skull capsules, even fossilised ones, their living expressions can only be witnessed in the behaviour and anatomy of living animals.

Here we have subjects where the single-strand, linear description I am attempting here, even supported by two-dimensional diagrams, seems ill-suited to explore such structures. Where Picasso found a witty short cut to a problem in self-expression, I turned to 3D modelling for some resolution of such conceptual problems, along the way taking some pleasure in their construction.

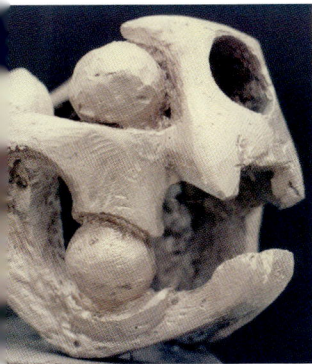

*Sculptural rendering of otter skulls: (left) Sea Otter (*Enhydra lutris*); (middle) North American River Otter (*Lontra canadensis*); (right) Giant Otter (*Pteronura brasiliensis*).*

A fascination with otters and their marvellously liquid movements led to my examination of their skulls. To my surprise, these diverged almost as much as rodents and road vehicles. In a fit of exploratory energy, I constructed three models. One otter, which foraged for eels, crabs and flatfish under pebbles, roots and rocks, resembled the Dassie Rat in having a relatively flattened head. By contrast, Giant Otters hunting big fish in Amazon waters have Lion-like maws, served by vertical muscles anchored onto deep, long skulls. Pacific Sea

Otters, diving for clams and other shellfish, have rotund heads with deep, well-muscled, nut-cracker mouths, fine exemplars of form determined by function.

My most memorable tutorial in skull morphology concerned the enlargement and transformation of a mere mandible, and my teacher for this unforeseen lesson in anatomy and behaviour was a young Hippopotamus. It admonished me with its own mandible and lips.

As an extremely stupid and fearless youth, I had joined a live-animal-catching team on a nocturnal hippo-hunt, during which our quarry, a more than 2 m-long juvenile, adopting its best anti-Lion strategy, had backed into an impenetrable thicket to face off a gallery of head-lamped tormentors. Armed with a very large purse-stringed hessian sack and an over-rated belief in my own agility, I tried to hood that unhappy, quivering hippo. Brushing the sack aside, it came at me, mouth agape like an erupting volcano, ramming its broad, horny lower lip into my groin. The impact hurled me backwards just before the hippo's upper mandible came down with a clash of ivories and a gush of saliva that soaked the front of my dungarees. Before I could get up, the hippo had backed even deeper into the thicket. I had escaped being disembowelled and/or decapitated by the sheer force of that lower jaw's impact. How come I escaped?

In the first place, hippos have transferred the task of harvesting grass and herbs from cropping teeth or tongues to broad, flat, horny lips – upper and lower. These shielded me from the prongs of still-growing incisors and canines. In adult hippos these fore-teeth operate somewhat like the tines on a stag's antlers when males, while immersed in water, test each other's weight and power by ramming together their hugely reinforced lower jaws. Human jaws are flimsy

Tracing of a tracing (by D. Coulson) of a rock drawing of man felled by Hippopotamus (several thousand years old).

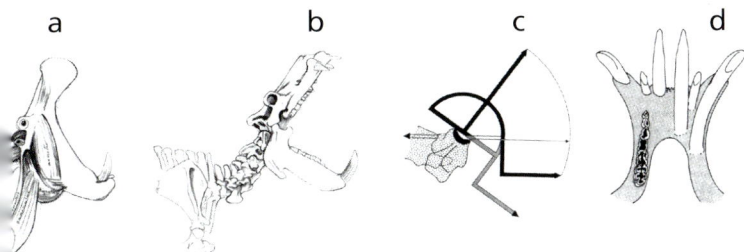

ABOVE: *Mandible clashing between Hippos.*
LEFT: *Hippo gaping: a) musculature; b) bones; c) diagram of the skull and jaw rotation; d) embedding of mandible teeth.*

a   b   c   d

hinges below our big, heavy heads so it is hard to reconcile the hippo's near inversion, in which it is a relatively light upper mandible that flips upwards (in spite of all the teeth, eyes, ears, nostrils and brain-box that it carries). Hippo mandibles have become hefty weight-testing rams, mostly employed while immersed in water. Whatever I learned that night was a lesson not to be recommended or repeated. Nonetheless, the experience helped inform a series of diagrammatic and anatomical drawings I made a few years later when a Park-culled hippo cadaver was the subject of my attention.

Among primates, the size and weight of lower jaws may vary a bit by diet but the depth and splay of primate mandibles can also reflect, indirectly, their ability to generate noise. African Guereza monkeys have deep mandibles that enclose the apparatus for their prolonged washboard-like *basso profondo* choruses, but I challenge any drummer to hit a beat anywhere close to the single, deep explosion of an adult male gentle monkey announcing his existence – 'Pyow!'

This voice demands your attention, even if you're a kilometre or more away, and it's pitched to carry through all the vegetable obstruction of a forest, woodland or thicket.

Portraits of 10 regional varieties of gentle and putty-nosed monkeys
(Cercopithecus (nictitans) *group spp.).*

An acoustic engineer would be hard-pressed to design a more assertive note, while a jazz musician might smile in approval at a pyowing monkey's calculation of seconds before it changes pitch with a deep, resonant 'hoooh', a call that seems designed to reassure and defuse tension. One 'pyow' is simply a statement: 'Me – here', serving as an alert, a rally, even a banal 'let's go'. A volley of 'pyow' calls is more serious, especially if accompanied by sharp, urgent hacks. I once glimpsed a Leopard sneaking away in the undergrowth after a consortium of gentle monkeys and mangabeys gathered above and subjected the big cat to decibel hell.

The 'pyower' is always a hefty (up to 12 kg) male with amplified vocal apparatus. Most frequently he is the dominant member in his group. The volume and force of his call are partly a function of his size and age, and the resultant bulking out of his larynx and vocal apparatus.

No less than a chorister who leaves the boys' choir as his Adam's apple keels and his voice 'breaks', a male gentle monkey must also await public recognition of his maturity with the 'breaking' of his voice and, eventually, social deference for his more imposing and noisier presence.

Gentle monkeys (including their spot-nosed branch) are among the more widespread of forest monkeys, so their 'pyow!' is probably one of the most familiar, certainly one of the most assertive, signals commonly heard in African forests.

Almost wholly arboreal and intensely tuned in to the auditory channel, gentle monkeys range over enormous stretches of tropical Africa wherever there are forests, even remnant ones. Just so long as it was forested, every narrow valley and every range of moist hills or uplands in tropical Africa could become inhabited by gentles (an alternative, biogeographic, name is the tongue-tripping 'Greater Periphery Monkey').

An early start, possibly as much as 9 million years ago, allowed them to spread over the tropics of an entire continent and along the littoral of its eastern seaboard. What explains their ability to maintain such a uniquely extensive range over millions of years?

Compared with other guenons, they are unusually versatile, being able to subsist for prolonged periods on leaves, supplemented by any other sort of edible plant or animal (for which they are champion foragers). They are also alert, fast in their responses and fecund (soon recouping sustained losses from eagles and other carnivores, people or epidemics). Their limitations derive from a relatively large size and a poor ability to survive for very long away from moist, shady forests. The finer, outer branches of trees cannot support their weight and this limitation might be linked to their strong preference for sticking to the shade – they actively avoid prolonged exposure to strong sunlight and generally avoid feeding on top of the canopy. Wherever forests have persisted, so have gentle monkeys. This combination of pan-African range, the passage of time and repeated fragmentation and isolation due to past changes in climate, has generated more than 20 taxa within what was once known simply as Gentle or Blue Monkey (*Cercopithecus mitis*). Today, their specific or subspecific ranking is in endless contention among taxonomists.

Face patterns in nine subspecies of Cercopithecus cephus *monkeys compared with a generalised guenon (bottom right).*

On the peripheries of Africa's equatorial heartland, the Congo basin, every period of drought in a dry continent has put great pressure upon the larger forms of forest-adapted inhabitants. At the two drought-prone extremes of their overall range, east and west, gentle monkeys survived by dwarfing.

In the vestigial forests of the Horn of Africa, the local gentle monkeys are nicknamed *scimmia nera* ('black apes') and their largest males probably weigh about half as much as others elsewhere. Their 'pyow' calls have yet to be recorded but can be predicted to lack the timbre of big-fella calls. In spite of their diminutive size, these desperately endangered little primates (also called 'Zammarano's White-throated Guenon') are generally included within the gentle monkey diaspora.

In the far western forests of Guinea, similar dwarfing had a quite different outcome. Here the contemporary descendants of those dwarfs are often known as 'Lesser Spot-nosed Monkeys'. 'Lesser Periphery Monkeys' is yet another designation, but I prefer to call all members of this most varied group the 'cephus red-tails'.

For a while, perhaps for a million years, those Lesser Spot-noses might have remained as much 'dwarfed gentles' as Zammarano's White-throated Guenon remains today. Then the advantages of small size triumphed, as those tiddly spot-noses gave rise to an extraordinary evolutionary radiation of small forest monkeys, all with brightly coloured or contrasting face patterns (*see* p. 219).

These monkeys, collectively known as cephus or red-tailed monkeys (and of which the Lesser Spot-nosed are but the westernmost population), became the subject of one of my earliest researches into the evolution of colour signal patterns. I chose them partly because they represented an exchange of channels – from noise to vision; from ear-blasting 'pyow!' to silent but ostentatious face-flagging.

Regardless of fluctuating climates or diets, smaller bodies are particularly favoured along forest margins or patches of regrowth around tree-falls, where small, light, outermost fronds and twigs bear berries, buds, leaflets and flowers that also attract swarms of catchable, edible insects.

Eventually, differences in foraging techniques and habitat preferences must have allowed gentles and cephus/red-tails to co-exist as separate taxa, though even today both groups still interbreed, albeit rarely. These hybrids are generally fertile and tend to breed back into their maternal population. Since this has probably been going on for millennia, wonderfully complex genetic jigsaw puzzles have been created. These have invited a succession of baffled reconstructions by more than one generation of geneticists.

The most obvious secondary cost of dwarfing and more extensive, exposed foraging is greater vulnerability to a wider range of raptors. Ancestral cephus/red-tails were probably under selective pressure for greater alertness and faster responses to escape eagles. Dispersing more widely throughout their food trees, they tend to forage as independent gleaners and this brings its own subtle evolutionary challenges.

Over most of their range cephus/red-tails have come to share forests with their gentle monkey precursor cousins. The overall body colouring of both lineages has remained relatively dull – it is only the facial mask and tail that have become exaggeratedly conspicuous in cephus/red-tails.

My friend and collaborator Jean-Pierre Gautier has shown that vocal repertoires retain a gentle monkey structure but all cephus/red-tails call at a higher
pitch, their throats no longer built to thunder out the ancestral 'pyow'.

Where home ranges are shared by gentle and cephus/red-tail groups, this has
an interesting consequence as cephus/red-tail troops seem to benefit from 'their'
gentle troop's dominant male's calls. The outcome can sound quite comical as a
cephus/red-tail male echoes the dominant male gentle's powerful 'pyows' with
a feeble copy, like an obsequious courtier or clown at court. It is as though gentles have remained in charge of the auditory channel while cephus/red-tails
have taken up the visual channel.

How could dwarfing (initially *within*) the most widespread and successful
of forest monkeys lead on to this switch in channels of communication? The
answer seems to lie in a fundamental change in male/female relationships.

Among the many expressions of male dominance in primate societies are
enlarged size, a louder and deeper voice, plus dominance displays and fights
among males. Dwarfing might have been proportional but overall reductions

Sketches of red-tailed Cercopithecus cephus *monkey.*

in size also tend to reduce differentials between males and females. Today, among cephus/red-tails, almost all sexual differences are minimal and I contend that this was ultimately a byproduct of greater scatter among foraging cephus/red-tails.

Among such troops, not only was it difficult for males to monitor female receptivity but, out on a filigree of branches, any hint of male aggression was easily evaded by females. In short, the successful male had to seduce rather than bully his way to female acceptance.

The evolutionary path to achieving this was extraordinarily oblique but governed by a psychological logic that applies as much to human as to a great many other societies. My grandmother, Bibi Minnie, often scolded me with 'don't stare – it's rude!', and deflected any implied impolitesse by her grandson by changing the subject and averting attention, usually towards the tea-pot.

Threatened primates often emphasise eye-aversion by moving their entire head from side to side, and it is this appeasing gesture that cephus/red-tails have adapted to signal social acceptability across both sexes and all ages. As Bibi Minnie might have put it, the approach of a courting male became 'less rude' if he kept averting his gaze. Once movements of the head had come to signal non-aggression, the entire facial disc served to divert attention away from those insolent eyes.

In all of nature (especially among highly visual birds and fish) sudden movements offer opportunities for the evolution of visual signals, These can become exclusive to a species or, eventually, to entire lineages.

A fellow primatologist once dismissed cephus red-tail twitches as mere 'facial tics', yet it is precisely their transitory flashing that commends such 'tics' as evolutionary raw material. Regardless of their behavioural origins or their anatomical position (face, fin, limb, ear or brow, anywhere subject to muscular control will do), involuntary twitches can be elaborated, even extended and coloured to serve as visual signals. Such signals can be exclusive to a species' private repertoire or serve some wider function.

For cephus/red-tails, facial colours might have become seductive, especially for reluctant females. For this Darwinian human, the challenge of trying to reveal the origins of those face patterns became all-consuming.

Return now to the timing of that newly emergent, far-western cephus/red-tail lineage's extension eastwards, out of its enclave of origin. One of those accidents of history intervened, to literally shape, diversify and determine their subsequent evolution. An Ice Age (which one of several remains to be determined) fragmented Africa's forests and, with it, created discrete populations of newly emergent cephus/red-tails. In my study of how a diversity of facial patterns could emerge from a common origin, I found one tiny relictual population of monkeys in the Niger Delta, the very locality where I supposed the ancestral cephus/red-tail lineage first began its eastward march. Named after Philip Sclater (a very great zoologist and biogeographer), Sclater's Monkey has

Distribution of the Cercopithecus cephus *group of monkeys in the central forest block: Sclater's Monkey at top left, Moustached Monkey at centre and various red-tails to the right (centre and east).*

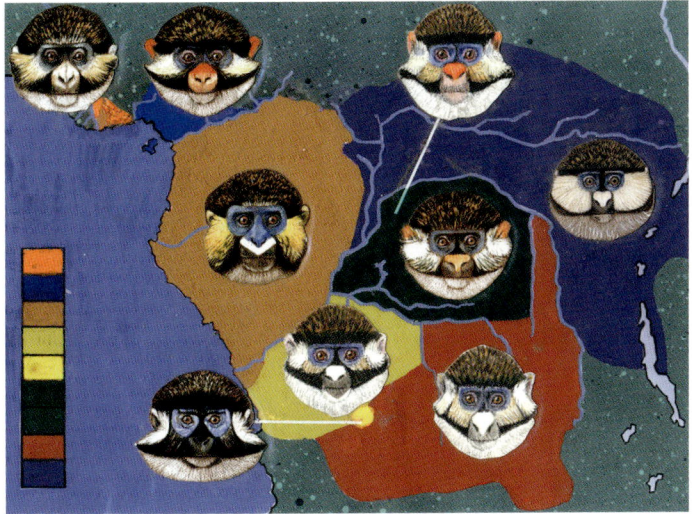

Field notebook sketches of red-nosed C. cephus *monkey.*

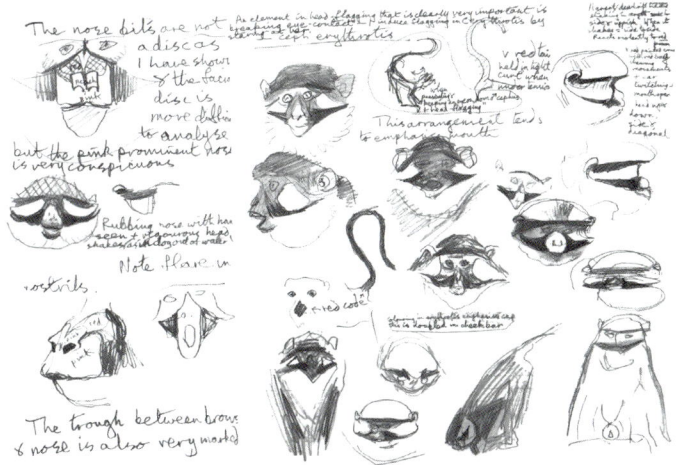

a complex facial pattern. Its significance is that the face patterns of *all* other cephus/red-tail populations can be derived from Sclater's somewhat gentle monkey-like face by various small alterations in colour or 'edging', to generate radically different visual 'facial flags'.

Cephus/red-tails north and east of the main Congo basin sport an incandescently white nose-spot, at the centre of a white facial disc bordered by a thin black margin that looks as though it was drawn by a pair of compasses.

The detailed genetics of this radiation await study, but among the many conclusions that have already emerged is the likelihood that these extraordinary descendants of gentle monkeys, first dwarfed in the far west tropics, became the most recent colonists of the vast forests of the Congo basin and central Africa.

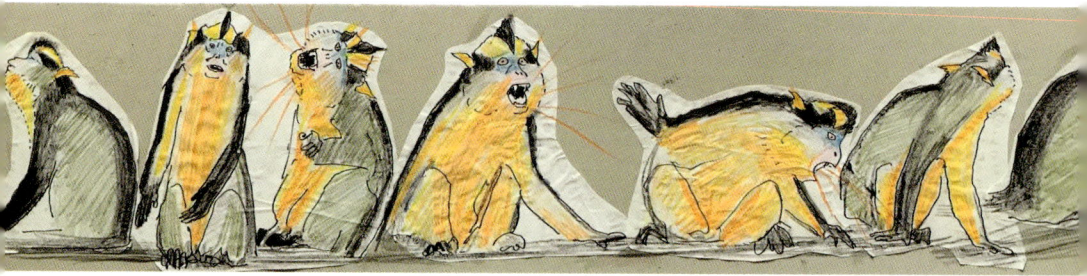

*A single male Crowned Guenon (*Cercopithecus pogonias*) calls (left to right).*
Each call presents a sunburst of orange crowned by a technicoloured mask.

I originally chose to study cephus/red-tails because study could be restricted to the manageable unit of a facial disc, but from the start I was confronted by a (likely earlier) radiation of similar-sized guenons, the mona monkeys, that co-exist with cephus/red-tails over most of central and western Africa.

What allows these similar groups to co-exist?

The most obvious separation lies in where and how the two use the forest. Cephus/red-tails, like their parental gentles, prefer forest edges and they seldom penetrate deeply into dense forest, instead foraging intensively within relatively narrow zones where the lower strata of the forest are preferred refuges from eagles. Monas, though, forage more widely and at higher levels of the canopy, deep within mature forest, catching quite a high proportion of animal matter with small, dexterous hands that have shorter fingers than those of cephus/red-tails. They are very accomplished and patient 'hiders' and respond to eagles by freezing in dense tree-top tangles.

Unlike cephus/red-tails, with their unambiguously far-west origins, monas seem to have their evolutionary origins in central, more permanently forested, equatorial Africa. Mona ritualised displays tend to involve swinging the entire forequarters from side to side, instead of just jerking their head. At such times, one mona species, the Crowned Guenon, flashes a high, banded, Mohican-like crest. Its closest relative, Wolfe's Guenon, has orange-plumed ears and a dark Zorro mask, and both (genetically close) cousins have also begun to diverge in their voices, the Crowned giving a short sharp hack while Wolfe's hack is more like a grunt. Vision is the only channel in which colour counts, as I set out to explore in cephus/red-tail monkeys.

The saga in which forest-adapted animals can afford to be colourful and conspicuous while their savannah cousins are khaki-drab finds its match among Africa's most ancient, extraordinary and fecund of bird families, the turacos. Dry-country turacos are called 'go-away birds', or 'crinifers', and wear grey or brown uniforms, while forest turacos sport vividly coloured plumage. Sexes can hardly ever be told apart, so much of their social behaviour remains poorly understood. Even so, most turacos are known to be monogamous and paired birds collaborate to raise their clutch of one or two young. Such a pair nested and

*Sonograms and 'viseograms' of Crowned Guenon (left) and Wolf's Monkey (*Cercopithecus wolfi*) (right).*

raised their clutch right outside our farm dining-room window, allowing us a front-seat view of their rearing.

Many tens of millions of years before stained glass was invented, forest turacos evolved translucent wing feathers that are flashed not only in flight but also in courtship. One bird, supposedly a male, apparently fully aware of the sun's position behind him, flies straight at his presumed inamorata. Blocking her way, head-on, the arriviste rears up and, like a traffic stop-sign, exposes two incandescent crimson discs with a gloriously symmetrical wing-flash. The flash is so

*Turaco (*Musophagidae*) drawings, 1955.*

Glimpses of turaco species feeding their young.

fast it is easily missed, especially by lazy human eyes. As a young naturalist I was so impressed by this display that I not only adopted several turaco species as enchanting aviary birds (including the biggest species, a Great Blue Turaco, though he was free-ranging, never caged) but I also designated the Rwenzori Turaco as our family emblem. In homage to their wing-flashes I made a stained-glass window which ornaments our kitchen to this day.

As with monkeys, telling aspects of turaco biology are concentrated in facial details of these avian dandies. Here, in coloured feathers, skin and beaks, turacos reveal that they are compulsive gift-givers, bringing fruit and flowers to mates or nestlings. Fresh fruits are sometimes served up by beak-tip tongs, while donors monitor the recipient through eye-ring lorgnettes that out-do the colour of the gift by contrasting vermilion eye-wattles against the cleanest and clearest of leaf-green plumage. Very often the donor has already swallowed or macerated fruit, flowers or leaves so it is the entire gape, not just the beak, to which the recipient must orient. To facilitate and guide partners into this extremely intimate open-mouth to open-mouth transfer, the donor's face wears sharply outlined tracts of black-and-white feathers that serve as pointers. This act of kiss-like giving is further embellished by lipstick-scarlet beaks, while the facial streaks, like white gloves on a royal banquet servitor, proclaim that such liberality has its own evolved elegance.

Birds, like mammals (even, perhaps, the earliest primates), had already evolved before Chicxulub, but both invite controversy about how far their radiations had proceeded before the meteor hit. For turacos, among the most primitive of all birds living today, some genetic studies suggest their ancestral lineage might have emerged as early as 75 million years ago, at a time in which their kind ranged far outside Africa. The great diversity of living turacos likely embraces an evolutionary 'Story of the Ages' that spans many tens of millions of years.

Turaco wingspreads.

A highly lucrative and cruel trade in these and other birds now threatens their very existence. Trappers for the contemporary zoo and cage-bird trades are stripping Senegal, Benin and parts of many other African countries, all effectively competing to exterminate turaco species, among many other orders of life that are all vivid expressions of the very process that brought humans into existence.

On the day that Tanzania celebrated its independence, President Julius Nyerere decreed that every caged bird would be set free, a release and ban on trade that should have been enacted throughout Africa up to this day. It is beyond dismaying to know that authorities in modern Africa, including Tanzania, now allow the export and potential extermination of much of its fauna, including turacos.

The officials granting such licence are seldom knowing criminals – they are just ignorant of the ecological fratricide they are permitting under their watch. Here, again, we desperately need pan-African education in the privileges and responsibilities of being custodians of the richest biota on planet Earth. We are among the first generations to explore such unexpected evolutionary histories behind the skulls of otters and humans, the jaws of hippos, the masks of monkeys and the kisses of turacos.

Rwenzori turaco (stained glass).

Spooning in the Mud.

BEHAVIOUR DRIVES MORPHOLOGY

In which the immaterial and ephemeral governs the material and durable.

'Oh! how hideous! Disgusting!' My neighbour had heard the ruckus that night and came over to find out what it had been about. A wild genet had seized a Hammer-headed Bat just as the bat had started to 'sing' in a tree beside my house. Somehow the little arboreal carnivore had managed to fatally wound the magnificent bat before it fell to the ground, but the genet was too timid to retrieve it. At dawn I had found the corpse that set off the neighbour's horrified response. 'Is a saxophone ugly?' I asked. Her horror was not without precedent, because this bat's scientific name is *Hypsignathus monstrosus*, which could be translated as 'tall-jawed monster'. As for my reference to a musical instrument, I was already familiar with this equatorial forest fruit bat, the males possessed of a voice to out-horn a foghorn any night. I was grateful to the genet for providing a specimen for me to dissect, because I wanted to compare its extraordinary vocal apparatus with that of its closest relatives, the appropriately named epauletted fruit bats.

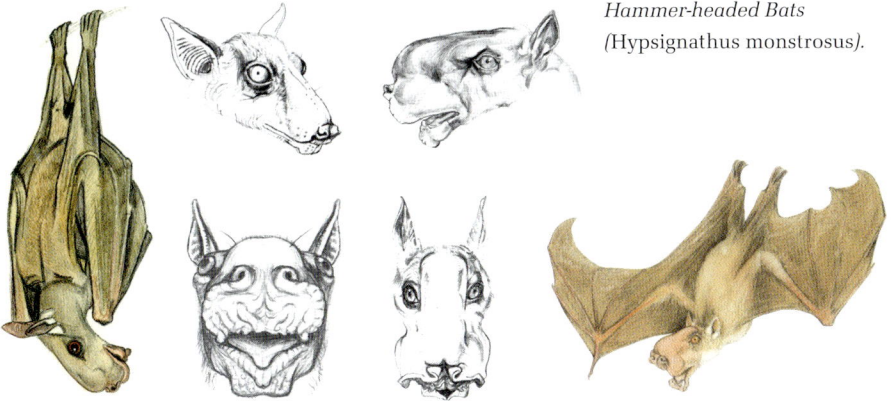

Hammer-headed Bats
(Hypsignathus monstrosus).

Why 'epauletted'? Because the males of most species wear deep glandular pockets on their shoulders, within which are hidden pompoms of the finest, silkiest and whitest-of-white hairs. Investing shoulders with such complex structures seems superficially quite quixotic, but the 'purpose' of these epaulettes is hinted at when males gather to out-sing one another, ornamenting each contralto phrase with a shrug and token flap of the wings. It is that shrug of the shoulders that exposes those incandescently white epaulettes. Females, drawn in like

*Straw-coloured Fruit Bat (*Eidolon helvum*). Sketches from Kampala roost.*

groupies by the magnet of a concert, flounce around on soft wings until particular songsters are so compelling that females alight beside their chosen maestro and invite coition. Some fly off to pick a fig instead, unimpressed. Call it sexual selection by epaulette? By bat-song? Or even by elastic power of diaphragm?

Fruit bats, left to right: African Rousette, Benguela, Golden, Dwarf, Singing (and singing portrait).

I pause here to stress how easily these common bats can be overlooked, especially during daylight, because, when they are not actively foraging or flying, they hang in groups, strung out like wilted, sienna-tinted leaves along a frond, or else alone under the umbrella of a drooping leaf.

Bat wings, powered by gigantic chest muscles during flight, fold around the body like a camouflaged blanket when at rest, even covering heads and shoulders. Even at night their calls are commonly dismissed by the bland and knowing as the 'croaks of mere frogs'!

We can actually reconstruct the logical sequence of adaptive innovations that culminate in a choir of male epauletted fruit bats. Start with the 'fruit' bit of fruit bat. Subordinate to the reproductive needs of trees, the bats need night-vision eyes and sucrose-sensitive noses to ferret out their food, without being foxed by unpredictable patterns of fruiting. This necessary nomadism makes keeping in touch with one another difficult, as well as posing problems for the bats' own reproductive needs.

Humans bell the faithful to prayer, bell their livestock to retrieve wanderers, and bell their ankles to infuse the stamp of animal feet with the rhythmic fever of a crowded dance-floor. There are a number of avian species that could deserve the name bell-birds, then there are frogs, even insects, with bell-like chimes – so the far-carrying calls of these bats have many precedents in the voices of other animals. Few, however, can maintain a chorus for up to three hours, as epauletted male choirs can. Nor can they reach the decibels of a male Hammer-headed Fruit Bat in full voice. Given the volume of sound, it is scarcely surprising that timid people find nights in a tropical forest frightening, particularly when Tree Hyraxes join the cacophony with screams that sound as if a baby is being strangled.

The evolution of a flying loud-speaker. The laryngeal structure in three fruit bat species: (left) Epomophorus; *(middle)* Epomops; *(right)* Hypsignathus.

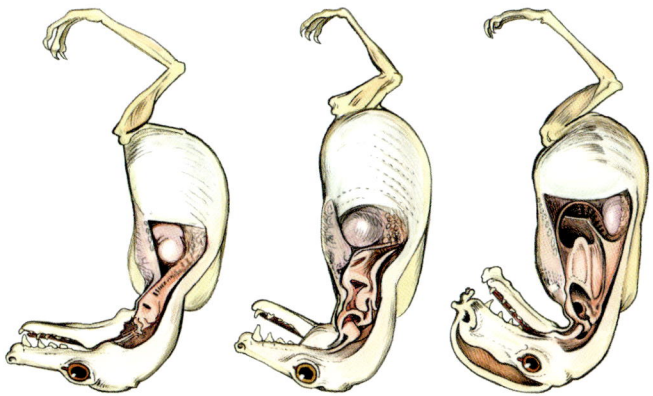

Earlier I inserted the word 'diaphragm', to draw attention to the deep, bellows-like force needed to expel each note from a male's chest. Pumping the lungs, flip-flopping the wings and expelling the call combine to flex shoulder-skin on the bat's *visually* most exposed surface (that is, for a viewer hovering just below the singer). It is also the site best suited to waft the fruity, musky fragrance of *L'homme Chiroptère* toward any female approaching from that nearby fruiting fig tree.

The following sequence illustrates how physiological and anatomical structures evolve, by natural selection and step by step, *in response* to a succession of behaviours.

Begin with wings, the result of a behavioural need, in an arboreal ancestor, to get from one tree to another. Just as distant, those ancestors mapped out their lives employing all their senses. At its simplest, our epauletted males shout to 'keep in

Epauletted fruit bats.

touch', but as competing individuals, males try to out-shout other males – hence the choirs, and the internal 'diaphragmic' effort put into each spasmodic call.

Being olfactorally challenged, humans are poorly placed to judge just how sexy any individual bat's musk might be. Nonetheless, we can infer that the particular fragrance inside every epaulette plays an active role in appealing to female bats. We can also infer that scent-guided frugivores are particularly susceptible to tracking ripe scent cues. That sensory susceptibility makes it likely that a concentration of scent glands on the shoulder became the second lure in the evolution of epaulettes.

'Blind as a bat' certainly has no relevance for wide-eyed, dog-faced fruit bats. Nonetheless, in any hierarchy of nocturnal sensory perception, vision is likely to come after sound and scent. The elongation, thinning and drastic bleaching of epaulette hairs and their implantation within an evertible pocket of shoulder skin were, almost certainly, the later or last refinements in the elaboration of a structure as extraordinarily eclectic and complex as a fruit bat epaulette.

Most important of all, each evolutionary increment emerged from the night-to-night behavioural needs and responses of a single lineage of fruit bats. A succession of peculiarly batty behaviours drove the evolution of a succession of uniquely batty morphological innovations.

Among bats and other mammals, as well as birds and fish, it is the males that emerge, over and over, as the more ostentatious sex. Why? Because how male blandishments impact on females matters. She judges him on how he looks, sounds, smells or feels to her. In matters of reproduction her appearance can be irrelevant because she is the selector. He is the one who must compete to be selected and how he manages his sensory impact on the selection panel is crucial in the paternity stakes.

For me and many others, life's central reward has been the survival and welfare of offspring. Their existence, like mine, emerged from the countless happenings of living, while their sex, like mine, was incidental to the many pairings of parents.

For a few of us, life as it can be lived right now has been blessed by opportunities unique to our time, but Covid-19 has reasserted some of the risks, dangers and constraints inherent in our animal origins. Those constraints govern the courtships whereby every one of us begins through the uniting of two separate cells. Consider some of the techniques of courtship in other beings.

The beaches of southwestern Africa once hosted many breeding colonies of Cape Fur Seals, in which huge, blubbery males weigh in at up to 350 kg while females average about 60 kg. Not surprisingly, some females become casualties of copulation. Beaches providing safe landings have been a precious resource for seals ever since some Nordic mammal took to a marine existence some 25 million years ago. Those northern ancestors of fur seals first colonised the south Atlantic some 5 or 6 million years ago. Probably sharing an ancestry with bears, seals differ from whales in needing *terra firma* to copulate, give birth and, in some species, to survive the first few months of life.

*Cape Fur Seal (*Arctocephalus pusillus*).*

The gross disproportion between fur seal sexes can be attributed to the scarcity of suitable beaches. Every male must defend and maintain whatever area of territory his size and aggression can claim. Bellowing and honking, every giant slug of a male rears up to buffet and bite his way to dominance. Among fur seals, male supremacy has a bloated body and bloodied face, but their evolution can be decoded in terms of constraints deriving directly from our planet's very diversity, the distribution of its opportunities and the behaviour that life itself demands.

If humans can judge male fur seals competing for as many copulations as possible as a process that has created monsters of ugliness, the very same sexual drive has created paragons of beauty, emerging from equally accessible histories.

I once had the opportunity to visit Papua New Guinea. My pilot brother-in-law flew me along the Huon peninsula where former coral reefs, aligned in long, long terraces, rise out of the sea to reach mountainous heights. Here the Pacific Ocean, over its own giant plate, meets the northward-cruising Australian Plate (a.k.a. 'Eastern Gondwana'), pushing up former sea-bed to form the new, great, mountainous island of Papua New Guinea.

This gigantic island, the product of Australia's bruising punch into the Pacific (and some of its surrounding islets and volcanoes), is sufficiently new for most of the predators and hazards of land life elsewhere to be absent. Papua's volcanoes have made the island exceptionally fertile, so fruit, flying animals and a few stray mammals have flourished there.

Among the earliest colonists were a few dowdy crows, the females of which suffered so few constraints they could feed, nest, raise their crowlets without needing to share the task with a mate, and live a sort of avian feminist dream. Their only constraint was a momentary need for sperm, and in choosing what male might be allowed the privilege of copulation they not only became exceptionally picky, they demanded dazzlement. It was not only the females that lived in an avian paradise. Males, free of all domestic chores, could devise and stage ever more elaborate dances, and develop the ever more inventive foreplay that females demanded before granting their favours. Immured on these fruitful, nearly predator-free islands, those crows became birds-of-paradise, taking choreographed contrasts in colour, shape, outline, movement and iridescent bling to extravagant extremes.

There can be few things more gloriously inventive, as 'original', as the plumage of birds-of-paradise, yet the evolutionary logic behind a singular place becoming a paradise for birds, even for birds as funereal as crows, has become accessible as never before. The logic has been worked out by friends and colleagues I have met and supped with.

Take that avian pole-dancer, the Wilson's Bird-of-paradise, a small crow that performs its histrionics around a beak-pruned stem sprouting from the forest floor and, perhaps, beneath stage-lights created by leaf pruning up in the canopy. On this miniature stage, the performance follows a practised routine consisting of ever more surprising flashes of contrasting and previously concealed colours.

*Wilson's Bird-of-paradise (*Cicinnurus republica*) a sculpture dedicated to David Attenborough and all the BoP gang.*

Hidden tracts of vermilion red on the bird's broad back and rump are followed by the opening of a perfect disc of dark metallic green: the bird's thin but very extensive ruff. Throughout the show, two delicate feather spirals quiver in front of the tail, subtly hinting at oversight by two empty, ghostly eyes. A bald, sky-blue pate is the only feature shared by our flashy pole-dancer and any dowdy female that condescends to be his audience. Even here his crown is several degrees brighter than hers. His act culminates in a single flash of searing yellow from a tract of canary-yellow feathers on the neck and, finally, revelation of a triangular yellow gape, so sudden and so surprising that even a human heart misses a beat or two. In homage to my friend and fellow lover of birds-of-paradise, David Attenborough, I once made an enlarged model of Waigeo's uniquely ravishing inhabitant.

Ex-African humans arriving in New Guinea, flightless, monochrome as crows, were so awed by the creativity they found on their land-falls that they devised

TOP LEFT: *In Kuria country, it's the girls that choose.* TOP RIGHT: *The most magnificent takes centre stage.* BOTTOM LEFT: *A Kuria warrior poses.* BOTTOM RIGHT: *Swaying and marching to the drums (photographs by Dorothy Kingdon, 1939).*

dances in deference to an evolved creativity in birds that they can never equal. A single gathering of male dandies displays the plumes of many thousands of birds. Even so, the efforts of male Papuans at one of their *sing-sing* dances, painted up and by contemporary standards almost as carefree as their bird models, can be pretty bewitching, not only for rival men and courted maidens, but for anyone lucky enough to attend.

I grew up on the margins of Serengeti, watching Kuria friends making themselves up for dances that sought to match the displays and ceremonies of fellow plains-dwellers. The men borrowed the plumes of Ostriches, they pronked like gazelles, they pranced like cranes and whydah birds while colobus monkey pelts provided the tasselled pinafores that swung to drumming that out-choired and out-syncopated any colobus chorus. Their faces became tableaux for vivid red patterns borrowed from bishop birds: 'Now! Am I not more beautiful?' Kuria maidens are susceptible to beautiful, beautified men.

Turning from days and nights filled with glorious noise and drama to more silent ones, I remember cool evenings during the early rains when a Barn Owl left its roost inside the cavity of an old stinkwood tree to fly through the coffee trees, more lightly and more silently than any moth or butterfly. The blunt pod of its disproportionately large face seemed to stare down through enlarged eyeglasses, as if the owl was scavenging for a lost coin. Perched, it was an inveterate peek-a-boo, keeping quite still while the shifts of its face to left and right were always horizontal and rather toy-like – the owl was trying to 'take a fix' on some hidden scuttle. Was it trying to look through the grass? Or was it trying to listen in spite of the cover – or was it both seeing and listening?

As an avid collector of road-kill corpses, it was not long before I picked up, skinned and mounted a dead Barn Owl to grace a shelf above my bed (to the

Silent Flight.

*Barn Owl (*Tyto alba*) road-kill (lithograph, 1954).*

horror of Bibi Minnie, my grandmother). For this ornithological taxidermist the contrast between a feathered and a featherless Barn Owl was memorable. The evolutionary meaning behind that mismatch has only been revealed by recent research. Barn Owls flying with their vision impaired can still find prey, yet those that have been shaved of their facial ruffs have the greatest difficulty, demonstrating that those discs really are receiver dishes that direct sonic waves toward the owls' ears, and have nothing to do with vision.

Without its facial feathering the owl resembles a pathetically skinny, naked and feeble vulture, but for the road-kill afficionado every one of its many hundreds of facial feathers is revealed as an individual triumph of creative sculpting.

It is as though every feather was competing for title as the most elegantly curved or rebated spoon, spatula or micro-tile. These feathers are densely packed but each one contributes to moulding an almost impervious, bulbous superstructure around that silly little vulturine skull.

Those spectacle-like discs, compounds of very numerous, individually moulded feathers, are, in fact, external 'ears' just as much as those soaring up from a Serval's crown, or sprouting, like sculpted mushrooms, from the sides of your face.

Barn Owl facial feathering with single spoon-like facial feather, enlarged.

Research has shown that Barn Owl facial feathering channels sound waves into apertures that are lower on the right side of the bird's skull (but tilt very slightly upwards) whereas the hole is higher on the left side but tilts downwards. This means that the relative loudness impacting on each ear is different. We now know that the discrepancies between left and right ears allow owl brains to compute '3D fixes' on shrew squeaks. Nocturnal shrews, being very numerous and extremely noisy, are the most obvious and easiest prey for Barn Owls in Africa (and I can confirm this from my consistently teasing out more shrew than mouse bones from numerous Barn Owl pellets). How I wish I had known about all these refined triumphs of engineering while I was a teenage naturalist. Nonetheless, Barn Owl behaviour and anatomy demonstrate how a single sense can dominate an animal's appearance, and define its niche.

Visual cues may have come last in the evolution of nocturnal bat epaulettes. But for mainly diurnal Painted Dogs or African Wild Dogs, *Lycaon pictus*, it seems that their piebald patterns derive from a specific detail of their parent/ offspring relationships. The consequences of that relationship go on to play a rather different role in the biology of adult packs. George Schaller thinks the strange, swirling blotches of black, white and yellow ochre plus white tufts on the end of tails help a scattered group of fast-moving hunters cohere as a pack.

Painting the African Wild Dog or Painted Dog – an evolutionary progression of pattern.

Painted Dogs.

True enough, but what sort of behaviours led to such a radical departure from the usual more subtle grey and brown camo-combats of wolves and jackals?

A Serengeti Painted Dog, its blotched, black belly ballooned with hot meat stripped off an Impala killed several kilometres away, is one of several trotting towards the old Aardvark burrow where a litter of pups is holed up. At the wheel of my veteran Land Rover, I follow, keeping my distance. This pack has been habituated by pioneer zoologist Willie D. Kuhme (to whom I and most subsequent scholars of canine biology must acknowledge deep indebtedness), so the dogs ignore the vehicle. On being besieged by tail-wagging puppies, each adult throws up its belly-load of meat, seeking to regurgitate its precious gift into the crowd's centre. It was provisioning parents within this particular pack that led Kuhme to describe what an Australian friend called 'Tall-Poppy Puppy Syndrome'.

Even the fiercest dog-hating human is capable of betraying some fondness for puppies, domestic, wild or Painted, so it is surprising to see adult Painted Dogs snap or bare their teeth at any tall-poppy puppy that detaches itself from its seething crowd of siblings. It would seem the adult feeder punishes any tendency to rise higher or to loiter on the margins, thereby ensuring that every pup has an equal share.

This egalitarian conditioning in puppyhood seems to carry over into adulthood, where a short ceremony called 'The Meet' initiates every set-off for a hunt. Each member of the pack rushes frenetically from fellow to fellow in what looks like a vague replay of den-mouth puppyhood. Every pack member alternates the postures of feeder to fed, quick-fire switches from dispensing senior to begging junior.

There is a visual dimension to 'The Meet' and its puppyhood precursor. I posit that the visual perceptions of ancestral parent dogs favoured anything about the behaviour and appearance of pups that served to dissolve pup outlines. Painted Dogs, adult or infantile, have coat patterns that reinforce, in almost cinematic fashion, the impression of fast movement and visual contrast while dissolving the outline and distinctness of any one individual within the pack. I think that the peculiar patterns of Painted Dogs were selected by the eyes of their beholders, beginning in the eyes of parents seeking to deliver fair shares among puppies.

The carnivores' ability to kill prey is based on teeth and temperament. Among Lions, huge clawed paws augment the clamp of fearsome jaws, and a hard-boned, muscular 120 kg lioness has enough killer instinct and heft to subdue and kill prey more than twice her own weight. Descended from solitary big cat ancestors, the power of a lioness reinforces her fierce intolerance of any unrelated outsider of her own species. Theoretically that should work against Lions becoming social yet it is its very perils that deter any maturing cub from leaving her maternal pride. If the same were to apply to males, Lion society would suffer chronic inbreeding. Instead, maturing young males effectively avoid mating with sisters and mothers by leaving their birth group to face the ferocity of foreign prides. Male Lions have the advantage in being heavier, and at 160 kg an adult male can take on all comers. If several strong young brothers team up, they may well succeed in killing or ousting the males of a neighbouring pride. Such confrontations elicit broadside strutting displays in which the males walk on tiptoe, lashing tails aloft, always staring down the opposition. It is this behaviour that seems to have favoured the evolution of Lion manes. Manes enlarge their owners' silhouette, deterring female attack and blunting the stranglehold

ABOVE The Lion's Maw *(terracotta sculpture).*
LEFT: *Lion symmetry.*

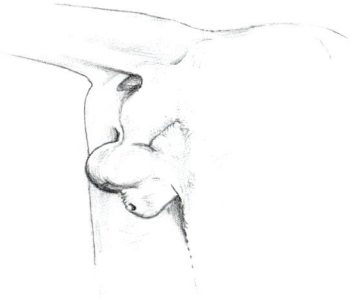

ABOVE: *Male Lion (*Panthera leo*) genitals.*

LEFT: *A male Lion struts his stuff.*

of a rival's teeth. This blunting might have favoured the evolution of manes but their visual effect is indeniable. Like cockerels and many other sexually dimorphic birds, here is an instance of a morphological embellishment evolving in response to some detail or details of specific behaviour.

The symmetry of Lion faces, their manes and beards, their rippling muscles and prominent testicles all serve to belittle even the shaggiest and most ferocious-looking of male humans. Numerous human cultures in search of symbols of dominance, especially male or martial dominance, have borrowed various male Lion attributes and behaviours. At the Ethiopian Court, trophies stolen from Lions were once obligatory details in the dowries of prospective bridegrooms. In Syria, an ancient bas-relief depicts a maiden standing behind her warrior groom as he slays Lions.

Pastiche of an Assyrian bas-relief.

*Comb Duck (*Sarkidiornis melanotos*) displays.*

In birds, as in Lions, sideways struts serve to present the viewer with an enlarged presence. The boldly pied Comb Duck has heightened its profile by raising enlarged scapular feathers and growing an enormous, flat carbuncle above its nostrils.

My father, who shot several for not-very-appetising *canard-à-l'orange*, called them 'knob-noses' and disappointed me with his disinterest in explaining the purpose in such a bizarre adornment. Indeed, from babyhood on, his various 'game-bird' trophies more often than not presented me with just such enigmas.

In some lake-side localities male Comb Ducks could be seen perched in bare or dead trees. Teddy's friend and colleague, ornithologist Hugh Elliot, labelled them 'lakeside lotharios', as they monitored small flocks of females under the jealous eyes of their 'harem masters', all grazing and a'dabbling down below.

Comb Ducks are mainly tropical birds that breed during the hot wet season. My fascination with these highly ornamental birds has endured. In contests between lords and lotharios, knobbed males stretch tall, clapping their main weapons (their wing wrists), and buffet, chest-to-chest, hissing like crazy cobras. Those same warriors, in courtship mode, defer to females with bows and coy curls of their freckled necks, revealing that their plumage is structured to present viewers with a geometry of curves and circles that culminates in that grossly contrived, soft and circular blade sprouting along the mid-line above beak and nostrils. That nose-knob has no meaning beyond the rituals of its species, beyond the hormonal drives and sexual behaviour of its owner and the observing females.

There are few behaviours more compelling than those that surround sex, most provocatively so among primates. Male Mandrills strut their rainbow-tinted boners, balls and buttocks so blatantly that I have seen a shocked lady whip out her zoo magazine to shield her fascinated, pop-eyed little girl from viewing anything 'so horribly vulgar, so damn tropical'.

While she is 'on heat', 'in season' or generally 'receptive', a female primate can also be pretty blatant. A she-Mandrill or she-guenon will turn her tail aside to allow full view, tactile and olfactory inspection, even penetration of her swollen, pink vagina lips. With what a human onlooker would call a saucy come-hither look, she has to twist her neck and shoulders round to monitor male responses to her nether regions.

There are two interesting exceptions where primate sex has become back-to-front. Once again, the behavioural needs of females (with some help from climate and altitude) seems to have set out to divert or augment male interest in back-end genitals to include full-frontal displays in the region of mammary glands – breasts.

Fossils tell us that more than one species of Gelada-like baboons once ranged widely in Africa. Today a relatively small, woolly-coated species survives on bleak, chilly highland cliffs in Ethiopia. The Geladas forage over the remnants of the grassy plateau, where their sinewy forefingers and thumbs pluck sparse grasses and herbs, shunting each harvest-leaf-by-harvest-leaf back into the care of rear fingers folded over the back of the palm. Every minute or so the products of this gleaning are passed to the mouth, palm-load by tiny palm-load. Yes, Gelada hands are extraordinarily adept but it is probably human agriculture that has helped push these baboons to such extremes of finger-food foraging. It is gleaning not only on the precipices of survival, but also close to the limits of temperature tolerance among the mainly equatorial primates of Africa. Frosts are a regular feature of the Ethiopian highlands and Geladas try to conserve warmth by shuffling along in a hunched, bunched squat that keeps genitals tucked away down below woolly cloaks. Like any baboon, female Geladas advertise their period of maximum fertility with raw pink skin around the target of a swollen and scented vagina. They also ornament the outer margins of their rear-ends with vesicles that look like pink beads strung around a pink doily. The female's sexual condition determines her scent as well as the colour and size of these vesicles.

Gelada: (left) female backside; (middle) female front; (right) frontal view of male yawning.

Unique to these highland Geladas is a mimicry of this elaborate visual and physiological structure *on the chest* of adult females (the chest vesicles even wax and wane in synchrony with the genital ones). Because their two nipples converge above the sternum (allowing both to be sucked at once) this has become pseudo-vagina in what looks like the transfer of a signal from rear end to front end. 'Up-end' might be a better term for the 'Life-on-the-Squat' that Geladas have adopted. Both sexes have hour-glass patterns on their chests, but only females mimic so faithfully the messages of their rear ends.

The physical details of what makes Geladas so distinctive are only explicable in terms of continuous, ever-changing adaptation to ever-changing environments. Embedded in a single, living Gelada is the history of how its ancestors, ancient and recent, responded to the challenges of their times. There is no enduring value in any specimen, any piece, any image of a Gelada. In the extraordinary lengths to which they have gone to survive on the edge of their kind's tolerances, Geladas have become messengers for the preciousness and the precariousness of life on the edge. In more than one respect we are all Geladas.

Earlier I mentioned a second back-to-front primate display. As every pubescent girl discovers, sometimes in exhilaration, often with dismay, her frontage suddenly excites the interest of every passing male. Here, in hopeful symmetry, is revealed one of the byproducts of bipedalism. Rearing up onto two legs has posed a conundrum not dissimilar to chilly days on the heights of Simen – a primary sexual signal, the vulva, is now hidden away; not beneath a dense cloak of wool but, in this case, between two hefty columnar legs.

Geladas, pushed to the edge of their species' ability to adapt, find an unlikely equivalent in yet another case of survival 'on the edge' – piscivorous genets.

Genets and oyans are cat-like arboreal or semi-arboreal members of the family Viverridae, and they inhabit all of Africa's grand mosaic of forests and woodlands, where some 17 species provide a measure of our continent's diversity of habitats. I have heard them referred to as 'rat-like Leopards' for their shade-adapted spotted, mottled coats and black-and-white banded 'barbers' pole' tails. Endowed with sharp claws, sharp teeth, and sharp noses, ears and eyes, each species seems to have specialised in its preferred prey and in its hunting technique for its choice of local prey – birds, lizards or mice, often buffered by fruit. All species of genets and oyans have spotted coats and banded tails and one or other species of genet occurs throughout Africa (excepting treeless deserts such as the Sahara or coastal dunes in Namibia).

Even so, one might think dabble-fishing from stream banks for very small fish should be well outside the range of any genet's adaptability. The Aquatic or Fishing Genet's shift into piscivory and its suppression of spots and a banded tail for a fox-red coat and bushy black tail calls for some explanation. Can this genet's adoption of new and unprecedented behaviours and habits explain anything about the way its anatomy and appearance have been transformed?

*Servaline Genet (*Genetta servalina*).*

Begin with the known borders of the Fishing Genet's overall range and the ecological peculiarities of its habitat, Limbali Forest, named for the dominant tree, a species of *Gilbertiodendron* which is almost the only tree over a vast area of the northeastern Congo basin. The brilliant biogeographer, meteorologist and palynologist, Jean Maley, together with his colleagues and Therese Hart, has long had a special interest in Limbali as a classic 'ecological island'. Excavating deep soil cores and analysing profiles (often revealed by road-cuttings), they have confirmed that central Africa's forests have had many and diverse responses to past climate changes. In the northeastern Congo basin, proximity to the Sahara and exposure to very seasonal boreal climates has meant that rains

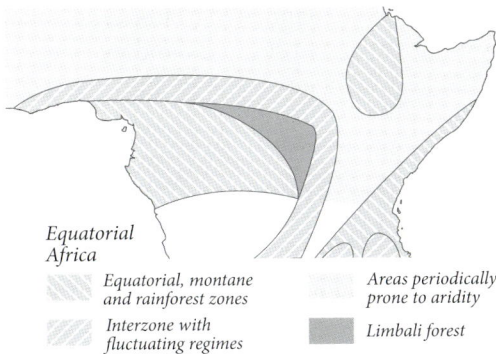

Equatorial Africa

Equatorial, montane and rainforest zones

Interzone with fluctuating regimes

Areas periodically prone to aridity

Limbali forest

LEFT: *the Limbali forest is sandwiched between broader bands of more and less aridity and humidity.*
RIGHT: *Fishing Genet (*Genetta piscivora*).*

arrive in deluges but last fewer months than they do in the rest of the basin. Very few equatorial forest trees are able to withstand this alternation between floods and droughts. Jean and his teams have revealed that today's Limbali tree cover is little more than 2,000 years old, following a major climatic crisis some 2,500 years ago which followed an earlier, still worse perturbation some 4,000 years ago. The last Ice Age was even worse.

The beginnings of such a forest are thought to consist of spaced-out trees growing on seasonally flooded but mineral-poor ground where palms, especially oil-palms, and dense growths of 5 m giant gingers and *Marantochloa* are the main fellow-colonists. This species-poor habitat creates dark, dense shade and the Limbali trees' crucial advantage is their ability to grow and propagate saplings in the total absence of direct sunlight. In as much as they are true shade-lovers, genets are logical colonists of such a habitat but there is no dapple in Limbali – the shade is unbroken so the advantages of a spotted coat are lost (many deep-forest animals are black and russet). Alternating floods with droughts, all under shade, is a challenge for most organisms, including most of a genet's normal choice of prey. One exception is very small fish that can weather both extremes in the rivulets and swamps that lace the Limbali region. The restriction of Fishing Genets to Limbali confirms how extreme their specialisation has had to be, but I also take their presence there as a measure of the adaptability of genets – it is as though some genet genie-in-the-genome took up all of non-desert Africa as a challenge to genet adaptability. In Limbali, the Fishing Genet has won the genie's dare.

Consider this fisher's dainty, short-clawed feet and spiny, fish-piercing teeth, but also examine the subtle, otter-like spread of its moustache of bristles. Natural selection begins with individual variations in behaviour, especially individual persistence under the shadows of an unpredictable world. Out there, in deep, deep shade, Fishing Genets and Limbali's trees are victors.

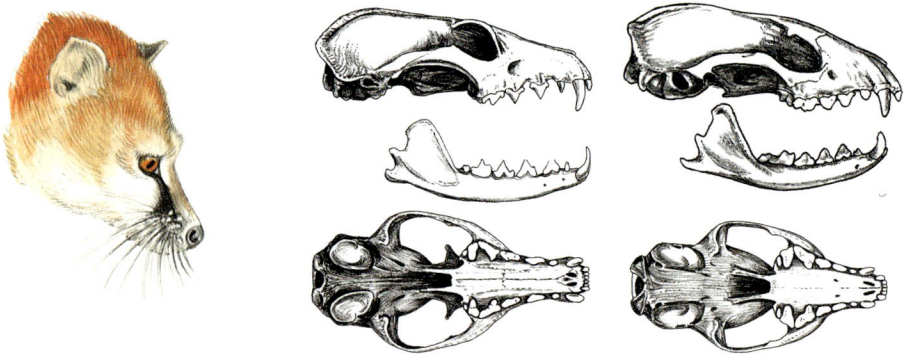

LEFT: *Fishing Genet moustache.* MIDDLE: *Fishing Genet dentition.*
RIGHT: *Common Genet (*Genetta genetta*).*

The perennial vicissitudes of Congo's northeastern corner and its frequent oscillations between dry woodlands and wet forests remind me, almost as an aside, of rare little monkeys, called Dryas Monkeys but I prefer Dryads – those quiet, shy spirits of classical mythology. For the Greeks, Dryads served in the retinue of Nature, guardians of many mysteries hiding deep in Greece's now long-felled forests. Dryads still inhabit the fast-falling forests of the Democratic Republic of Congo but I am among a growing militancy determined that Dryads and all that they both manifest and symbolise must and will survive the 21st century's forest-felling pirates and all their damnable auxillaries.

Age, sex and hormones account for varied coat colours which range from ill-defined greys, browns and off-whites to geometric masterpieces of design. The intensely black faces of all Dryads stare out, framed by white wimples of short fur. Judging by the few corpses that have been bought from hunters or photographed on village market stalls, adult males (probably charged up with hormones) wear a cabaret-cum-military costume in which all contrasts are maximised. Cheeks and throats are off-white in females, youngsters and in rare and faded museum skins. The faces of living adult males are framed by a bright orange that resembles spilled egg yolk. Black-sided limbs support a mahogany brown torso and back. Then it is as though jet-black trousers have been dropped to turn bare blue bottoms and the bluest of blue testicles into the show-stopper. In our imagined cabaret, white-lined tails end in black tufts that wave provocatively in the faces of their Dryad audience. It is no accident that Dryad males are among the most flamboyant of all guenon monkeys. Their habitat of dense tangled undergrowth certainly deters eagle and carnivore predators, but if males

Two extremes of Dryad facial colouring: (left) 'Salongo' morph/species?
(right) 'Inoko' morph/species?

are to make an impression they have to make it at very close quarters. Across all taxa, bright blue and black-and-white are the colour combinations that signal best at low light levels, so there is a visual logic in Dryad males advertising their balls with blue.

Dryad monkeys have intrigued but escaped the attentions of biologists for close on a century. Their numbers and total range remain unknown and it seems likely that Dryad avoidance of any exposure has helped keep them hidden from eagles, hunters, scientists and other busybodies. All specimens have been the outcome of chance in the lives of their finders. All of which brings me to the operations of chance in the very existence and evolution of these spectacularly colourful monkeys.

Therese Hart, with her husband John and a dedicated team of Congolese naturalists, all great pioneers of Natural History in Congo, has had previous mention. The Hart duo, Albert Lokasola and their colleagues have had the genome and molecular clocks of Dryads analysed. Dryads have been revealed to be remnants of an ancient stock close to the origins of today's savannah-dwelling Green and Vervet Monkeys, and their genes now add still further questions about primate evolution in Africa.

Sharing black faces and white wimples with Dryads, male savannah monkeys could scarcely look more different from their forest cousins. In particular, advertisement of *their* genitals is reduced to pathetic proportions – little accents of red, white and blue (which goes to show how any parade of male priapism can be shrunk by exposure to any approaching eagle!).

While the calibration of molecular clocks remains tentative, such clocks suggest that the Dryad and savannah lineages parted about 1.4 million years ago, but plenty more mysteries remain, hidden deeply within the lives of Congo's Dryads of the forest.

If intelligence has any part to play in human affairs, and if traces of fellow-feeling for fellow primates exist among humans, then the survival of Dryads within their Congolese fastnesses ought to take precedence over the so-called 'needs' of the timber, furniture and real-estate trades.

Seeking to invest monkeys with the dignity they deserve, I modelled their images on several occasions. I have marvelled at the many ways in which facial expressions and mask-like patterns serve quite subtle and specific social ends in the lives of these, my fellow primates, fellow mammals. One subject, a mountain monkey from Central Africa, took the form of a large, nearly spherical bronze head. It's form was invoked by the males' globular ruffs, out of which protrude assertive black muzzles. But inanimate objects are just that. The only ultimate value is the life that pulses through animal or plant tissues. The vicissitudes that living things have survived over millennia and more still shape their every physical detail.

Living individuals were once indivisible from the monumental skulls they left behind – bony accretions that are diagnostic of the many peculiar faculties that once created and inhabited them. In Africa, the wholesale slaying of all large wild animals has generated landscapes that can be seen as cemeteries. More abundant than their survivors, rhinoceros skulls incorporate flying buttresses that once arched between gargantuan necks and the fearsome weapons that once defined them – 'nose horners'. Horns that once perched above all the other activities that once clustered within every rhino's ponderous head. My 1960 skull sketch is but one of too many obituaries

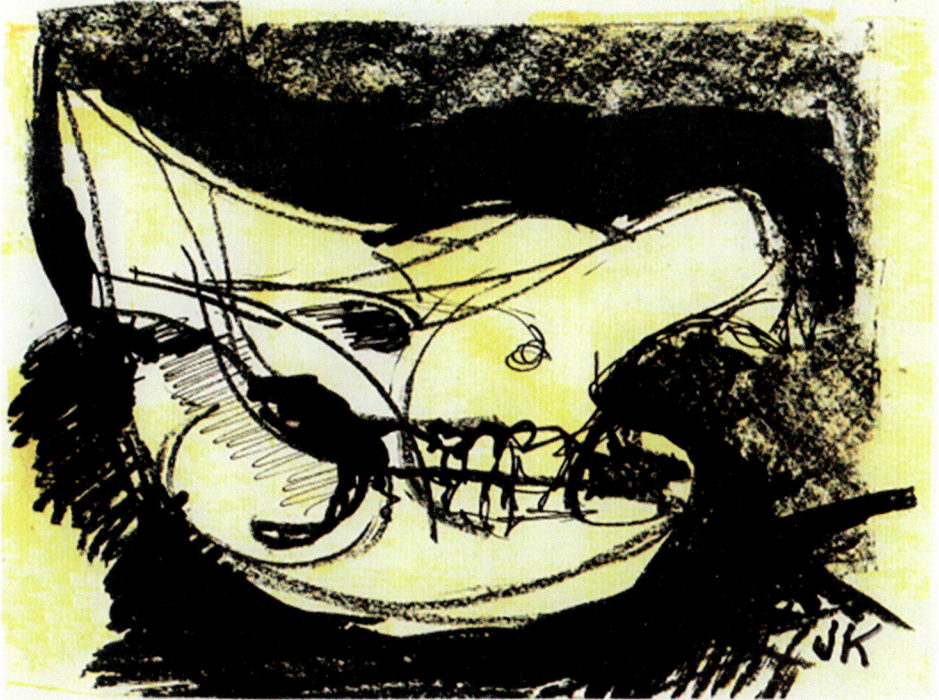

Grass Rhino skull (of an animal killed for its brain as a gastronomic delicacy), 1960.

ECOLOGICAL ELDERS

In which age has its advantages.

At less than 2 kg, talapoin monkeys are the smallest diurnal primates in Africa. My first acquaintance was finding one curled up, like a foetus, inside an alcohol-filled museum pickle jar. Then, when I went on to prepare a field guide to the mammals of our continent, I spent many hours in zoos, museums and labs on four continents where I watched quite a few captives and examined some dozens of skins and skulls.

I found that talapoin monkeys on each side of the Congo River were not only consistently different – they were *very* different. Those from south of the river wore an eye-suppressing black eye-mask, separated from mobile black ears by complicated little whorls of white whiskering. Hints of a crest served to emphasise their highwayman mien and their legs seemed quite long for an arboreal monkey.

By contrast, those from north of the river had rounded, infantile foreheads protruding above pale pink faces and equally pink, waggly ears, while their limbs seemed more compact and juvenile – even the texture and patterning of their coats was finer and softer.

Talapoin (Miopithecus) *monkeys north and south of the Congo River mouth:*
(left) Northern Talapoin (M. ogouensis); *(right) Southern Talapoin* (M. talapoin).

I combed through the scientific literature and, to my great surprise, there was but a single species described. So the descriptions and paintings in my field guide represented science's first recognition that the talapoins on each side of the Congo's mouth were distinct species.

OPPOSITE: Ecological Elders.

It is natural to look to the material, verifiable present to explain why two such species exist, so several colleagues were soon speculating as to which population was the founder and which was the crosser of that wide, deep, deep torrent.

Actually, talapoins are older than the Congo River's present outlet so it was the common ancestor of those two species that was eyewitness to one of the most stupendous dramas in African prehistory.

No human ever witnessed Congo waters make their first trickle over some faint depression along an ill-defined hogsback in the Kongo hills above today's Boma, but we can infer that this was where one among many eastward-flowing streams back-cut to a point so close to the source of a west-bound stream that some inland flood or temporary lake rose to a level where their joint waters suddenly flowed down into the Atlantic. Every subsequent flood took more and more water over that spillway, cutting an ever-deeper canal until its course became the thunderous wonder of the world it is today.

Fifty thousand cubic metres of Congo water cascade into the Atlantic every second. The erosive force of all that water is reinforced by loads of sand and soil that deposit 86 million tonnes of sediment onto the floor of the Atlantic every year, its nutrients feeding an abundance of sea life (that is currently being harvested by contemporary commercial hunter-gatherers, fishermen turned pirates).

The river is very far from finished in carving its way through the hard, rocky hills between Kimpoko and Luozi. All of this combines to suggest a relatively recent date for the Congo's present point of exit. Several lines of scientific sleuthing converge to suggest that 2 million years ago approximates to when the river broke through at this particular locality. Molecular clocks and phylogenies have arrived at that same date for a first divergence between the two talapoins, thereby helping confirm the validity of my field guide's insistence on two species. In spite of their now quite limited range along Africa's western equatorial littoral, talapoins have the important distinction of being little-modified survivors of the root-stock from which all Africa's arboreal 'long-tail' monkeys have evolved during the last 7 million years. They probably ranged over most of equatorial Africa during the earlier part of this very lengthy reign. What can be inferred from the talapoins' present range and biology? The sheer number of descendant species implies that an enhanced agility in trees might have been the clincher for both the ancestral talapoins and their diverse descendant kin.

One of the largest bodied of descendant species in the talapoin lineage is the Patas Monkey, yet Patas infants are strikingly talapoin-like. It is as though a single Patas life spans its own evolutionary trajectory.

The descendant radiation of monkeys has come to exploit a wide range of habitats and they have developed many refined and varied ecological strategies. For any ancestral population (usually less specialised than its descendants) there tend to be two main choices. One is to squeeze into some minor, under-exploited niche. For talapoins that could have been acheived by reducing body size (below that of any other known guenon).

Infant Patas Monkey
*(*Erythrocebus patas*).*

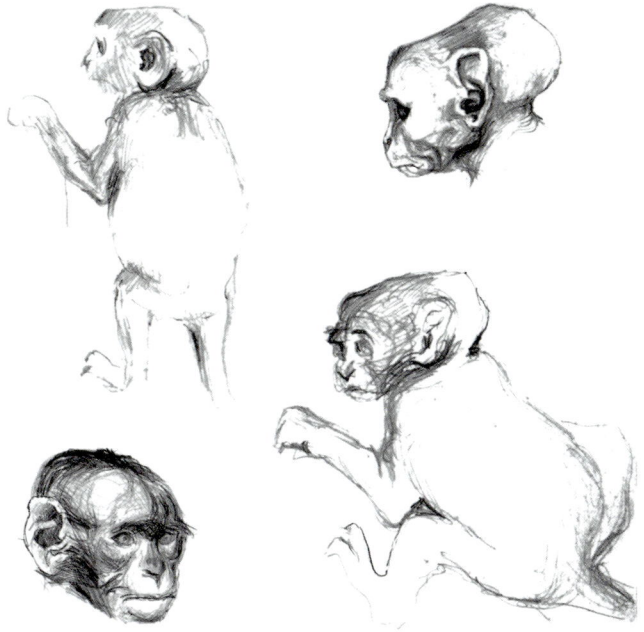

BELOW: *Portraits of seven*
African monkeys: a diversity
of face patterns.

The second option, for talapoins, suggests that their range has steadily (or by fits and starts) shrunk back to heartlands where some 7 million years of adaptation to local conditions or niche is so comprehensive as to have become unassailable.

Both options probably apply here. We are fortunate indeed to be the first generation ever to try to reconstruct how an abundance of primates has responded to gigantic events of Africa's geological history, such as the Congo seeking, losing, then seeking again to spill back into the oceans whence its waters came.

Only one third of a million years old, modern humans never witnessed any of these dramas. Compared to us newest of newcomers, talapoins are what I call 'ecological elders'. It is a title that challenges us to examine and try to understand natural history in the context of deep time, and requires a degree of empathy and fellowship for other beings. We have all, animals and plants, evolved subject to the same amoral selective forces that allow us to inhabit this small planet.

With 7 million years of survival behind it, the genome that inhabits every living talapoin should be regarded as a wonderful gift of chance (a gift of God for pre-Darwinians). For any analytical mind in search of a value system in this story, there are two main lines of argument in favour of paying intellectual, moral and scientific respect to our ecological elders.

The first is an obligation to try to learn as much as is currently possible about LIFE on this planet, integral to which is the biology and history of our elders, even their interactions with other animals and plants. This is a quest that includes how our elders relate, genetically and ecologically, to other members of their clades.

Our second field of enquiry must unearth whatever details we can find about the elders' habitats over time and space. The space they occupy today is probably sub-optimal and is certain to differ from some in the past. It remains essential to remember just how dynamic and frangible Africa's physical and ecological history has been.

Fossil secretary birds were probably stalking snakes some 30 million years ago, whereas the great mass of Kilimanjaro volcano is a mere million old! This prompted me to paint several pictures on the theme 'old birds – young mountain'. I could as well have titled them 'living birds – mineral mountain'.

As for the dynamics of past talapoin habitats, return to the still-eroding mouth of the Congo River where there is no escaping questions about where else Africa's greatest watershed might have found exits to the sea.

Old Birds, Young Mountain.

Answers must be premature, but the dynamics of Africa's unique geomorphology point to several possibilities, even several sequences.

We have already absorbed the need to visualise Pangaea and Gondwana as global shells of brittle rock that broke up, long ago, into separate continents. Today's rift valleys betray that, for Africa, the fragmentation continues. Beyond such generalities it is clear that entire land surfaces can lift or depress over vast continent-wide areas, tilting waters along new gradients, searching for new channels to follow.

Outflow of the Congo into the Atlantic is a product of very widespread uplift, beginning about 30 million years ago, all over eastern Africa. It was this blocking off of former eastward flows that forced another great river, the Nile, to drain northwards. Consistently blocked by rising ground in the east, waters' relentless quest for lower ground culminated in the Nile becoming the longest river on Earth. This also explains why the Congo, in all its more recent iterations, has had to flow westwards.

The first place to search for the Congo's exit previous to 2 million years ago is to look north. Less than 650 km north of Boma we have the answer. The Ogooué river flows into an exceptionally expansive delta after flowing through gorges so deep that typically marine fish still inhabit its depths. Far greater volumes of water carved those gorges than those that now drain off a very modest catchment. Emerging right on the equator, this was the Congo's previous exit before it burst through at Boma to cut the talapoins' range in two.

The Congo's present basin (see map a below) can be directly compared with the Ogooué/Congo's more than 2 million-year-old old basin, together with my crude sketch of hypothetical courses for its then-time tributaries. The map suggests that very modest raising or tilting of land surfaces had forced the Congo to reroute along gravity-governed but otherwise quite logical tree-like courses.

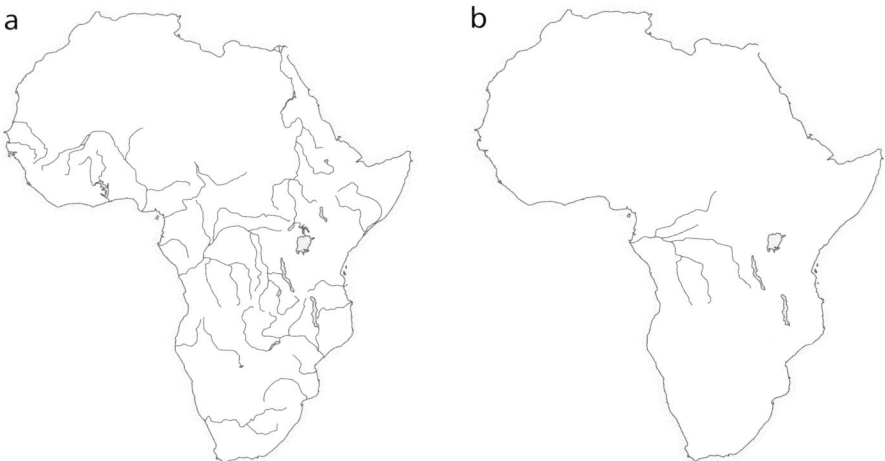

Congo River routes: a) present major river courses; b) hypothetical course of 'Ogooué/Congo' River 3–4 million years ago (Congo Exit II).

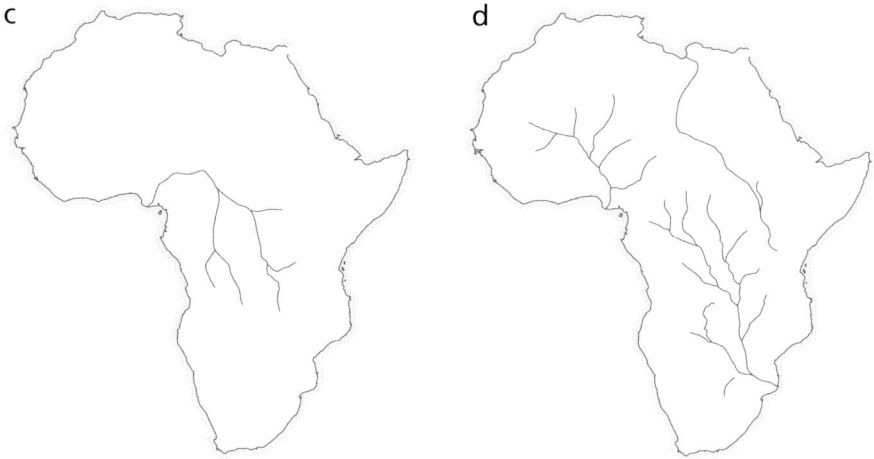

More possible Congo River routes. c) course of the Benue/Congo (Congo Exit III).
d) Zambezi/Congo (Congo Exit IV).

This cannot be the last word about Congo exits.

The broad trend of uplift that had pushed waters southward had already displaced still earlier out-flows of Congo waters. Before eastern uplift had begun to spread westward along the equator, there was no obstacle to flow to the northwest. However, Cameroon was already high enough to force waters to detour around its elevated mass so the Congo possibly gushed out in 'the armpit of Africa', or Gulf of Biafra.

Today modest and seasonal flows along the Benue valley have allowed it to silt up but, as Benue/Congo (Congo Exit III), that huge river probably arched westward to follow the course of today's Cross River and empty into today's Niger Delta. Even today, the Cross River, with its small catchment, demarcates boundaries for quite a few organisms, implying that it once had a more influential biogeographic role.

Now, reach back to some 140 million years ago, when long north–south uplift in Rump Gondwana created rifts that would eventually become the South Atlantic Ocean. This uplift along both margins of the proto-Atlantic/West Gondwana Rift would have tilted the entire Congo basin floor eastwards. Where waters flow from southeast to northwest today, it is possible that they formerly flowed, in an exact inversion, from the equatorial northwest to the far southeast.

In eastern Africa, uplift that had begun in the north began propagating southwards, step by step. As a direct result of this very extensive uplift, the Zambezi/Congo (Congo Exit IV) found its way into the Indian Ocean, probably along parts of the course of today's Zambezi valley, possibly still further south, creating a delta much too extensive to have been created by today's Limpopo River.

In a previous book, *Island Africa*, I summed up the Congo basin as an 'evolutionary whirlpool'. In less rational moments, this brief regression into the great river's extraordinary comings and goings can feel like trying to keep one's balance on a raft or punt that giddily pitches about over a whirlpool.

This regression into the Congo basin's distant past mainly serves to emphasise the deep time-frames in which we try to unearth evolutionary histories, and the many surprises that await us along the way. Some of these, such as reversing rivers, are quite disorienting.

Some of the more venerable of the Congo's elders have already had mention – lungfish, clawed and ghost frogs and, of course, a host of invertebrates, such as termites and spiders.

One inhabitant of tropical Africa, including the Congo basin, has to be the arch-elder of all terrestrial inhabitants because it represents one of the first

'Execrable scorpions' as experienced by execrable humans many millennia apart, possibly a painful experience for both artists: (above) image in a rock shelter on Chincherere Hill, Malawi (traced from a photo by William Pawek); (right) image drawn by Jonathan Kingdon.

aquatic organisms to crawl onto dry land. It rejoices in the appellation 'Execrable Scorpion' (*Lychas asper*), a name that my Bambuti friends certainly endorse because its stings are among the more painful and debilitating of the many hazards of Mbuti life in the Congo.

Scorpions were one of the hazards of my childhood. On one camping excursion a very small one got into my bedding and proved to be the scarlet pimpernel of scorpions, stinging me repeatedly whenever I had the temerity to lie on top of it, but so expert was it at scuttling into every fold and crevice of sheet or tentage that it survived my ire. In Dodoma, another fat, glossy, ebony scorpion inflicted a single, much more painful sting, but it was me that survived – not him. Like lungfish, scorpions evolved a mechanism for breathing air when, of all surviving animals, they were the first, like miniature lobsters, to crawl out of the water and survive on dry land. Scorpions are now known to have colonised the land about 400 million years ago and had arrived at their present appearance by 360 million years.

In spite of respecting their well-advertised tail-tip sting, I have always admired the geometric logic and symmetry of the scorpions' crevice-adapted body architecture. Envelope-thin flat one moment, they can arch in an instant into an agile, *noli-me-tangere*, 3D elegance. If scorpions pioneered life on land and then survived for 400 million years, maintaining roughly the same body plan throughout, then it is much more likely that the descendants of my camp-bed companion will survive more millions of centuries than will my own, much more fragile, kin.

Pondering the rich soils of his garden, Darwin showed that their fertility and that of other worm-friendly parts of the world were the product of earthworms digesting humus over millions of years. In tropical Africa it is mainly termites that have processed soils for at least the last 140 million years.

Before tarmac began to be used to surface African roads and before cement was widely available, hard, well weathered 'murram' or laterite was the most widely used mortar and bulk soil for roads. The chemistry of laterite conforms with the observation that iron and aluminium hydroxides get separated in termite guts from the phosphates that these insects need. If so, the most basic material all around me while I was growing up was the boundless product of termites. I grew up in a landscape of red hills, red termitaries, red bank-sides, the very roads, the very buildings, the plaster on our walls – it was all animal shit and had been accumulating since the Cretaceous – elder shit indeed.

Termitaries (or 'ant-hills' made by 'white ants', in the parlance of my childhood) were, indeed, scattered across the landscape. Seen from an aeroplane, parts of Africa are as freckled with termitaries as a guineafowl's plumage is freckled with white spots. After dark the termites emerge to forage for dead grass or wood, and tunnel-lets of still-damp mud snake out from the termitary, to become particularly dense and tangled wherever clumps of grass have begun to die back.

Pattern of termitaries, tiang and grassfires in the Sudd.

When Public Works Department (PWD) labourers sought to widen or straighten a road they often used their *jembes* and *pangas* (hoes and cutlasses) to hack away part or all of a termitary and spread it out over the road. Each night, tens of thousands of termites quarried and cement-mixed mud to re-build their city. Each morning there was a newly moulded shape covering up the PWD's blade-marked embankments and gutters. Sometimes the spire of a

Dewed Cobwebs at Dawn.

termitary might sprout in the very centre of a newly graded murram road, and occasionally the PWD might admit defeat and permit two lanes to develop on each side of the subversives' mandibular master-work.

A substantial proportion of so-called 'development' activities were (and still are) in a state of war with termites. As a child I admired the insects' industry, but also followed along in seeing such roadside war in terms of us against them. Back then almost all buildings, towns and road-works were as new as last night's termitary.

When I was taken to bombed-out Europe, no metaphor seemed more apt. The silhouettes of ruin were dead termitaries. The fever of rebuilding that followed had more than a semblance of the altruistic industry of 'white ants'. My East African home primed me for perspectives on London that metropolitan acquaintances found more than a little perplexing.

Ecological elders have sometimes pioneered new ways of living, their techniques or solutions so successful as to help explain why they have survived up to the present. They earn their title by virtue of retaining some ancient body plan.

Almost as ancient, and a lot more diverse, are spiders, animals that fascinated me as a child (I was sufficiently cautious never to try to catch or collect them but could watch them for hours). Acacia bushes shelter particularly handsome lynx spiders, *Oxyopes*, just one masterpiece built by an architect called Selectione Naturale, who has been actively building spiders for at least 400 million years.

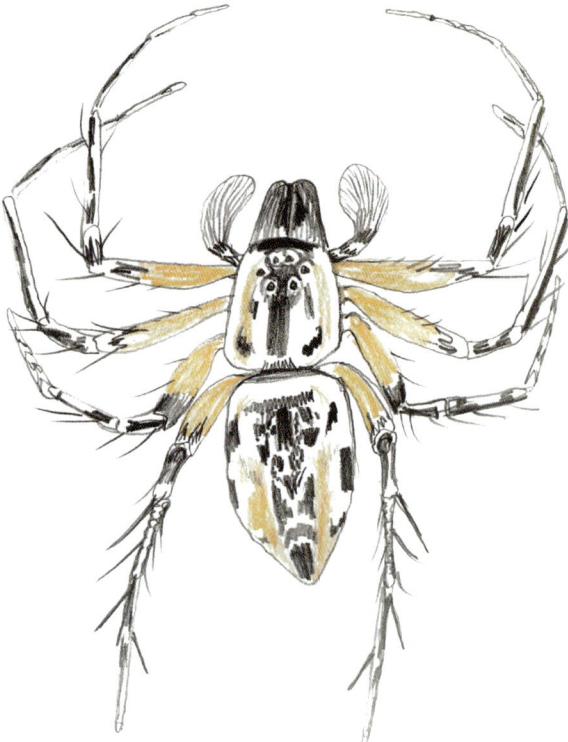

Acacia Lynx Spider, Mkomazi.

Living miniature chemistry labs, they secrete salivas that can dissolve almost any tissue and they synthesise a wondrous variety of silks – the stuff of webs, nets, elastics, glues, envelopes, purses, strum-strings, kite-strings, cocoons, fishing lines, trap-lines and trap-doors.

Both ancient and extraordinary, 'trap-door spiders' were sufficiently common for me to have seen males out and about during the rains, at which time they were presumably hunting for mates or for food, typically termites. Females are reputed to remain within or very close to their silk-lined burrows, wherein single individuals are known to live for decades. They have recognisable ancestors from 300-million-year-old fossils, and at least five trap-door families and an unknown number of species exist in East Africa (worldwide there are more than 100 spider families).

A friend once told me about being shown a large, cracked earthenware pot with a notch cut out from the overturned rim. Placed almost equidistant between two huts in a neatly swept village, this pot protected a trap-door spider burrow that was said to predate the village itself. The spider that came and went through its pot-notch was revered as a go-between, connecting dead, but dormant, ancestors deep underground and their living descendants above. It was seen as a being invested with the power of harbinger and messenger, and no villager would dream of molesting or harming such a sentient personality.

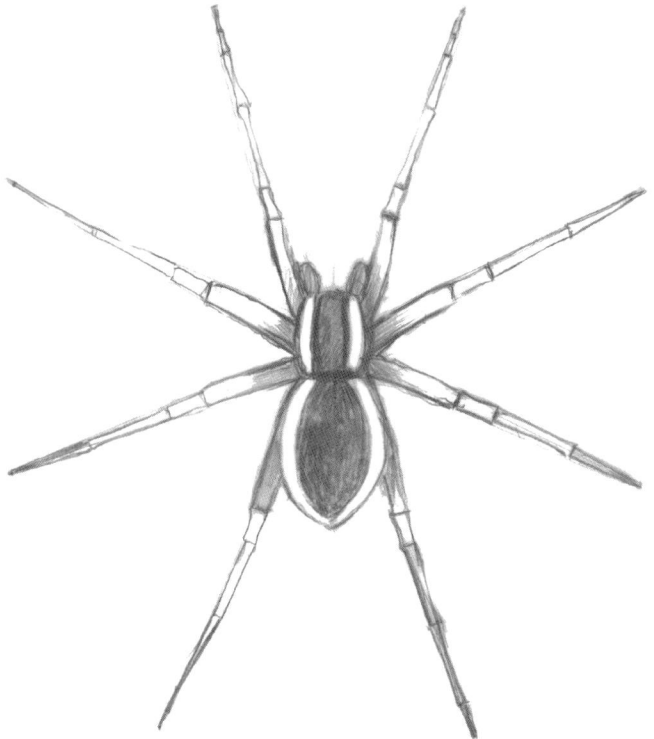

*Fishing spider (*Dolomedes *sp.).*

The 'paper quilt' or collage that heralds this chapter pays tribute to that pot-protected messenger, while hinting at our efforts to comprehend a planet inhabited by beings from eons that lie beyond our most earnest imaginings. A planet alive with elders.

During a *ngoma* in Tukuyu, one of the musicians once gave me a marvellous instrument called a *lipilenga*, which depended on vibration from a single spider's 'door' for its sound. A curve-stemmed calabash had been dried, both extremities of its emptied shell had been cut away and a spider-made, trap-door membrane had been glued over the cut stem. Pressing one's mouth to a round blow-hole cut into the outer curve of this stem allowed any sound, even words, to become distorted by the vibrating spider-web membrane, while the open body of the calabash, like a trumpet horn, hugely amplified the noise. The *vuvuzelas* blown at football matches seem to be constructed on similar principles, but lack the musical versatility of my trap-door *lipilenga* trumpet.

That spiders may play their own music was suggested to me by a friend as we watched a daddy-longlegs or vibrator spider (family Pholcidae) bouncing upside-down under its seemingly carelessly constructed webs. Was Daddy actually strumming his web guitar to attract tiny edible *dudus* to his concert? Or did the victims just happen to get snagged on those twanging gluey lines?

Malice – *wood, metal and hippo teeth sculpture.*

In the sedges and reeds around Mwanza Bay I watched spiders with bold white margins around their brown thoraxes and abdomens. They caught minnows and tadpoles by tickling the water surface. As prey came within reach the spider plunged, encircling the victim with all eight legs, simultaneously injecting a fast-acting venom-cum-digestive agent. A particularly strong line of web served as a hawser to link and return the fishing spider to its water-side perch.

In lakeside forests all around Lake Victoria there are spectacularly coloured, large, long-legged golden orb spiders (*Nephila*). I once found an exhausted munia (a small seed-eating bird) hanging helplessly, entangled in the yellow webbing that these spiders extrude. I also remember harvesting lengths of this yellow web, twiddling the strands into a twine and challenging other kids to break the thread with their fingers. This was impossible, yet a razor blade cut through it with ease.

Many small animals protect themselves by exuding poisonous or distasteful secretions, and the spiny kite spiders (*Gasterocantha*) seem to present their bright colours and spiny abdomen as a warning to predators larger than themselves while, at the same time, seeming to fool prey smaller than themselves that they are some component of a flower. I once made a sculpture based on one of these spiders which I entitled *Malice* because, to me, the hard spines around the kite spider's enlarged abdomen suggested a crab-like animal in some universal and size-free posture of threat – in spite of the headless animal behind this surreal disguise actually facing in the opposite direction!

Very few spiders are dangerous to humans and fewer still are aggressive, but among the exceptions are nasty little sac spiders (family Clubionidae) – I once had to rush a close friend to hospital after a sac spider bit his knuckle while he was gardening. The blackened, swollen skin around Ali's wound took weeks to heal.

A dewy dawn is perhaps the best moment to appreciate the ubiquity of spiders in Africa. Plains, swamps, rocky hillsides, forest and beach margins, even gardens, glitter under first light as dewdrops hang from every web (*see* p. 259). A closer look reveals spiders busy reeling in their nets. It is as though all the architects of New York gobbled up their skyscrapers for breakfast, only to rebuild them by suppertime.

This abundance and diversity of spider lifestyles evolved in the tropics and, however reassuring I found it to encounter garden orb spiders in northern Europe, the tropics host most of the more than 45,000 spider species, and some 117 families ranging from trap-door ambushers to fish-eaters.

I remember watching a hole-nesting King Baboon Spider in our rudimentary Singida 'garden', where we inhabited the *Boma*, a great, white German-built fort. The only East African tarantula, and the largest I ever saw, it had spine-armed digging pincers for jaws. It hesitated between retreat and advance at the mouth of its silk-lined hole. Its indecision seemed to oscillate between the deliberation of a traveller about to set off and the menace of an on-duty graveyard guardian.

ABOVE: *Aardwolves (*Proteles cristata*).* BELOW: *Springhares (*Pedetes *sp.).*

Its hole was but a few paces from a spreading jujube (*Zizyphus*) bush. Here there was a much larger burrow occasionally occupied by an Aardwolf, a delicate, long-legged animal with a broad, blunt face, huge black eyes, pointed semi-transparent ears and a sandy coat with thin black stripes.

I did not think of the Aardwolf as an 'elder' at the time but I have subsequently learned that this miniature, weak-toothed hyena pioneered its termite-eating skills some 10 million years ago, about the time that molecular clocks reveal its lineage parting from more conventional hyenas. This individual was obviously quite used to people and would sit among the desert roses at dusk or in the early morning, warily watching passers-by. At the sight of a dog it was instantly down its burrow – upstart dogs have become the bane of too many carnivore-elders, let alone ecological-elder-prey such as sengis or springhares.

Singida is built on higher ground between two soda lakes and, at the time our family lived there, sandy flats extended around these lakes, densely scattered with excavations – the work of numerous springhares. A couple of nights after my arrival on vacation from school I went lamping and shot two. Skinning and dissecting them for 'jugged hare' (improvised from Mrs Beeton's cook-book) revealed that they were more like long-tailed, very bouncy kangaroos and very different from true hares. Before allowing my quarry to be cooked I made drawings and some detailed studies. It was only later that I learned that they were descended from the very oldest stock of African rodents, their earliest African ancestors having arrived in Africa some 50 million years ago.

Mousebirds, *a composition in muted tones.*

This was not my first induction into the gang-crime of eating ecological elders. I am among the tens of thousands of young Africans who, armed with snares, catapults, nets and guns, keep up an ever-increasing barrage of missiles or traps to consume our elders. In my case Teddy was my role model. Remembering his own early induction into the then very male pleasures of shooting, he gave me a succession of guns. The first was an air-gun which came with the injunction to keep birds away from the kitchen garden and peach trees. The principal 'scoundrels', as Teddy called them, were Speckled Mousebirds. As a 10-year-old marksman I was no slouch and soon expressed gratitude to Teddy by preparing *Ragout de Coliou*, in which onions, carrots and tomatoes joined a half-dozen mousebirds or 'colies', swimming in thick gravy – all ingredients harvested from the same kitchen garden. Their naked, gutted cadavers revealed bright red legs, disproportionately large and muscular, attached to the sacral end of a uniquely S-shaped vertebral column. These were not 'perching' birds. Instead, they were 'hangers', even 'contortionist clinger' birds. Their 10-year-old assassin spent hours watching acrobatic travels by living colies through the canopies of our peach and guava trees.

Over the years I have seen every one of six very similar, equally peculiar coly species, at least one of them as precisely and beautifully patterned as a reef fish (together with birds-of-paradise, such fish are my ultimate bench-marks for perfection in evolved form and colour and a perennial source of models for ex-periments in pictorial imagery).

It has been disconcerting to learn, many decades later, that colies are the most archaic of ecological elders. Fossils that belong to an early coly lineage have been unearthed from beds more than 62 million years old in North America (only 3 million years after Chicxulub!). Later fossils from Eurasia and Africa combine to suggest that the coly lineage was among the very earliest to flourish in the meteor's aftermath. Like other survivors of our blasted planet, the class of theropod dinosaurs that we now call 'birds' quickly spread and diversified in Earth's newly emptied lands. Coly ancestors were among the very first theropods to move into new or empty ecological niches.

I have already stressed that it is one of the dynamics of evolution for rapidly adapting descendants of early arrivals on virgin lands to challenge more conservative, even ancestral, members of their own lineage. Other theropod lineages eventually boxed in or outcompeted the colies, but, of all the continents, only Africa has remained ecologically diverse enough to support such birds. Only Africa – but for how long?

Once the pleasures of watching behaviour overtook those of shooting, I developed a great fondness and admiration for colies. The term 'living fossil' could not be less appropriate for such vital personalities, as I discovered on raising an orphan, miraculously unhurt, that was brought in by the cat. This very fast-growing little bird was a true 'cuddler', especially in the early morning, when it would actively seek out my hand or search for a sun ray. Snuggling was consistent with the behaviour of its parent flock, which I often watched at dusk gathering on some multi-pronged stem that could support all of a dozen colies, so densely and so inextricably packed they were like a fluffy, globular Rubik cube.

Come dawn, the globe remained intact until the sun's rays were warm enough for each bird to unlatch and clamber up to some sun-lit support. There, spread-eagled and with half-opened wings, every one of them became a perfectly circular solar panel. Distributed through a small tree or shrub, they would have resembled flowers or fruit, were they not mouse-brown.

If we ask in what respects colies might be closer to their theropod ancestors than other birds, that abject dependence on absorbing and keeping warm is the closest I can get to an answer. Could a few coly-like theropods manage to huddle and cuddle up enough to survive many generations of nuclear winter? Having admired, cuddled, even eaten coly-theropods, I'd love to know.

In compensation there is, among Africa's endemic mammals, one lineage with a plausible Chicxulub-survival strategy. Along roadsides, in gardens and patches of remnant forest, burrowers split the soil just below the surface, creating a network of miniature embankments. The diggers, found all over southern Africa and a scatter of (mostly upland) localities south of the equator, are exquisite little golden moles, their velvety coats reflecting metallic glints from gold and copper to blue-black silver. Something like a golden mole could have survived Chicxulub. Their subterranean existence has even allowed some species to dispense with eyes.

Being blind to all that lives above their burrows is a trait shared by another South African animal. For people in pursuit of fortunes, South African soils are a feast, not of earthworms and beetle larvae, but of underground minerals. Their noses are adapted to sniffing out gold and diamonds, and mining towns mark out their network of burrows – Randfontein, Kimberley, Priesca.

Above ground the bait is wood-pulp, sugar, beans and meat, so vast plantations and vineyards overwhelm one of Earth's most precious living repositories of half-known and yet-to-be-revealed knowledge. South Africa is also Africa's most concentrated gathering of ecological elders in need of respect and space to survive. If only.

Society must reduce the importance of dead things, inert stuff, gold, sugar, gems, oil, disintegrating temples... and learn to value the ephemeral but *alive* fabric of a most ancient tapestry. We must value living, reproducing things, including our elders, as the only guarantors of a future.

Where Gerenuks Pass.

Primate Patterns *(a mural in Duke University Biology Department).*

14 IMPORT AND EXPORT OF PRIMATES

In which our cousins and predecessors reveal their travels.

Of all the twigs and branches of the banyan tree of Life on Earth, it is those of my own, human/primate lineage that I find the most compelling.

In the deep recesses of a real banyan tree I have locked eyes, from no more than a few centimetres away, with the huge, pale olive orbs of six 'ecological elders', more different from me than anything I could imagine. Those eyes, staring back at me out of deep shadow, belonged to six Sulawesi tarsiers that were about to emerge, at nightfall, to disperse through the shady undergrowth of the Tankoko forest, their stem-clinging leaps more like those of frogs than mammals. These were amongst the most thrilling hours of my life, particularly when I watched them return at dawn, one by one, their bellies replete with hand-caught crickets and cicadas, back to Home-Banyan.

Why the thrill? Because I brought some knowledge of my own evolutionary history to this encounter. I can even hazard a guess that it was some 66 million years ago that my ancestors parted company from theirs.

Higher primate class of 66 million years, reunion – six Sulawesi tarsiers (Tarsiidae).

I like to imagine our ancestral line choosing sunlight while their furtive, insect-eating ways were lit by the moon. If the arrival of the Chicxulub meteor was that marker, there was no escaping the thought that whatever history I might share with those Sulawesi tarsiers, it is not parochial. When this planet takes a hit from outer space the suffering and the consequences are global.

The fact that a variety of placental mammals had already maintained their distinctness from dinosaurs for more than 100 million years gets lost behind the melodrama of a 'reign of dinosaurs'. Modern genetics (and the logic of every organic structure being the product of serial, small and cumulative modifications) tells us that Chicxulub was less a moment of creation than a moment of release for the holocaust's survivors. Their subsequent elevation into ancestral status has been among the themes of previous pages.

Diurnal tarsier fossils are known, and the foveas, nostrils and genes of living tarsiers suggest a pre-Chicxulub ancestry shared with monkeys, apes and me, but not with lemurs, where the ancestral connection is earlier. Once lumped into a taxon called 'prosimians' ('before monkeys'), their fossil record demonstrates that at least one tarsier species reached Africa. Indeed, some form of proto-tarsier might even have given rise to all of Africa's anthropoid primates, so I like to think of our meeting that Tankoko night as a 'Higher primate class of 66 million years ago reunion'.

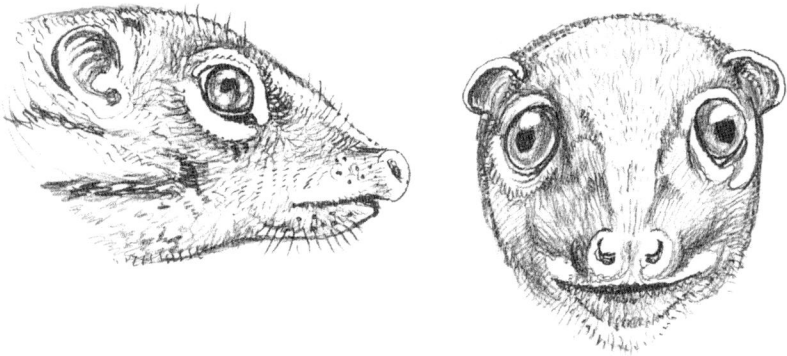

Profile and portrait of colugo (Cynocephalidae).

The last refuge for tarsiers, Southeast Asia hosts still more remote cousins – the colugos (often called 'flying lemurs' but more accurate would be 'gliding lemurs'). An Indonesian night can take you as close to flying back in a time machine as it is possible to get.

The roof of my forest bivouac in Sumatra resounded for much of the night with debris dropped by colugos feeding in the fruiting tree above. A torch revealed shy lemur-like animals, smudged all over with swirling speckled camouflage, their impossibly long thin limbs supporting supple membranes that they manoeuvred with gawky but careful, snag-avoiding movements. Wherever possible their bulging eyes and reflective lower eyelids evaded the beams of my torch as it probed the canopy, but even the faint and distant glimpses they allowed were as thrilling as my close-up encounters with tarsiers. Those blurry glimpses thrilled for similarly dynastic reasons, but geneticists have calculated that colugos diverged from primates about 86 million years ago.

This date should mark primate beginnings, because ancestors shared between two such highly divergent lines technically belong to neither. Were we able to examine animals immediately before or after such partings, such categories could be too trivial to register. So I like to think of those ill-lit ghosts up in the canopy as cousins whose passport to survival over 86 million years was to evolve into the most elegant of gliders. Meanwhile a ruthless, arrogant and greedy logging industry steals away the very fabric of my exquisite cousins' existence.

On present evidence we do not know how early the colugo lineage took to gliding but, once tall trees offered habitat, shelter and resources, we know that the evolution of gliding must have become one response to the problems of getting from tree to tree and that gliding long predated Chicxulub. In the case of birds and bats, gliding might have been an ancestral way-stop to flying, but if fruit bats, insectivorous bats and birds got there first, then competition with them may well have prevented cousin colugos from becoming the true flying lemurs that my imagination likes to toy with.

At this point, return to the beginnings of placental mammals, remembering that the primate lineage was but one early offshoot. The super-specialisations of living tarsiers and colugos are most likely byproducts of competition within what began as the then-largest and most complex community of mammals on Earth, centred on Southeast Asia.

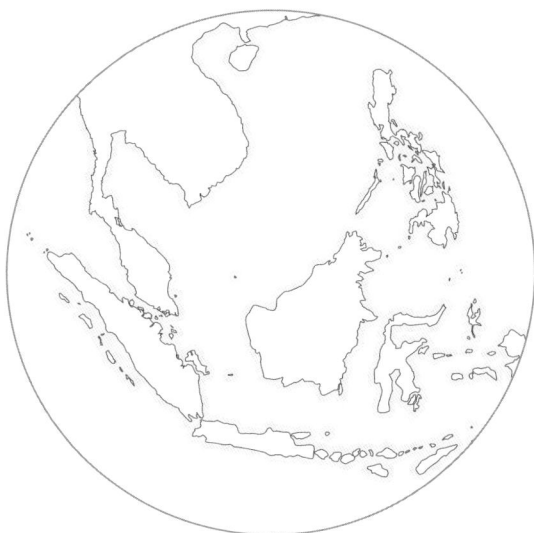

The region of the world most distant from the impact of Chicxulub.

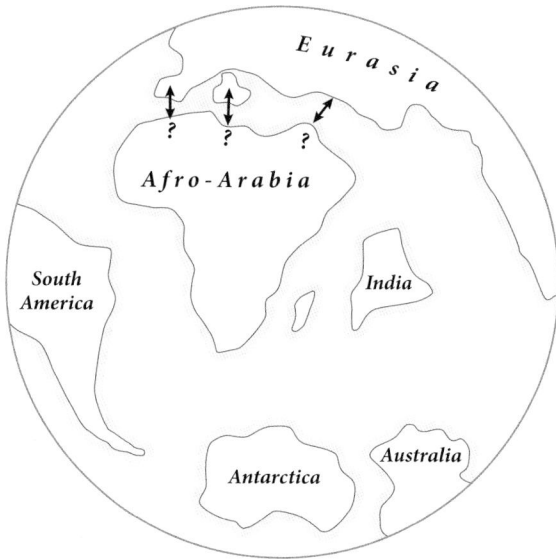

Afro-Arabia in relation to other land masses 63 million years ago; three possible points of entry or exchange are shown.

As the area furthest removed from Chicxulub, eastern Asia was probably the single most important refuge anywhere on Earth for mammalian survivors. Long before that, it was already the likeliest region of origin for placental mammals, then for primates and then for a divergence between 'prosimian' strepsirrhines and 'simian' haplorrhine primates.

We can safely assume that survivors of Chicxulub were small animals, but reliable tallies of survivors and of their diversity have yet to be made. The plant and animal communities that survived on each continent became the matrix upon which subsequent evolution has proceeded, but every traveller, every arrival from 'outside', challenged earlier inhabitants and perturbed whatever status quo might have existed before.

All the earliest comings and goings to and from Afro-Arabia were greatly inhibited by some 10 million years of continental isolation, because the broad Sea of Tethys lay between our continent and Eurasia. So it was left to a freak of geology to raise a temporary bridge, at least once, and perhaps more often. There are three potential connections, the shortest strait being near today's Gibraltar. A Libya–Italian link was considerably wider while a more 'tropical' connection has been posited by some geologists as an 'Iranian–Arabian corridor'.

Best guesses for the first primate arrivals in Afro-Arabia, both strepsirrhine and haplorrhine, are very rough, somewhere around 60 million years ago, and both have invoked an Iranian–Arabian bridge.

Other likely co-immigrant Eurasians were 'thryonomyid' and 'zegdumyid' rodents (savour these names with me, rolling their syllables over your tongue), both of which gave rise to very diverse contemporary rodent radiations that are especially interesting for close matches between skull shapes, sensory specialisation and behaviour.

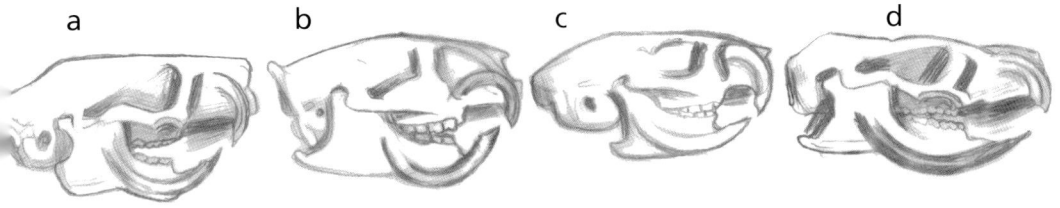

The persistence of functional geometry in living and extinct rodents: a) Anomalurus; *b)* Thryonomys; *c)* Carterodon; *d) extinct* zegdoumyid.

Any immigrant or import into Africa arrived with limitations shaped by the Eurasian community it came out of. In a largely 'empty' continent, the second largest on Earth, potentials for ecological success or speciation could be immense. A rough tally of contemporary African mammal species, all with immigrant origins, pans out at about 300 rodent species, 150 shrews, 100 primates, 70 antelopes, 30 mongooses, 11 mustelids, 10 hares, seven hedgehogs, four pangolins and one chevrotain. Were this a bird list, it could end with a single Asian partridge in an Uzungwa tree.

*Thomas's Galago (*Galagoides thomasi*).*

These tallies are the residue of some 60 million years of immigration, proliferation, attrition, specialisation, even super-specialisation, and extinction. They provide some measure of ecological success within a welcoming continent.

A very significant component in the advantages enjoyed by primates was their head-start as early arrivals on the second largest of all the continents, its vast spaces once blasted but now re-vegetated, and blessed with fewer competitors (and probably fewer predators).

Add to this the advantage of brains that were more versatile, certainly better adapted to tree-living than that of any of the afrotheres, Afro-Arabia's only endemic placental mammals.

Two early galagos are known from 40-million-year-old fossils, and today galagos are among Africa's marvellous array of strepsirrhine primates (yesterday's 'prosimians').

While on the subject of names and name-changing, it was the Swedish aristocrat, Carl Linne or Linnaeus, who turned naming into a branch of science. Thereafter, a thirst for naming plants and animals was triggered in most of the educated travellers who left the shores of Europe. One of the first regions of Africa reached by European mariners was the mouth of the Senegal River. Here, in about 1796, a small primate was collected and 'galago', its then local name (presumably in a dialect of Wolof), became the specimen's scientific and common name. It was a type of animal previously unknown in Europe.

*Zanzibar Galago (*Paragalago zanzibaricus*).*

Senegal Galagos were frequent household pets during my childhood, but we youngsters always called them *komba* or 'bush-babies', and we fed our pets on bananas, locusts and gobs of acacia gum, or we let them forage for insects brought in by pressure lamps.

Trying to get names to correspond more closely with genetic relationships had begun to gather force at the time I set out to prepare the first systematic inventory of the mammals of East Africa. Having kept several galago species over the years, I had noticed consistent differences between Senegal and Zanzibar Galagos, in spite of the authorities of the time lumping them as a single species.

Also, for four years I had naively kept a male Zanzibar Galago with a female Thomas's Galago without issue and without me realising that, in spite of their close resemblance, nearly 20 million years of genetic separation lay between them! Even so, every morning they crawled into their sleeping box to cuddle up, seemingly as lovey-dovey as Romeo and Juliet.

Eventually, my recognition that the Zanzibar Galago was a distinct species helped focus ears upon rich and varied galago vocabularies, and focus eyes upon previously ignored distinctions such as the shape of penises, in which some emerged not unlike dwarfed human members, while others resembled microscopically shrunken pineapples or cacti – ouch!

Galagos and their penises: (left) Greater Galago; (middle) Southern Galago;
(right) Zanzibar Galago.

Meanwhile, galagos and their allies have become a thrilling field of research led by the vision and energy of Simon Bearder, a colleague based in Oxford. It is thanks in large part to Bearder initiatives that more than 20 species of African strepsirrhines (sometimes called 'African lemurs') are now recognised, and their biology is being intensively studied.

The main reservoir of true lemurs, Madagascar, currently has between 50–100 surviving taxa, while the primates' land of origin, tropical Asia, only shelters two genera of lorises.

A plausible explanation for more than 100 taxa of 'streps' outside Asia yet only two genera in their supposedly ancestral Asian home is that the latter were constrained by the survival of many more pre-Chicxulub biota in Southeast Asia than anywhere else on Earth, whereas Africa and Madagascar were *terra nulla* for any primate arriviste. The picture for higher primates is less clear-cut, with Africa and Eurasia trading primate imports and exports over an enormously extended period of time. But what about South America?

Begin with considering what sort of African monkey rafted over a narrow Atlantic to found South America's single dynasty of primate colonists. Some cite the Oligo-mini-ape, *Oligopithecus savagei*. Fossils from both sides of the Atlantic were equally tiny around 32 million years ago and some were sufficiently similar to approximate to the sort of monkey and the approximate time (currently thought to have been between about 35 and 40 million years ago) when that

*Dwarf Marmoset (*Callibella humilis*). An aid to visualising the first primate immigrant*
to South America?

Madagascan pseudo-monkeys: Hadropithecus, *a short-faced diurnal lemur.*

random, unplanned but fateful crossing took place. Contemporary dwarf marmosets present us with a passable mimic of that fertile ancestor.

In the absence of strepsirrhines, those American primates even split into day and night monkeys, the nocturnal species becoming quite passable mimics of Madagascan night-lemurs.

On Madagascar an opposite swap took place in which one lineage of nocturnal lemurs moved into an unoccupied day-time niche to become *Hadropithecus*, the 'pseudo-monkeys' (sadly, these fascinating animals became extinct quite recently – only skulls and bones remain, so my sketch is an extrapolated reconstruction).

Within Africa the oldest fossils, mostly from northern or southwestern Africa, can only hint at what was going on in the equatorial regions (in which both primates and destructive decay, then as now, flourish best).

For me the outlines of primate prehistory became tangible and particularly stimulating at the hands of my good friends Bill Bishop and Sonia Cole. Bill was Director and Curator of Uganda's National Museum and it was as a Trustee of the museum that I joined them on an expedition to northeastern Uganda, where I marvelled at Bill and Sonia's ability to pick out a fragment of broken bone hiding in a sea of dusty gravel, no matter how abraded or goat-hoof-shattered that little bit might be. Our base was a Karamoja settlement near Mount Moroto, where an elder with white wisps on his chin came up to me, tugged my then quite luxuriant beard and, with undisguised disgust, announced, 'It's like dead grass!'

With the express purpose of providing material for a giant mural I was making for the Uganda Museum, and incidental to deepening my interest in palaeontology, I was delighted when Bill gave me the plaster cast of a skull numbered 7290, with the odd name, *Proconsul* – it still graces my desk.

Beginning as one of those insider-joke names, *Proconsul* implied 'before chimps', because 1930s tabloid media had elevated a stunt-performing Chimpanzee named 'Consul' to popular celebrity status. On the Lake Victoria island of Rusinga, Mary Leakey had unearthed an almost complete skeleton that soon became a star in the firmament of popular science.

Mary and Louis Leakey, with financial backing from the globally read *National Geographic* magazine, went on to excite a flurry of interest in ape and human evolution that has not abated. For this, Bill Bishop, Sonia Cole and a whole bevy of other scientists, myself among them, were deeply grateful. In spite of name changes, Mary's *Proconsul africanus* has come to be recognised as an 18-million-year 'dental ape' and thereby from her bag of bones hangs a much larger story.

Belonging to the larger category of Proconsulidae, or dental apes, Mary's specimen was among the earliest to mark out what has become an enormous trove of African fossils. Proconsuls are thought to have given rise to rather mixed bags of 'pithecuses' (in Greek *pithecus* meant 'monkey') – *Afro, Helio, Nachola, Equatorius* and, later, *Griphopithecus* and *Kenyapithecus*. Their fossil bones fill many drawers in many museums and their description and placement within primate trees cover many pages of many scientific papers. Some authorities have represented these exclusively African radiations as the broad highway of human evolution.

In common with many colleagues, I regard this fecund procession of primate trees, each flourishing during Africa's millions of years of isolation, as fascinating radiations that can be fancied as evolutionary experiments in becoming more or less human. Nonetheless, every mini-radiation manifests a diversion that led, in the end, to extinction of that lineage and its descendants. Except for one.

Bill Bishop had invited me for a preview of what had begun as random bits of bone and teeth. On his desk in the Uganda Museum he assembled these into a nearly complete palate, attached to an incomplete and battered, big, male face. He showed off these bits of broken muzzle as if they were the crown jewels of some medieval monarchy, and of course for us that ape was much, much more interesting, valuable and significant than any such juvenile baubles.

Bill and our mutual colleague and friend, anatomist David Allbrooke, published a description of Bill's find (it was 1963). There was not much to compare it with and its 21-million-year age was close to guess-work (the real age is probably somewhere around 18 million years). At that time the fossil record was extremely patchy, bones and teeth popping up from all over the eastern half and northern littoral of Africa, their ages in mostly guessed-at millions of years.

The two top authorities and intellectual patrons of the time, Louis Leakey in Kenya and Le Gros Clark in Oxford, had recently assigned some fragmentary fossils from Kenya to a newly found, large dental ape which they named *Proconsul major*. This was the rather hierarchical atmosphere within which Bill and David assigned their nearly whole muzzle to *Proconsul major*. A couple of years later a brilliant young palaeontologist, David Pilbeam, visited us in Kampala, expressly to examine Bill's *Proconsul major* specimen minutely and critically. He had been studying *Dryopithecus* fossil apes from later European sites and concluded that what had come to be called 'Bishop's ape' could more accurately be seen as a precursor to the dryopithecine lineage. This introduced an entirely new and very controversial dimension to the discussion – comings and goings of primates between Eurasia and Africa: what Bill called the 'import/export trade'.

At this juncture, Louis S. B. Leakey, with ever an eye to the sort of public debates he loved, weighed in with his gut opinion that 'Bishop's ape' was directly ancestral to gorillas. Because African apes and humans share common roots, L. S. B.'s opinion reinforced the (still current) idea that human-Afro-ape evolution is an unbroken African story.

At issue is a broader realisation and need for documenting the transformative effects of intercontinental exchanges. New opportunities, even built-in constraints, are always without precedent for both immigrant and for the established ecological communities into which the newcomer comes. When that new arrival is a primate more advanced than any resident mammal, the consequences are not predictable.

Bill's excavation site at Moroto has continued to deliver more fragments of the same individual that Bill was first to find, the once-alive licker, taster and owner of the very palate over which we shared excitement back in 1963. These bits can be read as a hotch-potch in which component parts of bodies adapt in similar fashion to similar circumstances. Fragments of its shoulder bones suggest Bishop's ape was an accomplished climber, albeit a heavyweight one; its thick thighbones resembled those of a modern ape. Most significant of all, its short-backed ape-like vertebrae proved to be fundamentally different from the long-backed, more monkey-like *Proconsul*.

This meant that *P. major* had to be ditched as a name for Bishop's ape. It has been replaced by the more explicit and appropriate *Morotopithecus bishopi*, but in 1985 another fossil of younger but comparable age, *Afropithecus turkanensis*, was found in the same region and certainly belongs to the same lineage.

Both types are now called afropithecines, reinforcing the idea that African apes gave rise to hominins directly, without involving any excursions to Eurasia. There have even been suggestions that this enlarged collection of *Morotopithecus bishopi* bits represents a biped directly ancestral to humans.

I follow David Pilbeam in seeing Bishop's ape as our best model of the sort of ape Africa exported to Eurasia. Now that the genes of Asiatic orang-utans have been found to cluster tightly among later hominids, the idea of a 'Returnee Radiation' has come in from the cold, and a 10-million-year Eurasian immigrant ancestry for the gorilla-chimp-human lineage has become much the more likely reconstruction of hominid evolution. Having exported early apes (perhaps as early as 20 million years ago), Africa reimported their progeny, an advanced dryopithecine, about 10 million years later, to become a much more recent pre-chimp than was old *Proconsul*.

Like the evolution of super-smart capuchin brains in South America, ex-African apes followed a similar trend as they flourished and proliferated all over the length and breadth of tropical and subtropical Eurasia. In spite of being just about the brightest animals on Earth at that time, these ex-Africans left behind contemporaries that were embedded within an already very extensive community of higher primates. This can be predicted to have put very considerable constraints on ape innovation back in Africa.

In Africa, the fate of apes smaller than their rapidly evolving monkey competitors can be inferred from 12.5-million-year-old *Simiolus minutus*, a talapoin-sized ape whose lineage probably succumbed to competition from the leaf-eating monkeys that were just then evolving, probably in equatorial regions of Africa. Given that tiny marmosets can behave like triumphant wrestlers, I like to imagine this little fellow, known only from a few teeth, strutting his stuff in the spirit of 'apes will be apes'.

In Eurasia, the apes had the higher primate niche all to themselves and radiated accordingly. Once established in Asia, apes seem to have diverged into the smaller gibbon lineage and larger, orang-like or 'pongid' lineage. In the absence of any competing primates, pongids gave rise to a particularly widespread, successful and diverse radiation of western apes called dryopithecine apes.

Just how diverse those ape pioneers became remains to be deduced, because a few gibbons, siamangs and three species of orang-utans are mere shadows of Asia's 'age of the apes,' which included a true Goliath ape.

It is to one of those glorious little asides of deep-time natural history that a 'mere rodent' offers us some measure of the abundance and diversity of

*'Silver-back' Mini Ape (*Semiolus minutus*),
all astrut.*
INSET: *Lion Tamarin.*

Jonathan beside a Gigantopithecus *(I wish).*

apes in ancient Asia. Over millions of years, countless generations of porcupines have garnered skeletons, dragging bones back to their burrows or caves where they can safely gnaw them away.

I once found an elephant skull with contours modified by porcupines, but it was way too big for them to drag home. Porcupines need masses of calcium to reinforce their armour of quills, but the dentine and enamel of ape molar teeth are even harder than porcupine incisors so piles of button-like ape molars accumulated within the retreats of Asian porcupines. Imagine what a small proportion of ape teeth are likely to have ended up in porcupine refuges. Then learn that our knowledge of such accumulations of teeth comes from chance encounters by farmers, the majority of whom just hoed them back into their fields. Then learn that but two or three generations of archaeologists have purchased a few thousand ape teeth, all sold as 'dragons' teeth'. These belonging to several species, including many from a real giant of an ape. Now, remember that the archaeologists could only visit a mere handful of thousands of Chinese apothecary stores, where many generations of gullible customers over many centuries have been buying, grinding up and consuming countless such 'dragons' teeth'. Over the several millions of years in which apes were the only higher primates in Asia, that uncluttered primate community evolved a 3 m Goliath ape weighing in at 500 kg. This enormous ape, *Gigantopithecus blacki*, closely resembled a scaled-up orang-utan and flourished over extensive areas of tropical forests in

Southeast Asia and China for at least 3 million years, probably a lot longer. The cause or causes of its extinction some 100,000 years ago are debated, but my intuitions point to human hunters and their eventual penetration of forests that had become (literally) marginal under Ice Age climates. But oh! how my teenage naturalist self longs to have met my gigantic cousin. If only we could have foraged together for wild asparagus!

We now know that the two earliest lineages of higher primates diverged about 30 million years ago. One led on to larger, short-backed, tailless, slower and marginally bigger-brained apes (simplistically demeaned as 'intensive gleaners'). The other led to smaller, long-backed, long-tailed monkeys (sometimes lumped together as 'expansive foragers').

The former made slow, careful movements using strong, clamp-like hands and feet on ever-more dexterous joints, to lever an undifferentiated lump of a body through a lattice of branches.

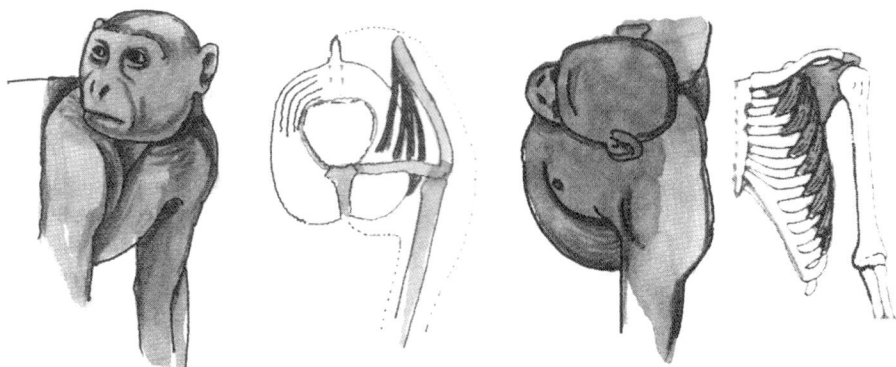

Shoulder and X-ray (sort of) of monkey (left) and ape (right).

By contrast, monkeys have long, flexible backs, narrow shoulder blades flanking deep, relatively narrow chests, thereby allowing an easy quadrupedal gait that could take them all over wide areas of forest or open woodland. This distinction is crucial for our understanding of ape emergence, and for their reversed fortunes with the birth of our own lineage.

I remember the perils of baboon existence being dramatised for me and my companions one June, when the rains were over and strong winds were drying out soils and grasses. Smoke dissipating over the horizon betrayed the first fires of the season. Shortly after dawn on the edge of an open floodplain we suddenly encountered three baboons dashing towards a single, leafless tree, a lean, athletic lioness hot on their racing hands and heels. They just made it up onto spindly branches where, gasping and panting from their exertions, they looked down on their equally puffed, tail-lashing pursuer. No ape ever ran that fast.

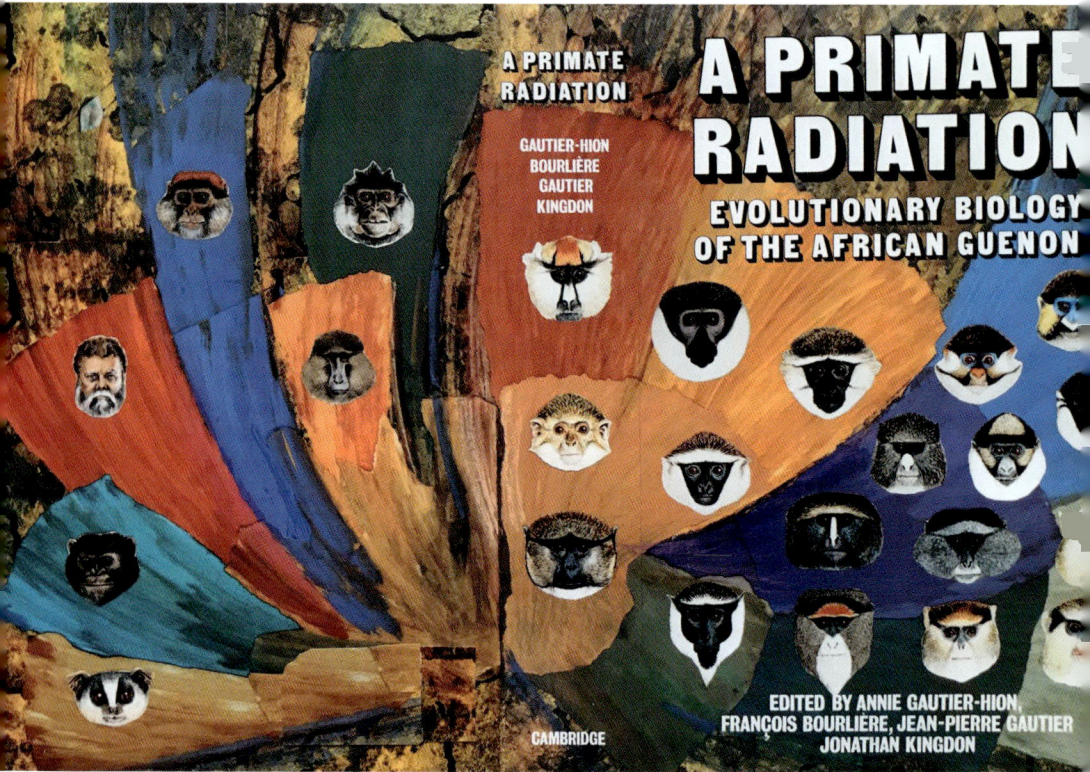

A Primate Radiation *(dust jacket).*

Regional morphs of Angola Colobus (Colobus angolensis).

The advantages of fast movements are numerous, facilitating escape from both predators and rivals, encompassing a larger home range, and moving rapidly through the trees or over ground from one food patch to another. For monkeys, running and leaping through forest canopies, long tails are as much of an asset as the pole of a tightrope walker.

The radiations of monkeys in Africa are astounding. I have scarcely touched upon the colobus monkeys, one of Africa's most widespread exports to Eurasia, nor macaques, another successful emigrant lineage. Earlier I hinted at the face-painting origins of cephus monkeys, one small branch of the particularly attractive, talapoin-descended and widespread 'long-tailed' cercopithecine monkeys. I reminisced on baboons, their motor-car skulls and the refreshing vulgarity of male Mandrills, while neglecting their derivative descendants, the mannerless mangabeys, wild, unruly mop-heads, their turkey-gobbling part of every healthy forest's orchestra.

*Grey-cheeked Mangabey (*Lophocebus albigena*).*

As for the colobids, they too fill dawns, dusks and other moments in need of advertisement with croaking, rumbling choruses of extraordinary volume. Their pied pelage has been likened to the restrained elegance with which early Protestants responded, several centuries ago, to the colourful excesses of a decadent Europe. I see their noisy treetop ceremonies as a splendid assertion of presence in a green and crowded environment.

Angola Colobus.

Now, return to the 1960s, while I was assembling my illustrated *Atlas of Evolution in Africa* (heartily endorsed by Bill Bishop and David Allbrooke). I set out to depict (and dissect) the adaptive anatomy of a representative range of African primates. This was part of an effort to present every mammalian radiation in the context of whatever perspectives of deep time were available in the 1960s.

The finders and describers of fossil hominids sometimes vie with one another to present their fossil (usually its disjointed fragments) as directly on the line of human ancestry. I prefer to assume that perhaps no such fossil, no such 'missing link', has survived, anywhere, yet the fact that every fossil samples populations that came and went over deep stretches of time, and across diverse regions of Africa and Eurasia, makes every fossil precious on many, many levels.

A bold pattern and some dates are now emerging from an ever-expanding treasury of both African and Eurasian fossils. Africa's 18–20-million-year-old export of something resembling Bishop's ape was followed by African leaf-eating monkeys that got to Eurasia about 14.5 million years ago, where they generated a very extensive radiation of Asiatic colobids. Today there are many species of leaf-eating monkeys or langurs all across Asia (most of them on the cusp of being exterminated by neo-colonial pirate logging companies). All derive from a single African species that made it into Asia.

Three wise monkeys (as seen in a Kyoto Museum).

About 6.5 million years ago yet another African species, this time a macaque, followed, generating a similar diaspora all across Asia. In Kyoto I was shown beautifully carved images of three macaques that, hands over eyes, ears and lips, *saw* no evil, *heard* no evil and *spoke* no evil. That good maestro carver could not have known that his Japanese Macaques were ex-African primates; for him they were cartoons of all-too-human closure of the senses. I like to think that the artist was begging us *not* to be blind, deaf and mute in the face of our own mischief-making.

Prehistoric exports and imports of biota on a grand scale have enriched both continents, enhancing biological complexity in more ways than we can ever begin to comprehend.

The very earliest Eurasian export of a haplorrhine primate pair into Africa might have been small enough to nestle in your palm, but their continent-hopping

progeny changed the entire trajectory of life on Earth because, after more accidents of timing than we can ever know, one of their many descendant lineages
became *Homo sapiens* and so generated the Anthropocene, the era our species
must now most urgently examine and reform.

The text on apes that follows ends with the tale of a gorilla that was persuaded
to trust the humans who named him 'Rugabo'. A year or two after I met Rugabo,
he was decapitated by armed gangsters, and a head that once seemed to smile in
trusting fellowship was sold, pickled, to some passing foreigner.

In angry disgust at people's absence of empathy for our cousins, I painted
a 'pietà' and a couple of pastiche 'depositions', dedicating them to Rugabo in
outrage and pity for the sad fates of too many great apes at the hands of loggers, poachers, smugglers and perverted city gourmet diners. It is criminal that
the way things are going, decapitated gorillas and chimps will soon join all our
shared ancestors as candidates for oblivion. Victims of a culture of death –
ecocide. The principal assassins include international logging companies with
headquarters and lobbies in the world's most influential cities.

Chimpanzees, gorillas and orang-utans hold jurisdiction over all our most
recent genealogies. With respect to the shallows of human origins genealogists
must submit to the authority of living ape genes. Lifeless geology, crumbling
fossils and all the armories of contemporary science must bow before the living
authenticity of these prophets – the elders of all our origins.

Gorilla sketches, 1964.

SLOW BUT SMART

In which apes outwit the competition.

A few hundred metres above Mongiro Hot Springs, in the northwestern lee of Rwenzori, the 'Mountains of the Moon', stood several giant fig trees. Their produce would have filled the shelves of a dozen superstores several times over, and for a month or more. Fruiting between trees was staggered but the ripening of a single tree's enormous crop was fast. Just as well, because no sooner was the fractious, flap-flitting fruit-bat clientele dispersing at dawn than early-bird barbets, green pigeons with empty crops and hungry red-tail monkeys arrived. Once the sun was up, a second wave of fig-eaters displaced the dawn debutants. Honking hornbills, assertive baboons and mangabeys and an assortment of fruit-eating birds cluttered the tree's crown, tutting and scolding in a chorus of mostly contented, greedy diners.

Finally the Chimpanzees arrived, indifferent to whether the figs were ripe or not, until bellies were full enough to allow more selective harvesting. By this time the ripest of the day's fruits had all been creamed off and an army of dispersers were crapping tiny fig seeds onto, into, under and all over Bwamba/Rwenzori, the single most densely packed, super-diverse patchwork of forests on Earth.

Late-breakfasting Bwamba chimps exemplify several of the advantages enjoyed by apes. First, they are close to the size and weight limits for arboreal animals, so the timetables of lesser beings are partially determined by the need for other animals to get places early, to give way, sometimes on threat of becoming someone else's breakfast.

Chimps are pretty gross feeders – not specialists like leaf-eating colobus monkeys or nut-eating squirrels. Nor are they fast, like red-tail monkeys, aerobatic rollers or bee-eaters, animals that live life in the fast lane, absorbing sensory information and making innumerable decisions in splits of seconds.

Instead, chimps learn, accumulate and share knowledge over extended lifetimes. Compared to other primates they are relatively slow and deliberate, they mature slowly and live longer, absorbing knowledge from an early age from well-tested mentors, or creating it through youthful experimentation. Who knows what detail might be encoded in their lung-emptying hoots?

OPPOSITE: Deposition *(a pastiche).*

Hooting Chimpanzee.

Where other species rely on built-in *savoir-faire*, apes adapt to local circumstances.

Whatever complex of merits allows chimps to breakfast late in Bwamba, their primacy among primates needs to be seen in the context of deep, slowed-up time. Every adaptation is the product of time, often drawn-out, but ultimately measurable time, hence my seemingly endless repetition of millions of years, after millions of years, for which I make no apology.

Fig-tree visitors: Chimpanzee and hornbill.

Consider first what all primates share, from Tankoko tarsiers, Senegal Galagos and Congo talapoins to Geladas and chimpanzees. All share round heads and alert faces, vertebral columns, ribs, four limbs ending in hands or feet that are modified 'hands', revealing that primates, from their earliest beginnings, all shared an arboreal history. Today there is no shortage of arboreal mammals but primate ancestors probably triumphed in this niche close to the beginnings of placental evolution, giving them an important head-start as brainy mammals.

Primates have been remarkably consistent in adapting, over some 100 million years, to an astonishing range of arboreal niches. The hands of even the most terrestrial of primates betray their inability to shed this primitive vestige of long-lost ancestral, arboreal lives. As an inveterate sandal-wearer I am grateful for my inheritance of toes and shudder at the very thought of having a satyr's paired hooves instead. I am grateful to be a primate.

The ability to find pathways through three-dimensional space, to judge distances, to discriminate between ripe and unripe, edible and inedible, all depend on good sight so no primate has ever abandoned vision and visual acuity as a senior sense. Primate reliance on vision is well expressed in the size and sculpting of eye sockets and in the relative size of those portions of the brain dedicated to processing visual information. Externally, visual acuity is manifest in alert, attentive eyes.

Some leaf-eared nocturnal primates, such as big-eyed, big-eared galagos, also have excellent hearing and can locate their prey by sound. Primates communicate with a babel of calls, using transmitters that vary from bony voice-boxes to bellows-like vocal sacs, while receptors have taken the form of relatively conventional external ear pinnae and skull ossicles that vary but never overtake eye sockets in size.

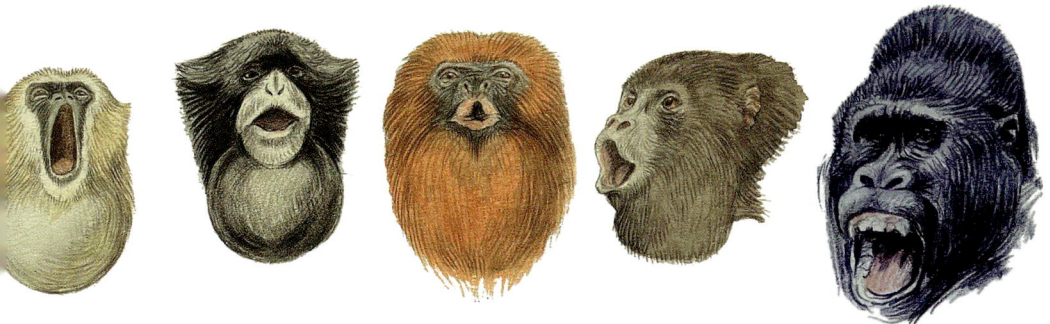

Primates are vocal primadonnas.

Primates scream, whisper, sing or gossip and they exploit the multiple percussions of animal existence as vehicles for information. Sounds need interpretation and can express social and emotional needs, from an infant's cry to an aroused male's furious thrash through the vegetation.

Because primate evolution has, from the start, evolved in a three-dimensional lattice, sounds are made and interpreted in space. The potential for developing sound-based codes exists in any animal, but climbers, gliders, flyers and swimmers have potentials, including musical potentials, that are denied to heavily terrestrial animals operating in but two dimensions – just ground that goes east to west, north to south.

LEFT: Weaning. RIGHT: Mammal Madonna.

Galagos and other lemuroids with prominent noses follow scent trails to find their way along mazes of dense undergrowth or through the canopy, even on moonless nights. Higher primates relegate scent to minor roles in their biology; nonetheless, we are not so specialised that we eschew sniffing a perfume or a ripe mango, a palm toddy or a newborn baby's scalp. Smell and taste can overlap, as evidenced in a connoisseur sampling wine or a chef taking her live or TV audience through the preparation of her dish, sniff by sniff, lip to tongue.

An infant primate clinging to its mother, or her proffering a breast, expresses the sense we most take for granted – touch, the most direct, most physical of senses and the one most concentrated in lips, fingertips, breasts and sexual organs – all as closely co-ordinated and linked up with brains as any other sense. Remembering that human feet worked as second-rank hands for the greater part of our evolutionary history, it becomes less surprising to see feet enlisted by footballers, skaters, be-boppers, and ballet and swing dancers, as they demonstrate spectacular feet-feats or convert limb movements into creative art forms.

For most animals, touch is a more diffuse sense, yet many courting fish and birds delicately brush one another with a fin or feather (usually a pair of specially evolved plumes or hair-like bristles). Like a lover's caress, such extrusions have been evolved to entrance a sense. Even the most flighty of ewe antelopes can, during oestrus, be seduced into a dreamy-looking trance by being love-thumped ('*laufschlag!*') on the belly by clumsy foreleg kicks from a randy ram. Notwithstanding the evolution of a multitude of magic wand caresses, touches, love-bites, slaps and tickles, for most animals these are mere incidentals to social survival and the pursuit of a living.

Touch and all other sensory channels have evolved in concert with brains. We think of brains themselves as the ultimate measure of 'brainyness' but brains began as processing and exchange centres for information pouring in from multiple sensory receptors. With regard to spiders, my friend Fritz Volrath has shown that each leg has its own autonomous information/action exchange centre – eight self-contained mini-satellite brains, each capable of weaving a web of startling complexity. Watching footballer Megan Rapinoe, I found myself wondering whether she possessed a pair of spider legs on steroids, so great was the speed and accuracy of her evasions and goal-winning kicks!

It is no surprise to find that regions of the brain linked to particular sensors reflect the relative primacy of that sense in the life and survival of any one species. The odour-processing apparatus and brain of a termite-feasting Aardvark are sufficiently enlarged to influence the contours of its skull and face, outlines that reach cartoon proportions in a tottering newborn Aardvark.

No less cartoonish are the too-heavy-headed neonates of our own species. To learn how very late that brain-heavy head developed in the evolution of our

Newborn Aardvark with prominent turbinal swellings between snout and brain.

species is surprising but, on further reflection, there is the imperious logic of natural selection behind it. Begin with the selection necessary to enlarge birth canals in mothers. Any mismatch in the relative breadth of a mother's pelvis, of flexion at her pubic symphysis, or of the malleability of her baby's cranium, can result in death for child, mother or both.

Once born, helpless modern human neonates present mothers or parents with many anxieties and liabilities. Project that helplessness back into prehistoric times and imagine what hazards might have applied, even for new-born hominins and mothers that probably suffered their birth pangs and gave birth in the midst of busy encampment life, while walking around with a toddler in tow or while squatting over a melon.

An enlarged brain in the earliest primates was, in part, a byproduct of wide-eyed arborealism. It is also a fact that the ratios of primate brains relative to their bodies are, in general, larger than in most other mammals. I say most, but bats, whales and a few others *can* compare.

Take Sperm Whales. Not only do they boast, by far, the biggest brains on Earth but, like apes among primates, their lineage has specialised in being relatively brainier than any other, already brainy whale. We now know that those giant brains encompass that vital emotion we call empathy as well as love, exuberance and grief. In so far as our still clumsy sciences can discern, Sperm Whales have structured vocabularies as complex as anything human, while some of their sonic repertoires and compositions might have impressed a Mozart or Duke Ellington.

I remember as a teenager being given a magnificently big-brained Honey Badger skull. Gazing at it and clothing it in my mind with the trundling, muscular form of a live, supremely self-confident badger (usually seen from a discreet distance and never approached directly), I longed to raise a baby one just to learn how much I and the badger might share or have in common. I divined that it would be so much more rewarding as a companion than our only half-wild cat, or our midget-brained dogs.

That longing to raise a Honey Badger was preceded by my infatuation, at age four, with a super-intelligent, big-brained capuchin monkey, his barrel organ and the be-hatted, moustachioed hurdy-gurdy man he had in tow. Our family were on leave during the last months before World War II broke out and an Italian organ-grinder (and what Teddy called his 'Darwinian assistant') plied the streets of Exeter and Exmouth. I adored the monkey, the Italian and all their tunes (in that order). 'I love coffee, I love tea', 'Funiculum, Funiculi', and a twangy version of Verdi provoked some passers-by into paying the monkey (whose main job it was to solicit and harvest coins) to stop and go away.

Not me. For a few weeks that marvellous capuchin and his friendly attendant became my ultimate role-models for how life should be lived.

My mother delighted in recounting how, a few weeks later, I earnestly approached the Roman immigration officer on our arrival by sea-plane on Lake Bracciano. 'Italian officials came aboard. Jonathan said politely to the first,

"Please, have you got a monkey?" There was a suppressed titter all round but he realised that no insult was intended and an international incident was averted. Jonathan must have remembered that Italian organ-grinder and his monkey last summer.' Perhaps my delight at meeting a friendly, gentle, highly interactive monkey on barren English streets had me supposing that a monkey shadowed every Italian as his daemon. No such luck.

'Darwinian assistants': capuchin monkeys.

Wild capuchins have been observed to carry nuts to drying-out spots then, several days later, return with a stone or stones that have been specially selected for size and shape. Squatting above each nut, they hold the stone above their heads in both hands then whack downwards with enough force to crack open the now brittle-dry nut-shells. During the wet season, when mosquitoes abound, the capuchins find millipedes which exude an acrid secretion. Rubbing their bodies with squirming, oozing millipedes results in their mosquito tormentors being demonstrably repelled, and flying off to find less intelligent sources of blood.

Capuchins in the New World can stand in for what apes signify in the Old World because some 38 million years ago, the chance rafting of a single pair or party of tiny African primates gave rise to a South American, continent-wide radiation that is now reduced to some 150 species. Within that spectrum, the capuchins are, by many orders of magnitude, the brightest. They continue the primate tradition of selection for superior braininess, culminating in capuchin brain/body ratios that are but miniaturised parallels for those of modern humans.

Contrary to popular ideas about monkeys being primitive while humans are advanced, it is our human/ape lineage that is relatively ancient compared with baboons, colobus monkeys, macaques, guenons and even capuchins. All of these are defined by highly specialised characteristics that have evolved relatively recently. While *Homo sapiens* is the newest kid on the block, our hominin lineage aligned two evolutionary traits very early on. One strand of selection was for bigger brains, the other was the conservative trait of remaining 'generalised' (a necessarily ambiguous term). Apes, like those Bwamba chimps, eschewed any extreme of adaptation, remaining generalised in diet and in any trait that might have restricted them to some minor niche or immured them in some remote region. Chimps recently ranged right through the central and west African rainforest belt, having maintained that pre-eminence for several million years. Chimp genes reveal that some contemporary regional populations differentiated more than 2 million years ago. If they were hominins, taxonomists would separate them into species.

I have always enjoyed and participated in community dances, and reserve a special empathy for chimp equivalents where rhythmic noise and vigorous movement are involved. The components of drumming and singing depend on motor patterns that scarcely differ for man and ape. During those conditions of excitement in which such performances occur, I sometimes feel myself imagining traces of my distant ape ancestry.

It is difficult to recognise animal origins in some of our responses to visual, aural and olfactory stimuli, particularly when we know even less and most of us are even less interested in the behaviours and customs of mammals than in those of foreign fellow folk from far away.

As the Mongiro chimps exemplified, less selective bulk feeding, large bodies and social cohesion ensure priority of access to food, while decibels of noise and drawing together in massed numbers can actually intimidate most potential predators and competitors. The obvious cost of being restricted to thicker supports, or more frequent travel on the ground, suggests that apes combined tree- and ground-dwelling earlier than other primate lineages. Wherever heavy apes found it was easier to hang from or swing between branches, there was a shift away from hind-limb to forelimb dominance. This was no trivial change in emphasis and was linked to an absolute and irretrievable loss of a tail. Curled up in a vestige known as the coccyx, both ape and human tails are amalgamated into a wholly internal structure. As well as being diagnostic of the divergence between apes and monkeys, taillessness became the diagnostic detail of an ape-specific limitation on their later history. Monkeys out-ran apes and people until they, in turn, were out-run by arrows and bullets.

The anatomy of ape limbs has been sufficiently plastic to allow the arms and shoulders of chimps and gorillas to accommodate to levering heavy bodies over the ground or to haul them up to the top of tall trees. It is this heavy-duty haulage that makes a striking contrast with the relatively light work that typifies human arm work today. For people and cultures familiar with primates on a daily basis, our similarities (especially those with apes) demand some acknowledgement of affinity. The challenges that this posed and poses for thoughtful people in the tropics has been interpreted in many different ways. Batwa, Efe and Bornean hunters have long recognised that apes are kin. Being banished to cold northern Europe woke my 11-year-old self to how deeply primitive and wilfully ignorant northern primate-deprived cultures can be. Yet their historic 'discovery' of apes in the tropics eventually gave rise to new branches of science, even the new discipline of primatology.

Gorilla faces, gorilla moods.

A 22-year-old European, raised to be the queen of her extraordinarily obsequious nation, was taken to see a captive orang-utan. She described this poor little ape as 'frightfully and painfully and disagreeably human', a response echoed by many a contemporary zoo-goer. Classical notions of the ideal, of costumed courtiers obedient to crass religious iconography or to elite and popular, advertisement-driven, tabloid cultures – all combine to endorse and prolong primitive rejection of the tropical world from which we all ultimately derive. Only some education in biology, in rudiments of genetic literacy, or exposure to the concepts and cultures of some non-agricultural and ape-savvy peoples, can teach us that human genes can and have inhabited bodies and faces very different from the ones presented as 'normal' by primate-ignorant cultures. African scientists and artists must now take up our responsibility to educate a world all too deeply misled by pedants from intellectual wastelands.

Instead of the tens of millions of years that were formerly assumed for an ape-human divergence, Vincent Saritch and Ian Wilson used proteins as surrogate genes, to estimate an ape-human divergence at 7–5 million years ago. Pioneer geneticist Naoyuki Takahata and his colleagues made an even bigger splash when they announced in 1995 that, contrary to all appearances, chimpanzees were more closely allied to humans than either chimps or humans are to gorillas. Virtually all subsequent studies have corroborated that relative relationship, but many people, even scientists, still find it difficult to believe.

The refusal to respect animals as related beings has come to be called 'speciesism', an epithet closer to racism than one might think, and I can say this from personal experience. One morning, while I was buying fruit in Mbeya's open market, a gang of youths swanked past and one called out, '*Tazama huyu nyani zeru, mweupe kama usaha, ninajaa na karaha,*' ('Look at that albino baboon, white like pus, how disgusting.') A market woman turned to me. 'Don't listen to them, brother, they're just hooligans.' The taunt immediately brought to mind the image of an albino baboon, painted in 1770 by George Stubbs. There was

Arms aloft: Common Chimpanzee and human compared.

no way that hooligan could have understood where my laugh back at him came from. Anyway, 'albino baboon' can join many other supposedly less than complimentary names that say more about the limitations and prejudices of the namer than they do about the named.

Every human can, indeed, be disgusting, including me, but I pay attention to facial expressions as much as any primate. I also take some comfort in combining my apish facial anatomy with the artefacts of speech, text and imagery, to share my thoughts and emotions with as many as consent to listen, read or translate.

No other ape species can do that, so I am among those who must speak up for our cousins' right to exist. Our progress in this respect was made manifest when the president of ape-friendly Uganda actually described gorillas as 'cousins'.

Indeed, my face *is* ape-like but that is not what makes me revolting. My unruly hairiness intersperses moustache, beard, brows and sideburns with naked nose, cheeks, ears and forehead, a distribution of facial hair almost identical with that of a female Sumatran Orang-utan. As for ape babies, they evoke the strongest and most emotional responses from women: 'Oh! how adorable – oh! come to me, baby.'

OPPOSITE: *Bonobos (*Pan paniscus*).*

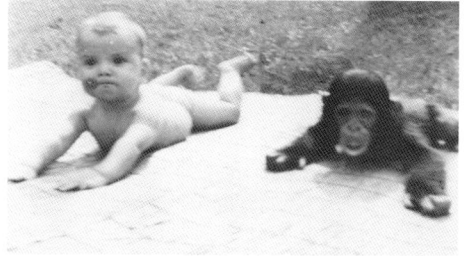

Babies on the mat: Human and Chimpanzee compared.

As for apes, perhaps they see us as aberrant, comical, certainly disagreeable versions of themselves. Here are sentient cousins whom we *must* acknowledge, not reject. Their current mistreatment and disinheritance should never have been countenanced, yet it proceeds unabated. It has become a moral duty to stop the systematic extermination of our cousins.

In the raw racism of our current barbaric era, facial expressions and facial hair easily become bridge or barrier, as can the relative reflectance of any individual's skin pigment. In our generally over-dressed cultures we have concentrated on signalling age, sex, status and race, as well as mood and temperament, in one fragment of our bodies. Some of us even try to hide this last vestige of natural openness behind masks, shades and veils.

I remember my surprise at seeing my generally august father inviting the laughter of villagers by telling a story from his undergraduate days in Oxford. In picturesque but fluent Kiswahili, Teddy opened his yarn with, '*Kulee*, Oxford, *iko Mto*, *wanamwita* River Cherwell'. To translate and continue, 'The river (*mto*) Cherwell, which flows through Oxford, has two secluded stretches reserved for naked bathing. One for ladies is called "Dames' Delight", while its male equivalent is known as "Parsons' Pleasure". One hot summer's day, two professors taking their pleasure in the Parsons' stretch found themselves subject to embarrassing laughter from an intruding boat-load of young undergraduettes. Both

LEFT: *Female Sumatran Orang-utan.*
RIGHT: *Male* Homo sapiens.

professors scampered for their towels, one covering his loins while the more eminent of the two wrapped his own grizzled head. "Why?" asked his companion. From depths of towelling came the muffled reply, "In Oxford I am known by my face!"' Teddy's audience, familiar with European coyness over their genitals, found his story, and his style of telling it, hilariously funny.

Like Oxford professors, gorillas are known by their faces – they could just as well be said to know one another by their noses (probably their aromas too), and one gorilla bridged a bit of the gap between himself and humans with his own sly take on noses.

Here is my recollection of a story, probably best told in Kiswahili or Kinyaruanda, but English will have to do. It concerns gorillas beside a mountain stream, flowing off Sabinio volcano. Here, even within a single family, highly distinctive faces with uniquely shaped and crinkled noses stare out from undifferentiated woolly masses of black hair.

I began this exploration of origins with a Leopard's killing of Ikimuga, a Mountain Gorilla that lived on the saddle between Mounts Muhavura and Sabinio. Here is my memory of another visit, nearly 30 years later, to another group of Sabinio gorillas.

The group's alpha male was Rugabo, whom I have mentioned before. On our arrival, he gathered up some giant fennel stems and, ostentatiously turning his back on us, sat hunched over his snack, pointedly ignoring us.

Mike Catsis, our infinitely patient guide, had spent several years 'habituating' Rugabo and his group (which included, for sure, some relatives of Ikimuga). Mike was imparting disciplined and systematic skills to the scouts and guides that protected these precious, remnant gorillas and their habitat.

At this point I should remark on Mike's physiognomy. His tanned face, narrow by gorilla standards, was fronted by a not remotely gorilline (indeed, a prominently Mediterranean) nose.

Given the significance of nose-shape in gorilla society, it was perhaps not surprising that Rugabo eventually could not restrain his fascination with Mike's nose. With a slight jostle as he sat down beside this persistent but still foreign observer, Rugabo reached out his enormous fist to take Mike's nose between index and third finger. After brushing it very gently, he then brought his knuckles up to his own nose for a brief sniff.

As the two primates looked momentarily into each other's faces, Mike thought he saw the corners of Rugabo's mouth turn up in the shadow of a smile – ha!

16 THE BOREAL, LATITUDINAL REALM

In which humpanzees reach a parting of the ways, one playing the safe and well-tested, the other facing a surfeit of risks.

My earliest introduction to biogeography began with the Orange Ground-thrushes that used to sing all along the valley below our house in Tukuyu. With hind-hearing I can reminisce that their melodies, varied phrases and trills were not unlike those of American Robins or European Blackbirds. As a scientifically minded 16-year-old I collected a specimen which now inhabits a drawer in the British Museum of Natural History. When I took its carefully documented and labelled skin to my mentor, Mrs Hall, in 'The Bird Room' in Cromwell Road, she introduced me to Captain Grant who took me off down a maze of corridors between gigantic wardrobes, all full of bird skins. Homing in on a small skyscraper of layered drawers, he stopped at one labelled 'African Thrushes'.

'For those of us trying to name and understand the relationships between populations, these ground-thrushes pose an awful lot of questions,' said Captain Grant. He went on to show me how 'my' bird, *Geokichla gurneyi*, occurred in mainly montane forests all the way up the eastern side of Africa from the eastern Cape to Mount Kenya. 'We don't have any specimens from Tukuyu, so your skin provides a valuable connection and record from a locality lying between our Kenya and Nyasaland examples – the Museum would be grateful if you were prepared to donate it?' Of course, I was thrilled! My donation and others would join skins collected by Charles Darwin and Alfred Wallace, more than 100 years earlier, and induct me into the brotherhood of 'collectors', the naturalists on whom the whole structure of biological classification, nomenclature and scholarship then depended.

Captain Grant then pulled out another drawer in which examples of the birds' closest relative, *G. crossleyi*, were laid out in neat rows. Unfolding a sketch-map that he was preparing for a field-guide-cum-handbook, he went on, 'Now, look how these thrushes, clearly the same sort of forest-floor bird, range from the eastern Congo across to Cameroon.' I was being invited to visualise my specimen and a coterie of its relatives as contemporary inhabitants of my home continent.

Captain Grant then parodied these thrushes as if they were a human family, in which one branch had taken ox-wagons to settle way out west while another shipped to the far east. 'Yet a third weathered troubled times but were content to stick with the homely south, all settling down over a million years or so – of course!'

OPPOSITE ABOVE: IDs Isolated as Disks: *facial and shoulder patterns of five African ground thrushes* (Geokichla *spp.).*

OPPOSITE BELOW: *Specimen of Gurney's Ground Thrush, collected and drawn by Jonathan in 1951.*

Over less than an hour, and with no trace of condescension towards a mere schoolboy, Mrs Hall and Captain Grant put my specimen into the context of Ice Age/hot-house exchanges on a small planet, one detail in the natural history of Africa's birds, one of countless inhabitants. Tacit to our meeting was the

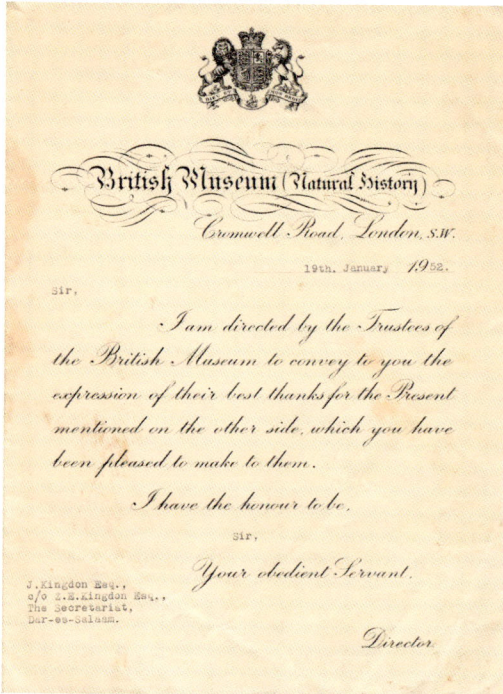

British Museum certificate (1951).

BELOW: *Opening pages of* Homo sapiens *profile in* East African Mammals *(1971).*

BELOW OPPOSITE: *Continuum of* Homo sapiens *profile.*

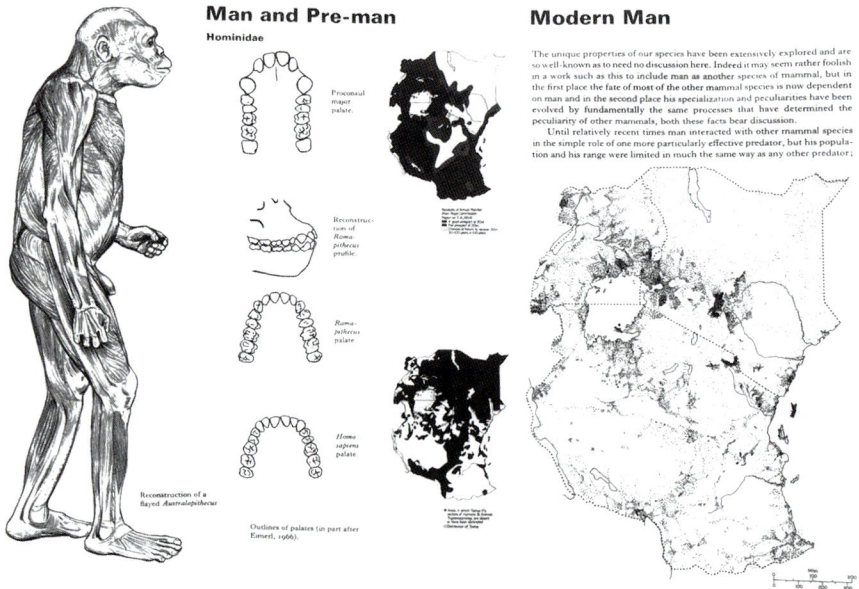

Man and Pre-man
Hominidae

Modern Man

The unique properties of our species have been extensively explored and are so well-known as to need no discussion here. Indeed it may seem rather foolish in a work such as this to include man as another species of mammal, but in the first place the fate of most of the other mammal species is now dependent on man and in the second place his specialization and peculiarities have been evolved by fundamentally the same processes that have determined the peculiarity of other mammals, both these facts bear discussion.

Until relatively recent times man interacted with other mammal species in the simple role of one more particularly effective predator, but his population and his range were limited in much the same way as any other predator;

assumption that evolution was the most endlessly interesting and open-to-scrutiny subject in all of human civilisation.

For schoolboy me their open-hearted welcome served to dissolve any awe I once might have had towards the hallowed halls of a great museum. Instead, the museum staff presented themselves as being like librarians in a public library; ordinary but kindly people, in charge of very unordinary resources. We pored over a single bird family's existence in evolutionary deep time – the only context in which the birds (even the entire Museum's collections) made any sense at all. As Mrs Hall ushered me back into the public galleries, she presented me with a handsome certificate of thanks, and these words: 'Remember, questions are much more important than answers.'

Those animated discussions in the Bird Room were precursors for much later explorations into the biogeography of human evolution. They were my introduction to the informal and intrinsically social pleasure of joining one or more like-minded people in watching animals behave, examining their build, their structure, their provenances, always searching for pattern and meaning in the speciation of plants, people and other animals. As in any society, there are myopic, hierarchically minded and jealous scientists, driven by small ambitions, but it has been my good fortune to have shared, throughout, the sense of an intellectual sister- and brotherhood that was first enjoyed in The Bird Room in Cromwell Road. This has been extended and deepened by my colleagues and friends in Makerere, Nairobi, Mweka, Wits, Cape Town, Paris, CSIRO, Kyoto, Los Angeles, Chicago, Duke University, and in Oxford's Zoology and Biological Anthropology departments as well as its Ruskin School of Drawing.

East African Mammals: An Atlas of Evolution in Africa was my first effort, begun in my 20s, to combine mapping evolution within my own home region

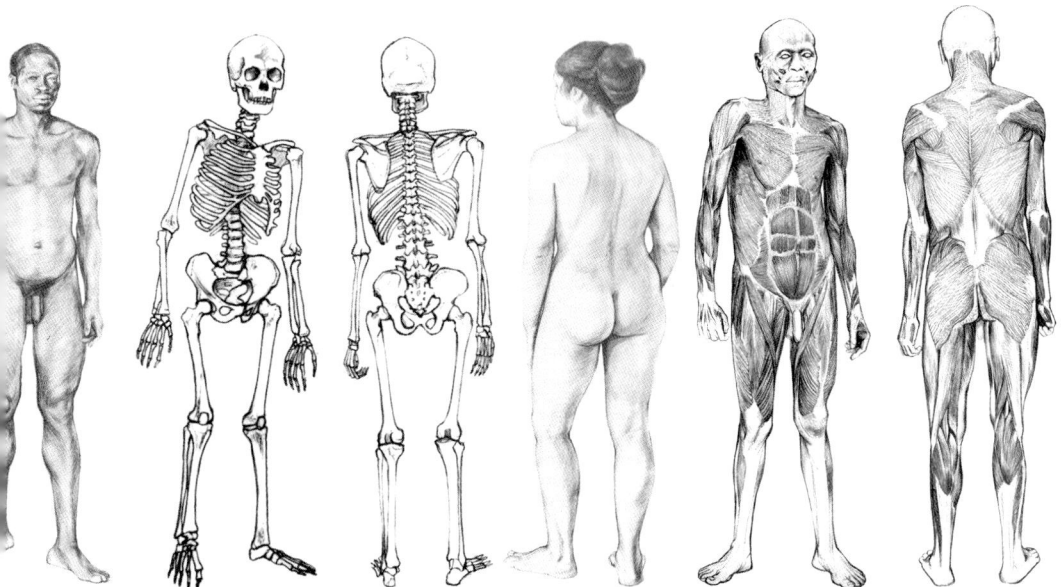

with a continental perspective. Opening my *Atlas* with a profile of the primate *Homo sapiens*, I saw no reason to suppose that human evolution had bypassed or overridden the forces that had shaped the behaviour, morphology and evolution of any other biota.

Why should human natural history depart from the gross biogeographic patterns that we seek in the biology of other animals and plants? My studies and these pages have been predicated upon such questions, as was *Island Africa* (a series of exploratory essays in continental island biogeography) and the more recent *Lowly Origin: where, when and why our ancestors first stood up*.

As an intrinsic part of my search for pattern in animals and plants (not uninfluenced by those school-day sessions in The Bird Room), I once drew attention to significant differences between the forest biota of southeastern forests and those of the centre–west. To my surprise, this biotic boundary came to be dubbed the 'Kingdon Line', and subsequent research has further emphasised the scale of disjunct distributions among many taxa of plants and animals. These have suggested a gross bifurcation of populations north and south of the equator. Furthermore, these divisions correspond with Africa's very shape – latitudinal north of the equator, longitudinal to the south and east. They are effectively separate realms that have been deeply influential in the evolution of animals and plants. In evolutionary terms they are two archipelagos within a single continental mass, each subdividing into lesser, ecological 'islands'. Populations living north of the equator are subordinate to boreal climates and invaders, and also subject to a fluctuating Sahara that imposes its own whims and disciplines upon neighbourhoods to its immediate south. During cold, arid periods, tongues of desert have probed deeply southwards, serving to subdivide a realm that is more than 5,000 km across, its borders and its several, elongated botanical bands fluctuating with every climatic pulse.

The composition of centre–west communities has been strongly influenced by the fact that virtually all immigration from Eurasia came in from the north or northeast. Their first encounters south of the Sahara were with a continent-wide east-west boreal belt. Initially, this realm was ecologically impoverished.

Perhaps the single greatest distinction between the two realms was founded 66 million years ago, when virtually all life must have been exterminated in northern Afro-Arabia (as then was) because this region lay wedged between Chicxulub, Nadir and the dreadful eruptions of Island India's Deccan Traps.

This has left a post-catastrophic legacy in which some species and groups of species, even quite complex communities, are absent to this day, whereas entire ecosystems, especially plants, survived in the far south. That longitudinal south–east realm will be the main focus of the next chapter. For now, my main concern is with the boreal centre–west realm, a broad sub-Saharan stretch of latitudinal strips of relatively flat, low habitats where all life is adapted to ever more rain or water the further away it gets from the Sahara. From open grassy plains, through savannahs and woodlands to forests where every substrate is a trunk, a branch or a root, where trees grow taller and ever denser as they approach the equator – this is all good habitat for primates (or *was*, before loggers got there).

One or more of these latitudinal strips would have been particularly welcoming for an incoming species of Eurasian dryopithecine woodland-adapted ape. This ape (and it is more likely it was a single species, not more) must concern us here because, in my view, it is to that 'returning' ape that contemporary African apes and humans owe their existence.

Within Africa, fossils of African Miocene dental apes (those descendants of Bishop's ape that never left their home continent) eventually declined in both numbers and species, just as the predominantly equatorial cercopithecid monkeys increased and diversified.

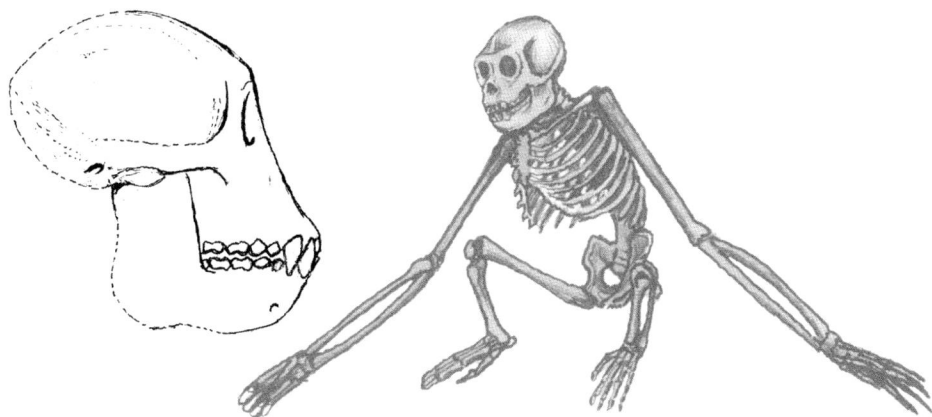

LEFT: Anoiapithecus *skull.* RIGHT: Oreopithecus bambolii *skeleton.*

Meanwhile, their ape cousins in monkey-free Eurasia became exceptionally numerous, an ecological and demographic success that is hard to imagine today but is testament to the adaptability of apes and to the breadth of niches that were open to them *in the absence of monkeys.* They were varied and adaptable, some had big canine teeth and prognathous muzzles, and they came in different sizes, sometimes differing in size by sex, sometimes not.

One, *Anoiapithecus brevirostris*, from Spain, had a very short, human-like face. Another, *Oreopithecus bambolii*, a long-armed swamp-ape from Italy, had greatly reduced the height of its pelvis and modified its feet into 'platforms'. In Africa, similar modifications in anatomy would reappear much later under very different circumstances.

All these variants on a specifically ape body plan help make later 'experiments in becoming human' more understandable. As was explored in an earlier chapter, ecological challenges generate behavioural responses that then lead on to modifications in morphology. Palaeontologists in pursuit of human evolution are charged with interpreting just such complex sequences.

The Eurasian ape that made, for us, that fateful crossing (probably around 10.5 million years ago) was among the more versatile and 'generalised', with a body-build and hands that were relatively unspecialised. The most likely type of dryopithecine apes to have found their way to Africa are called 'Greek apes', the *Graecopithecus* species. Another possible invader was *Pierolapithecus*.

.Very similar remains of an equally fragmentary and somewhat later fossil ape, *Samburupithecus*, in East Africa, provide as clear a proof as we are likely to get of the grand Africa-to-Eurasia, Eurasia-back-to-Africa safaris that lie hidden in the history and in the genes of humans and our cousin apes.

The picture has been complicated by persistence in the fossil record of a few conservative dental apes (such as *Nacholipithecus, Griphopithecus* and the somewhat short-faced *Kenyapithecus*). Such 'dental ape' descendants stood their ground in some regions for some time, but eventually left the field to their immigrant cousins.

Darwin's hunch that Africa was the most likely place of origin for humans may have been one of the earliest scientific insights into human origins, but he could have had no inkling as to how genetics and geneticists would reshape the way we think about ancestors and ancestry.

The distinctions that mark us off from other animals are generally assumed to be a large ratio of our makeup. Not so – by far the greatest part of a human is animal and much of what is left was specifically adapted for life in African environments.

What has been consistently underplayed is the significance of hands and fingers. I maintain that the whole, grand spectacle of human evolution hangs on the emancipation of hands.

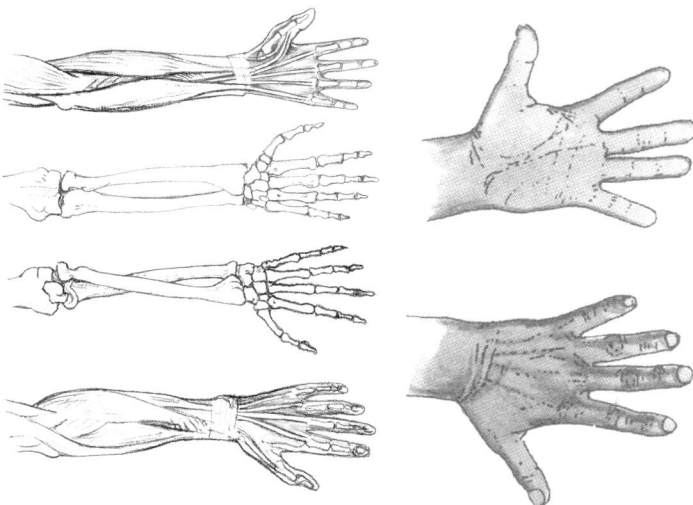

The Emancipation of Hands.

With typical clarity, Charles Darwin anticipated this insight, but in getting the sequence wrong he inadvertently misled subsequent generations of biologists. Early Darwinians imagined the hands' emancipation FOLLOWING uprightness. Instead, it was hand and finger-foraging that DROVE the adoption of uprightness. To understand how this happened, we need the prehistoric entities that a patchy fossil record can only hint at but which science and logic assure us really did exist. The ecological setting for that primary differentiation remains almost as imaginative as their names, but there are clues.

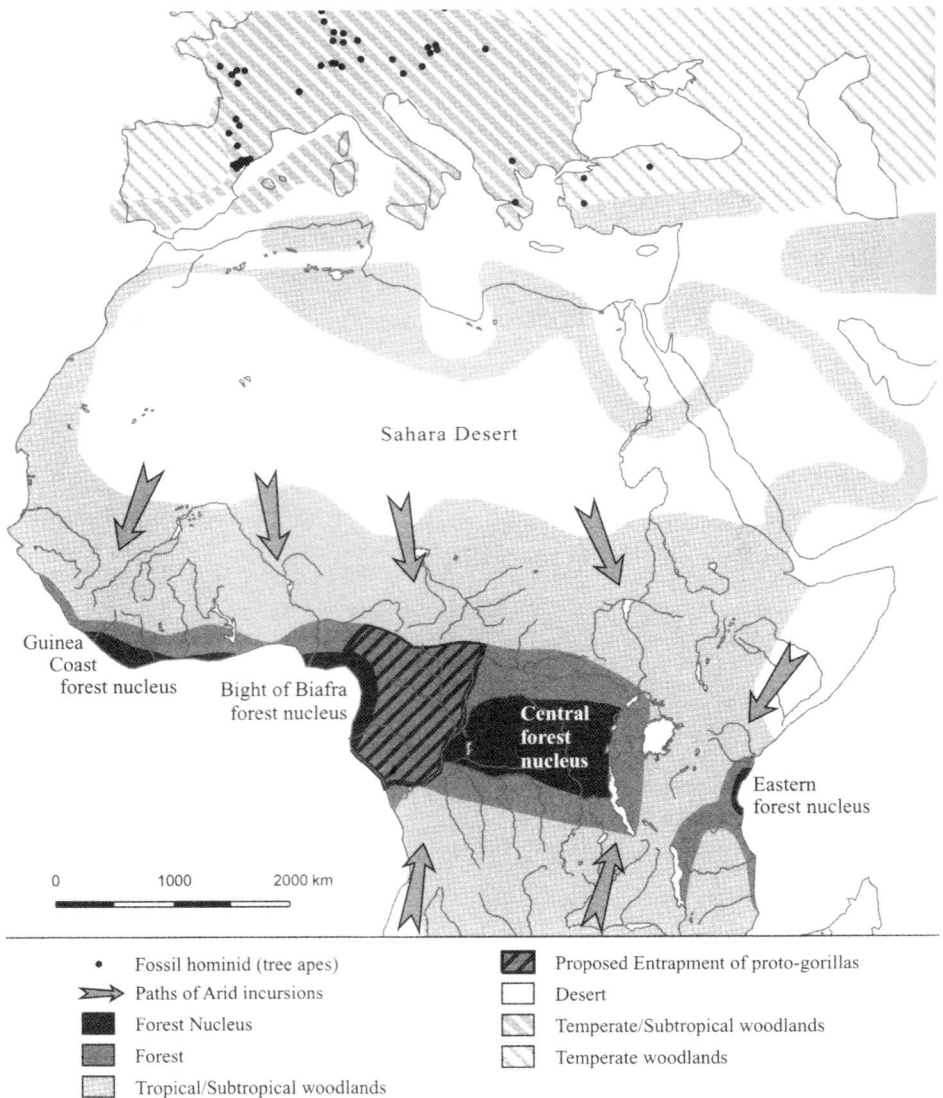

Miocene apes get to Africa (map borrowed from Lowly Origin, 2003*). (Note: the region between today's Congo and Ogooué exits probably lies south of the proto-gorilla range.)*

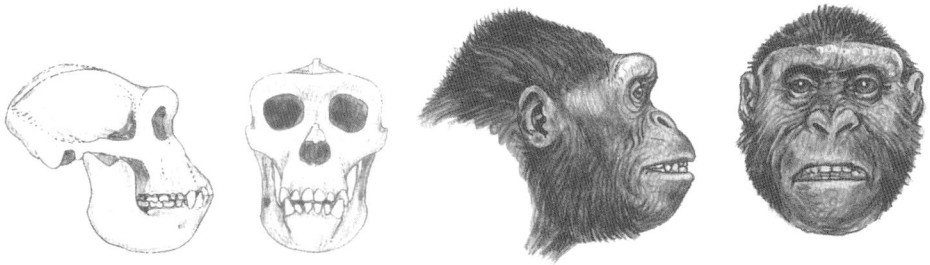

Stand-in for a dryoporilla? (left) Reconstructed Sahelanthropus *skull, profile and frontal views;*
(right) reconstructions of profile and frontal views of Sahelanthropus.

Because gorillas and chimps are contemporary species and have been separate from us and from each other for quite a few million years, I will call their/
our Miocene precursors 'dryoporillas' and 'humpanzees' and propose to explore
some of the conundrums that primary divergence poses. To imagine what a dryoporilla might have looked like, there is an enigmatic skull from Chad, named
Sahelanthropus, to hint at its appearance.

Returning to the immigration of a *Graecopithecus/Samburupithecus*-like ape
sometime after 10.5 million years ago, there may well have been a prolonged
period in which that single species adapted to its new home in quite a restricted
area or areas. Eventually one or more periods would have been sufficiently
benign for the incoming apes to spread very widely north of the equator. Once
populations had become widely distributed, speciation would have begun with
genetic drift, isolation and periodic changes in the boundaries of major vegetation zones, all subject to oscillating climates.

The contemporary distribution of gorillas suggests that a distinct dryoporilla
population came to dominate the moist and most favourable centre. At this
point remember that the huge Congo River, the greatest barrier to southward
movement, especially for apes, probably flowed out along the lower course of
today's Ogooué, substantially narrowing equatorial forest habitats north of the
river and barring dryoporillas from getting very far south (which probably remained 'dental ape' territory for quite some time).

Today, gorillas live in super-wet habitats – not so the earliest dryoporillas. Numerous anomalies within Africa's forest fauna and flora confirm that dry corridors have repeatedly introduced non-forest biota deep into the rainforest zone.
Again and again, forests have expanded from a string of persistent equatorial
'wet spots' or refuges. Every expansion effectively engulfs non-forest elements
living within such corridors. This has forced a wide variety of species either to
adapt or to die out. Conversely, drought periods would have forced forest biota,
on pain of extinction, to somehow adapt to life in degraded scrub or woodland.

Each time forests advanced after prolonged dry periods, the hinterlands of
Cameroon and Gabon seem to have acted as a persistent, engulfing 'trap' for

non-forest organisms. The abundance of gorillas in this region, and their persistence in the face of climatic changes in the past, is consistent with this being the locality in which they have lived longest and to which they have, over very long periods, become best adapted.

Gorillas remain the only living model we shall ever have of their dryoporilla predecessor. To walk beside a habituated gorilla troop as they forage, then to rest with them in the shade, can narrow a bit of the immense distances of time and evolution between us, if only in our own lonely, all-too-human imaginings. Sadly, the fact remains that we have, over the centuries, proved to be their deadliest competitor and enemy. Gorillas now inhabit a minuscule fraction of the ground where they once ranged, not in the far distant past but within historic times.

Assuming that geographic, ecological and temporal separation underlay the very earliest divergence between dryoporillas and humpanzees, we have yet to deduce a region of origin for the latter but we can surmise that dryoporilla ranges often became less extensive than the territories available to humpanzees. Extremes of every major fluctuation of climate would have favoured one or the other population, causing their overall ranges to expand or retreat. Both lineages could have repeatedly expanded, only to lose territory, over and over.

Then there must have come a point where separation became genetic and ecological, rather than purely geographic. Eventually this might have allowed the two incipient species to occupy adjacent bands of vegetation, most likely somewhere along the width of sub-Saharan equatorial Africa.

Dryoporillas were most likely to have preferred broad sections of wet, equatorial habitats (but with the wide Congo River as their own southern boundary). Humpanzees probably predominated in drier but well-wooded landscapes that, during benign extremes, might have stretched from Senegal to the Indian ocean.

It was as the least specialised/most generalised of all apes that humpanzees could occupy such a gigantic range, but it would have taken just one of a succession of Ice Ages to decisively isolate their most easterly population. It is at this point that we can begin to anticipate the eventual split between future chimpanzees and future hominins.

During the Cold War (about the time I got interested in the biogeography of human origins) rift valleys were proposed as notional 'Berlin Walls' between eastern and western biota – obvious enough for Lake Tanganyika (which was long thought to mark the eastern border of chimp distribution). As I had lived through droughts in the middle of the great diagonal rain-shadow east of the rift that runs from northeast to southwest Africa, as well as having studied forests on the Indian Ocean littoral, I came to think that periodic droughts in this region would have been quite enough to deter eastward expansion of most central African forest communities. Likewise, the same rain shadow would have inhibited westward flow for most eastern forest biota. That dry corridor could be, and, I maintained, *was* the very cause that separated ancestral chimps from ancestral humans.

My own first awareness of chimps hard up against their eastern limits was indirect, but no less memorable for that. My mother's albums, together with her sketches and photographs, have lain around in an old Zanzibar chest for as long as I can remember. The 1933 album recorded a walking safari, with porters carrying their tentage, that took her and my father west of Kasulu in today's Tanzania, towards Lake Tanganyika and the border with Burundi. One day, as their little caravan trudged over the green, grassy hills laced with forested valleys, they approached an enormous fig tree and Dorothy recorded their disturbing a troop of feeding 'baboons'. However, the accompanying sketch unambiguously depicted chimpanzees.

Much-degraded sketch of Chimpanzees west of Kasulu by Dorothy Kingdon, 1933.

As a precocious juvenile naturalist, I took her to task on the mismatch of caption and sketch. She remembered them resembling 'big, black fruit' falling out of the tree in their haste to get away to the safety of the nearby valley forest. New to the region and to its fauna, it was Teddy who had identified them as baboons, perhaps because neither of them could have anticipated chimpanzees east of the lake at that time. Our family friend, the local Game Ranger, George Rushby, was equally surprised when he encountered chimps south of Dorothy's encounter, and well within his custodial 'range'. George subsequently wrote a short article about these unexpected populations of chimps, which caught the attention of Louis Leakey, and led on, some years later, to Jane Goodall's celebrated studies on chimps living here, on the outermost eastern margins of their range, wedged between the lake and the settled pastoral countryside. Today, the remaining chimps have retreated to the lakeshore itself, to Gombe National Park, a last enclave where they are surrounded by settlement and cultivation.

Chimps and chimp ancestors have probably been fleeing from humans and their bipedal hominin ancestors for several million years, so their fleeing a party of humans is not merely a motif. Rather, it is symbolic of the whole nature of competition between primates – between them and us. It is a confrontation that has innumerable dimensions and consequences for human evolution.

Returning to what might have allowed the earliest hominin to bud off from its humpanzee ancestral population, it must have begun with some sort of sustained barrier between 'the-rest-to-the-west' and the far east. This barrier still exists as an arid corridor that was and still is the main interruption of what would otherwise be a pan-African belt of equatorial forests.

A tiny population of far-eastern humpanzees can be visualised occupying relatively stable habitat along Indian Ocean shores and on nearby mountain slopes. Here, refuge could be found even during the very driest climatic cycles. That earliest population probably flourished best in a small wedge of territory that lay between the Usambara mountains and the Pemba channel, a refuge blessed with perennial streams and, for omnivores, rich, year-round vegetable and animal resources.

All along this coast, proximity to the warm Indian Ocean has allowed coastal and littoral montane forests to escape the rigours of super-arid periods, while the eastern faces of mountains have provided year-round water and, in some cases, discrete, permanent and cooler extensions of forest habitat inland.

Forested rivers and streams that flow off neighbouring mountains and hills link these extensive and durable refuge areas with the humid Indian Ocean littoral. Together, the coastal forests and 'eastern arc' mountains represent a very elongated, almost snake-like ecological island wherein organisms have long been divorced from other forest areas of Africa. One measure of the eastern arc forests' isolation is that as many as 80 per cent of all invertebrates here are

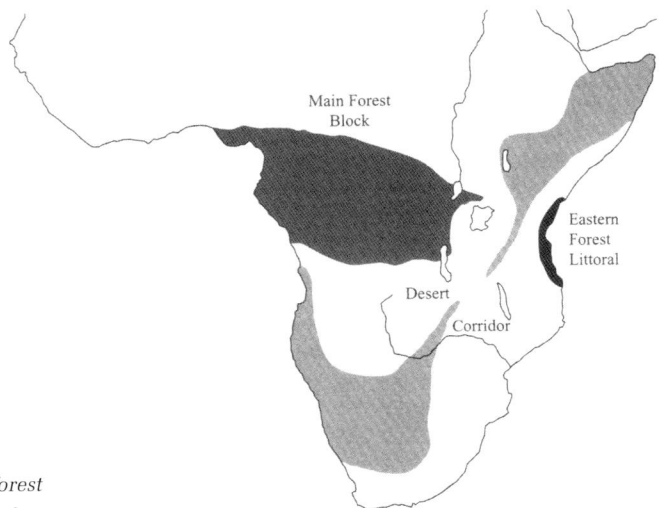

Arid corridor between main forest
block and eastern forest littoral.

endemic, as are 118 species of vertebrates. Of the larger trees, 274 species are endemic. Such huge numbers of uniquely endemic animals and plants confirm both the difficulty of crossing the Kingdon Line but also, crucially, significant contact with that great reservoir of pre-Chicxulub species and communities way down south. Biota living north or south of the equator must adapt to different climates, rainfall patterns and breeding seasons.

This finds expression in caves in Gabon. Here, leaf-nosed bats from north of the equator breed during one half of the year, while populations of the same species fly in for the other six months. Both populations maintain their regional identities, but if you extend such separation to entire populations – north versus south or east versus west – you have a potent mechanism for speciation.

Eastern forests also grow in diverse soils at diverse altitudes – clay marshes at every delta, interspersed between fast-draining shoreline beaches and coraline rags, all along a 4,000 km coastline. The seaward sides of nearby mountains have wetter, cooler slopes, sometimes climbing up to quite substantial altitudes. Eastern forests were not only more riverine, but their waters acted as magnets for thirsty animals and their seasonal waterflow and overspills concentrated or distributed nutrients. In such localities, environments are often consistently richer and more nutritionally diverse than the averages recorded further west. An enlarged dietary base is likely to have allowed groups of two to 12 individuals, dominated by a single male, to occupy defended home ranges.

Squatting Chimpanzees.

Today, no contemporary ape species inhabits these eastern forests, nor were they likely to have done so for millions of years (for living species, never). If this was the geographic and ecological nursery for hominins, we can turn to the habitats of their closest living relatives for guidance as to what might have been special about that homeland. Chimps living in western and central African forests have been the subject of multiple studies, and these provide appropriate models for examining environments from an ape perspective.

Along our East African littoral I envisage the emergence of apes more lightly built than either modern gorillas or chimps, but much less skilled climbers than orangs, true conservers of that ancient but ill-defined ape distinction of 'intelligent generalist'. Now examine today's vestigial East African coastal forests from that late Miocene ape's perspectives. How might they have responded to a habitat that, even today, differs from the high forests of central and western Africa (Table 1 overleaf). As eastern humpanzees suffered ever-longer isolation from their mainstream namesakes, what might this ever-more-isolated, ever-more-distinctive population be named?

I was inordinately pleased when, in 1994, a 4.4-million-year-old fossil from Ethiopia was invested with a hybrid Latin/Afar name that translated into 'rootstock ground ape' – *Ardipithecus ramidus*. Since then, older (5.8-million-year-old) remains have been called *A. kadabba* and fragments of a still-older, 6-million-year-old likely ground ape with very similar teeth have been named *Orrorin tugenensis*. It is possible that all three fossil taxa represent newly emergent populations of ground apes. If so, ground apes and their more immediate descendants began to diversify by region, living, surviving and evolving over at least a couple of million years.

Sketches of imagined squatting ground ape, with far right the supposed proportions while standing.

Table 1. Evolutionary implications, for ground apes, of isolation in East Africa's littoral forest.

East African Littoral Forests (Compared to Central Africa)	Evolutionary/Behavioural Responses and Implications for 'East African Ground Apes'
Isolation by Somali Arid Corridor	Genetic separation
Small overall extent	Small ape population
Long north/south extent	Extended 'linear' populations
Lower canopies	Less volume of arboreal habitat
Shorter trunks	Easier climbing
Lower rainfall averages	Drier feeding conditions
Drier microclimate	Food decomposition slower
More seasonal	Fruit diet less continuous
More seasonally adapted plants	More underground storage organs
More erratic seasonality	Fruiting less reliable
Fewer superabundant fruiters	Fruit quantity diminished
More deciduous plants	Seasonal differences on forest floor
Some leaves shed year-round	Leaf-litter continuous
More ground-level plants and animals	More activity on forest floor
More resources at or near ground level	More ground feeding
More small animals at ground level	More incentive to hunt/handle prey
Great diversity of terrestrial animals	More diverse hunt/handling skills
Diversity of prey behaviour	More flexible responses to obtaining food
Many small, scattered food items	More time spent foraging
Many food items small and diverse	Finer dexterity and coordination
Fewer seasonally superabundant foods	Foraging less clumped
Riverine habitats linear	Foraging more linear
Nutrients concentrated near river banks	Resources mainly in the vicinity of banks
Forest/woodland galleries sometimes narrow	Home ranges sometimes sharp-edged
Consistent resources in confined range	Home range well known
Undergrowth seasonally/locally dense	Vocal contacts episodic
More competition on floor	More competitive encounters
Predators more common	More alert; faster regrouping
Diverse large mammal encounters more frequent	More flexible group displays/threats
Unpredictable encounters more frequent	Frequent individual scanning and responses

The postures of primates have been debated since 1872 when the intact skeleton of a long-armed, gibbon-like fossil ape, *Oreopithecus bambolii*, from the mountains of Italy, was first described. Even then, it was remarked upon how the lower back and reduced hips resembled those of humans. In spite of the ape being clearly arboreal, here was the first convergent model for 'seated feeding' as an adaptive trait. Yet interest wilted as fashions in science moved on and away.

Ardipithecus ramidus: (left) skeleton; (middle) imagined portrait; (right) whole figure imagined.

One way of asking how ground apes might have responded to new environments is to examine the feeding strategies of living species. For example, when contemporary chimps are under duress from a poor fruit season, they break up into smaller foraging units that scour the environment more thoroughly, while trying to maintain their frugivorous preferences for as long as possible. By contrast, gorillas respond to the same pressure by maintaining their groupings, but diversifying and enlarging the range of their foods to include previously ignored or less digestible plants. Another variant, better suited to Eastern forests, would have been to include more animal and underground foods and spend more time and effort foraging for significantly smaller but still nutritionally rewarding items.

The micro-animal resources of eastern forests are very substantial and can be predicted to have attracted the attention of apes with omnivorous tastes. The leaf-litter fauna is abundant, consisting most notably of termites, ants, snails, land-crabs and many other small invertebrates. More than 100 species of reptiles and amphibians occur, include 36 lizards and at least 14 frog species, while shrews, rodents, sengis, mongooses and small antelopes are among the most common mammals. Forest-floor birds such as guineafowls, spurfowls, ibises, crakes, pigeons, pittas, robins and thrushes are particularly common, indirectly confirming the abundance of forest-floor cryptic fauna. Several typical forest foods, such as mushrooms and seed-sprouts, don't fossilise but are known to be important components in forest diets.

Field biologist Richard Wrangham has estimated that forest chimpanzees in Uganda spend 72 per cent of their total feeding time aloft, eating tree fruits, and only 17 per cent on the ground. Such ratios would have been impossible for ground apes in the east, because fruiting is more seasonal and less predictable. However, the forest floor was alive with potential edibles.

Living in small home ranges rich in forest-floor foods that require appropriate floor-foraging techniques must have induced substantial evolutionary change. I call the required change 'squat-foraging' as adopted by ground apes (my critics call it 'the bum-shuffling hypothesis') and I regard this as a necessary precursor condition before hominins could become erect. If eastern apes alternated between climbing relatively short shoot-budding or fruiting trees and feeding on the forest floor, a vital statistic would be the relative amounts of time spent between these two activities, and how much daily movement was necessary. The spatial distribution, the relative nutritional value and the size of food items would have significantly influenced the speed, versatility and energy demands of their daily movements. Foragers would have spent as much (probably a lot more) of their time searching leaf litter and excavating as they did foraging in fecund fruit-bearing trees, or moving from one feeding site to another. Quadrupedalism would never have been abandoned if substantial distances had to be covered, especially if such journeys involved exposure to predators. Easy refuge in trees and small home ranges were both essential prerequisites for habitual squat-foraging. That evolving ground apes needed a relatively secure, rich and well-known home environment rather than an extensive, insecure and demanding one is, in my view, a crucial precondition to becoming bipedal. An enormous fuss has been made of tool-use by chimpanzees, yet they best display cases of the spirit being willing while the body (without the right agents – fine-fingered hands) is weak. Whether they are foraging or socialising, modern apes constantly interrupt their progress with pauses to manipulate a food item, handle (clumsily) a temporary tool, touch or hug a companion or make a gesture. These operations require that they squat, lie down or stand on two legs. If so, a case can be made that squat-foraging was less about initiating entirely new behaviours, and more a matter of removing constraints on many pre-existent talents, traits and try-outs.

Remembering that 'behaviour drives morphology', the hands, head and shoulders of squatting ground apes would have had to swing easily from side to side as the forager investigated terrain within the arc of its arms' reach, and it was in this respect that apes would have had well-established assets. Releasing both forearms from weight-bearing and refining finger dexterity were the two evolutionary

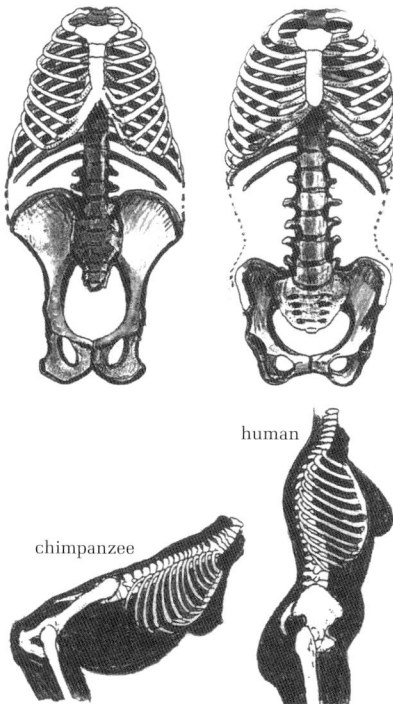

human

chimpanzee

Remoulding the single structure of an ape torso into the waisted human torso.

innovations that accompanied the pre-bipedal acquisition of an erect back. Initially, their hind legs and feet had to modify in less obvious but equally transformative ways.

Because the broad and elongated pelvis makes up almost half the length of an ape trunk, any pelvic retraction or shortening had to be compensated for through a combined lengthening and strengthening of the lumbar region of the vertebral column, which effectively means acquiring a strong, flexible 'waist' while retaining a broad chest and still broader shoulders. With each lumbar twist, the pulling back of one shoulder brought the head and neck back to the squatting body's more vertical centre of gravity, a less obvious effect of greater flexibility at the waist.

An oblique, cantilevered articulation of the skull would have been ill-suited to the sustained rotary panning of the head (or head and shoulders) that would accompany systematic litter-searching or even the monitoring of surroundings from a squatting position. By altering the head's balance on the neck, this was a foraging style that would have favoured shifting the spinal column's attachment to the skull away from the rear of the skull, towards a more balanced, forward position, below its central underside.

While all this swivelling and rebalancing was going on above, what was happening below the waist? The single most significant requirement was balance, and a stable base for those busy foraging arms and hands. Only easily adjustable legs and platform-like feet could provide a necessarily firm yet shuffle-mobile base.

Returning, for a moment, to quadrupedal humpanzees, their feet were likely to have been well adapted to fast four-footed travel between scattered food trees (as with the fig-guzzling chimps witnessed by Dorothy and Teddy). Contemporary chimps have only made quite minor modifications of this format (such as lengthening and arching of the digits), thereby accommodating to having heavy bodies but still climbing in tall trees. Remember, any such modifications long post-dated humpanzee anatomy which must have been a lot less specialised than in modern chimps *or* modern humans.

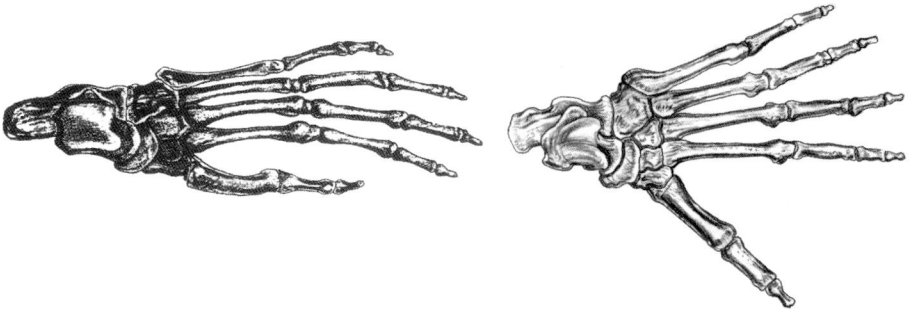

Foot bones: (left) Common Chimpanzee quadruped walk/climb skeleton;
(right) Ardipithecus *platform structure.*

This was context for the thrill I felt in 2009 on seeing casts of the foot bones of a 4.4-million-year-old *Ardipithecus* fossil, because the foot corresponded almost exactly with my much earlier, once hopelessly hopeful reconstructions. Predictably intermediate between chimp and human feet, they were modified in small but very significant ways. For a start they were FLAT, with rather rigid toes and no trace of an arch. The big-toe-cum-thumb and the outermost digit or 'pinkie toe' were larger and more reinforced than in any other ape, human or hominin, extant or extinct. These two toes formed the outer borders of a very broad bony arc – ideal for a firm pedestal.

In concert with three other sinew-strong digits on each foot and a hefty pair of buttocks, they spoke to a super-stable tripod for a new foraging technique. The describers of *Ardipithecus* feet, fixated on locomotion as the only role for a foot, came up with the rather lame function of serving 'careful climbing' – incidentally probably true, but not a very innovative beginning for the lineage that led to you and me.

A prime limitation on actively evolving bipeds would have been their dependence on gallery forests, rivers and streams. Part of this limitation might have lain in the physiology of a shade- and moisture-dependent animal, but the main restriction would have been the need for exceptionally rich and concentrated food sources. Improvements in walking would have been conditional on the ground ape's foraging technique; the manual sorting that so effectively scoured the floor could have been equally applicable to reaching up and out to sources of food that were above and beyond the radial cone of a squatting ape's reach. Initially, a prime incentive might have been the many bushes that grow in glades, on forest edges and in thickets, which produce great numbers of small berries such as jujubes (*Zizyphus*), wild raisins (*Grewia*) and capers (*Capparis*), and even the cherry-like pulp around wild coffee berries.

Senses and brains govern foraging in interesting ways. Take Marsh Mongooses, which go through shallows, hunkering down on well-spread hindlegs and tail, heads up, blindly scrabbling their thin, elegant fingers though the muddy water, every digit sensitive to the shape or movement of a great variety of aquatic or semi-aquatic prey. African Clawless Otters use shorter, thicker, stronger fingers to catch crabs or extricate aestivating fish, frogs or worms from shallow shelters in shallow waters. Unlike the mongoose, these otters are quick to see and pounce upon a fleeing crab or frog. Even so, carnivore eyes are consistently subordinate to carnivore noses, ears and whiskers as vehicles for finding their way about, for detecting food or fellows.

Of African primates, sharp-eyed Barbary Macaques and squirrel galagos are the most habitual of ground foragers, putting already versatile primate hands and fast responses at the service of primate visual supremacy. Even here, ground-grubbing could have verged upon becoming a specialist cul-de-sac.

What took this lowly foraging technique way beyond any such precedents? In terms of senses and diet I am proposing an enlarged menu, in which the humblest of foods becomes acceptable. This was linked to confined ranges within which adaptation to seasonal change could be accommodated without much season-dependent movement.

The emancipation of hands became the single most significant novelty and legacy of becoming a ground ape. The difference between ground apes and other foragers was a dissolution between hands as mere support and foraging agents and hands as true extensions of strategic minds, actual tools in their own right. It is crucial to identify manual emancipation as preceding upright stance because it was not 'getting up on two legs' but putting hands at the service of brains that was the primary driving force that ultimately led on to the evolution of modern humans.

This has found comic confirmation in the 'Penfield Homunculus'. This is a graphic that charts ratios of the brain devoted to the sense of touch. Hands, lips, tongue and genitals (in that order) demonstrate what grotesquely 'touchy-feely' people we have become.

The biogeographic expressions of these developments are relatively simple, but if my reconstruction is correct, the hominin radiation began with an equatorial stock of apes on an eastern coast, which became steadily more adapted to ground-living. Eventually, this foundational population began to spread along the coast, followed by upriver dispersals from a

Penfield homunculus.

4,000 km coastal base. Relatively narrow habitats along a range of this length might have taken many millennia to colonise, with break-up into sub-populations a recurrent likelihood.

This is a powerful conceptual model for understanding subsequent events in the evolution of hominins, especially 'bushiness' in their genetic family tree.

Perhaps the first, certainly one of the earliest, expansions would have been north of the equator, up into Ethiopia. The core geographic feature of northeastern Africa is the high, deeply bisected Ethiopian Dome. Two large rivers, the Webb Shebelle and the Juba, drain its eastern face and both exit together at the equator. Although these rivers flow off a giant dome, I call this the 'Ethiopian basin' to align it with other watershed regions. During peak Ice Ages, Ethiopia resembled a remote outpost of Eurasia (hills and mountains west of the Red Sea provided the long and tenuous link). Extensive glaciers and tundras formed at the highest altitudes, and narrow, foothill forests became degraded, impoverished in species diversity, cold and dry.

Colonisations upriver from the coast might have been inhibited at such times. By contrast, warmer, wetter periods supported more extensive forests, across a wider span of altitudes, and these would have held more attractions for primates. Indeed, the fact that fossil *Ardipithecus* species recur in Ethiopia over more than a million years suggests that, overall, biotic impoverishment was particularly favourable for ground apes, and especially first arrivals, with plenty of time to adapt with minimal interference from competitors or predators.

Later hominins and a few hardy monkeys might have shared such advantages as cooler habitats with fewer competing primates, probably fewer competitors, fewer predators and fewer diseases, in a sterilised post-glacial landscape. For a slow, frequently terrestrial ape, any mitigation of such constraints probably compensated for a fickle climate and a more limited choice of foods.

Two species of ground apes seem to have maintained a near-monopoly of this outpost for well over a million years. During glaciations, their range and numbers may well have shrunk drastically, but with the onset of more benign periods they had a head-start on other primates.

If ground apes were common within a very limited spectrum of species, it becomes less surprising to find them fossilised – the very abundance of which

Kingdon and Jengo at rock shelter site.

being some measure of their primacy in a community of mammals less diverse than in richer, more central parts of Africa. Other examples of phylogenetic hangovers in Ethiopia corroborate my suspicion that ground apes and other early hominins could survive here a lot longer and in greater numbers than they did anywhere else.

Finding evidence for 'platform feet' here, and then again in South Africa, provides strong support for my argument that ground apes came to inhabit the full length of east-coast forests. Identifying Africa's eastern littoral as the cradle of hominin evolution, and its even more localised wedge of equatorial coastline opposite Pemba Island as its choicest spot, invites anyone interested in human origins to ask about the potential for finding fossils.

I first visited the Amboni caves, near Tanga, as a boy and I remain hopeful that, one day, someone will find the skeleton of the first ground ape (or perhaps the last humpanzee) curled up in some corner of Amboni or, perhaps, at the bottom of a coraline sink-hole.

Some years ago I explored some caves and Rock Art sites in nearby Mkomazi with my old friend, colleague and former student, now Professor and Digo Elder, Elias Jengo. Together we earned the not very flattering title '*Wazee wawili wa Mpango*' (two old cave-men).

Elias was born in Tanga, and his local homeland was once inhabited by our shared ground ape ancestor, so we could reflect together on one of the tragedies of recent history. This tract of land, this nursery for the human race and all their hominin cousins, became an early victim of colonial industrial agriculture. This land, one of our most symbolic of birthplaces, has become endless fields of cane, not so long ago worked by slaves, and a landscape of vast sisal plantations, grown from seeds of henneken, stolen from Mexico by colonial scoundrels. Ever since, generations of Digo and Sambaa subsistence farmers have had their lives and livelihoods blighted by estate plantation peonage.

'They're all elephants!' exclaimed Elias. 'Elephants with heavy feet, their offices in London, in Paris, Beijing, Amsterdam, Moscow, Delhi, New York, Frankfurt – you name it – the only ones from Africa have tusks. If we don't control them, they stamp all over us, even in our remotest villages.'

Elias Jengo.

17 THE AUSTRAL, LONGITUDINAL REALM

In which the birthplace of Homo *and her descendants is located.*

Eyes scampering over views below our portholes, butts nestled in the green up-holstered seats of a sea-plane, my family and I were flying from Dar es Salaam to Durban. This journey of nearly 3,000 km was mostly flown along a coastline fringed by coconut and cashew plantations, fishing villages, mangroves and remnant patches of coastal forest.

In a matter of hours we travelled down a coast that had taken hundreds, more likely thousands, of generations for our ancestral ground apes to inhabit from end to end.

Of course, I had no inkling of such an immemorial drama at the time. It has emerged from my subsequent pursuit of monkeys, squirrels, sunbirds and innumer-able other biota down below those seashore clouds, within those vestigial forests.

While many a Darwinian has longed to find fossil 'missing links', too many have neglected the fact that our ancestors lived under the same constraints, sub-ject to the same ecological and geographic realities as any other mammal. Every species has a geographic locality of origin. For some it may be some lonely atoll, but over the 7 million or so years that the human/hominid lineage has evolved, multiple forms have lived in multiple localities, and belonged to diverse ecolog-ical communities-of-origin. Most of those communities remain open to study, if only in vestigial form, but they really can and do help reveal the dynamics and patterns under which we evolved as a single, triumphant species.

It is not my intention to join in the contentious fray that surrounds a far-from-adequate array of hominin fossils. Rather, I want to retrace possible peregrina-tions by our ancients down our coasts and along more inland highways. This is the main purpose for this chapter.

In the previous chapter I examined latitudes north of the equator with a Dar-winian biogeographer's eye, recalling the origins of African apes, the transfor-mation of eastern humpanzees into ground apes, and some special features of eastern Africa's sole region north of the equator – Ethiopia.

Now, remembering how the shores of the Indian Ocean glided by beneath my porthole, I turn that same eye down Africa's longitudes south of the equator. What follows is my reconstruction of how and why ground apes would have been prone to generating divergent descendant populations in different inland basins.

OPPOSITE: Lolui Memoir, *1966.*

It is also a story of red herrings, of how and why palaeontologists have sometimes been misled by a succession of side-lines. Every one of an ever-more-crowded collection of hominin fossils offers glimpses of what I like to call accidental, incidental 'experiments': experiments in natural selection that preceded our own emergence as the only, lonely and deeply intolerant, survivor.

I do not mean this in any sense to be a parade of triumphal predestination. Rather, early hominins responded to the vast, regional expanses and the spectra of habitats in our continent by adapting and diversifying in opportunistic fashion, as if they were yet another radiation of antelopes, mongooses or weaver-birds.

Return now to that humpanzee-descended population that came to inhabit the equatorial east coast. If the droughts that isolated them in the first place were sustained over long periods (possibly serving to fracture east-coast forests too), then the settlement and evolution of a founding population of ground apes could have been quite localised. I have already identified Tanga's hinterland as the most likely region for any such population to endure any such vicissi-tudes. Once the structures associated with erect posture (pelvis, backbone and skull base) had found some sort of functional stability and the animals could combine their ground-combing niche with incipient or actual bipedalism, that minuscule eastern ground ape population would have expanded north, south, even westward up the Pangani and other large valleys, out of their constricted equatorial haven. Their peripatetic descendants then responded to the physical, ecological and climatic peculiarities of whatever latitude their expansions took them to. All this occurred over several million years.

Thanks to an abundance of fossils and their dated excavation sites, we know that early hominins had reached central Kenya by 5.7–6 million years ago, and to the Ethiopian massif by about 5.8 million years ago. Later, the descendants of

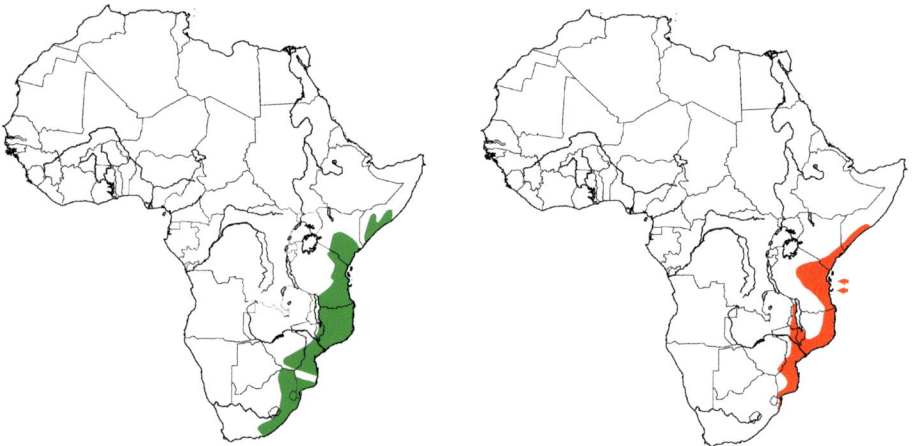

LEFT: *Distribution of gentle monkey* Cercopithecus albogularis *cluster.*
RIGHT: *Distribution of Red-bellied Bush Squirrel.*

some such hominins adapted to the high veldt in South Africa. What has had inadequate attention are the east-coast littoral forests, which provide the only consistent connecting link. That it is a connecting link is well demonstrated by numerous species of mammals, birds, plants and other biota that range from the Horn of Africa to its Cape of Good Hope. My favourite mammalian examples are White-throated Guenons and Red-bellied Bush Squirrels. Many organisms are exclusive to the littoral itself (some forest birds maintain a single gene-string along the entire coastline). Others subdivide into three or four sub-populations while others range quite far inland. Forest-adapted species sometimes inhabit the many inland montane outcroppings between Ethiopia and the Drakensberg. Not one fossil hominin has been found from anywhere along that high-road, but then not one palaeontologist, so far, has really tried to find them!

Remember that luck, chance, serendipity or whatever you like to call it has governed most of the great discoveries of human evolution in Africa. Louis called it 'Leakey's Luck'. Olduvai's fossil beds were discovered by a clumsy butterfly collector, Hans Reck, who, in his single-minded pursuit of a *Papilio*, tripped up and fell into the gorge (or so the story goes).

Turn, then, to biogeography for what evolutionary patterns might have had to apply.

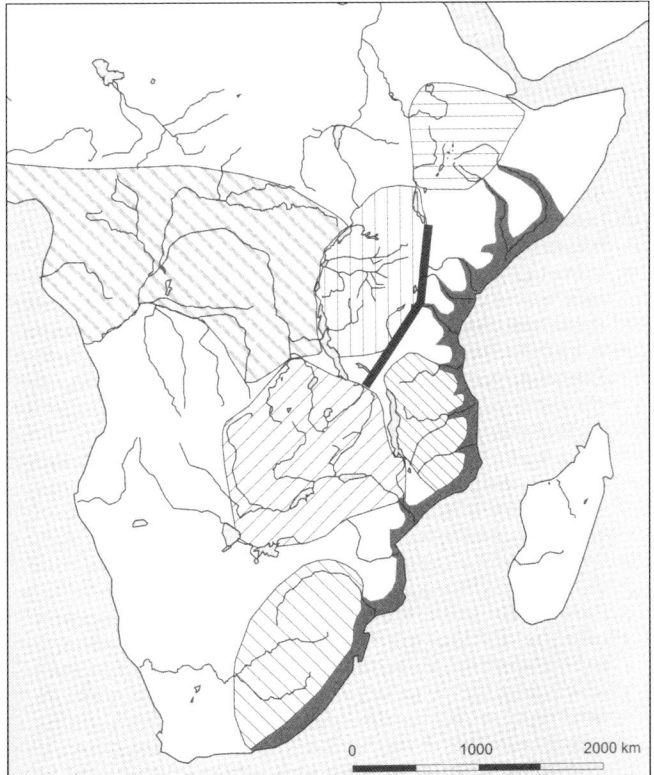

Map to illustrate basin evolution. Indian Ocean littoral forest mosaic and five basin regions described in text.

For any animal or plant that lived along that 4,000 km long coastal strip, its narrowness implies both diminished population sizes and much greater susceptibility to fragmentation. Thus, smaller gene pools could have been as influential as the narrowness and isolation of the habitat in encouraging sub-populations to bud off along such a long strip. If those narrow coastal forests formed the primary tap-root from which subsequent hominin populations could derive, each river basin provided the adventitious roots along which larger populations could follow and eventually expand up every rootlet stream.

Between the Web Shebelle in Somalia and the Limpopo River in South Africa there are some 12 major rivers that run down to the east coast from deep in the interior. All of these would have offered arteries of expansion to coast-based populations of animals. Differing greatly in length and depth, especially over time, they cluster into four or five distinct biogeographic divisions.

Hypothetical extensions in the range of ground apes would have involved separate, and very different, domains. Adaptation to such coastal stretches and/or their inland basins were probably staged – earlier near the equator, later towards the south. This poses unsettling possibilities – over extended periods did the hominin family tree branch, and did each branch develop slightly different mosaics of characteristics?

The fossil record that exists *can* be reconciled with this early basin-by-basin branching, but it raises the possibility that many hominins, once confidently stitched into the human genealogical chart by their discoverers, may belong to just such offshoots. Every one of them could be called an evolutionary experiment. Only one of them can be directly ancestral. If we have been misled, then that, alone, justifies exploration of how and why we might have been misled – so, why?

Fossils began as 'curios' or the paraphernalia of magicians, quacks or Oriental apothecaries, until the systematic study of evolution invested them with genealogical significance. The practice of comparative anatomy is several centuries old, but Darwin transformed the curious into a discipline relevant to his own and to all our origins.

It tends to be forgotten how recent the practice of palaeontology is, and how the study of human evolution began with the traditional and anatomical authority of doctors, the first to be challenged by the realities of our animal architecture and our resemblance to apes. 'Curious resemblances' were first explored in the surgeries and laboratories of medics.

All physical anthropologists are well versed in functional anatomy, and most can interpret every detail of human anatomy in terms of function, but even they can sometimes find it difficult to envisage now-extinct ways of life and living, let alone reconstruct past habitats from today's ecological ruinations.

Physical anthropologists have become hugely dependent upon those rare concatenations of geology that deliver precious outcroppings of fossil bones. Almost from their very beginnings, these treasure troves of palaeontology have

generated a near global audience of what have come to be called 'origins groupies' – I am one of them.

Like cineastes that look to Hollywood for material, content and style in the di-
rectors, producers and actors, we read scientific papers such as *Science*, *Nature*
and the *National Geographic*, and we watch the players, contemporary human
and ancient fossil, fitting or being fitted into whatever family tree or theory already (currently or temporarily) exists.

The study of human origins has long been constrained by all-too-tiny troves
of fossils in the hands of all-too-few scientists. The cameras dwell upon the
skulls of fossil heroes resting on the palms of 'star fossickers' while we, the
audience, hold our breath and, every now and then, wriggle in our seats as
some iconoclast smashes up some long and hallowed theory, but the theatre
has seldom strayed far from the fossils and their local provenances. From very
sparse beginnings, the numbers of fossils and their beds have proliferated
splendidly, but palaeontology and the study of human origins has remained
pathetically underfunded and neglected in our currently myopic, barbaric, so-
called civilisation.

As late as 1971, an influential anthropologist could maintain that humans
evolved over tens of millions of years along a straight line of descent from a
14-million-year-old Indian *Ramapithecus* to a contemporary Michigan professor.
Such super-simplification was the direct outcome of all-too-few fossils (and too
few sceptics) but the cultural quirks of palaeo-divas have played a part too. There
is also an unspoken commitment to house-keeping that wants to tidy up all-too-
messy scenes. I am as guilty as anyone and know whence that impulse comes.

Fossils of any organism cannot tell us much about their region of origin, which
can be quite distant from where a fossil is unearthed. Likewise, today's habitats
can only be a rough guide to those of the past, especially for ecological communities that were partly 'proboscifacts' – the products of sustained browsing of whole
landscapes by millions of elephants, over millennia and more, not mere centuries.

We have already seen how many immigrant species have found their way into
Africa from other continents, contradicting the popular supposition that Cheetahs and zebras, for example, are classically 'African' animals. They are not. From
a deeply prehistoric perspective they are recent American immigrants, now very
far from that distant continent wherein their spots and stripes were spawned.

All animals have their moment in the sun. Some are hugely prolific, most are
mobile (even limpets move), so there are plenty of similar displacements across
and *within* our continent.

Fossils also tend to sample species long (sometimes very long) after their
actual emergence. It is the particularly successful, widespread, numerous and
enduring species that are the most likely to appear in the fossil record. Some
hominin species seem to have persisted a lot longer than others, and such fossil
elders are a major source of confusion. In the minds of some of us, they beg to
be tidied up.

It is theoretically possible that as many as four bipedal or incipiently bipedal lineages could have emerged in four well-separated regions, immediately inland of our east coast. Ultimately, each might have derived from a shared ground ape parentage, each with its origins in differently timed expansions inland from widely disparate coastal enclaves.

Maintaining long-term genetic continuity over a 4,000 km, narrow coastal distribution would have been near impossible (especially during climatic shifts that introduced eco-gaps between lengths of coastal forests). The existence of ground apes as a single gene pool along the length of Africa's eastern coast could only have been of relatively short duration (nonetheless, there are contemporary species of plants and animals that *do* survive over the entire length of this ostensibly continuous and very narrow range).

All Eastern ground ape populations probably became capable of fully balanced and habitual standing and spasmodic walking, but derivative bipedal, manual and climbing skills would have varied in the rate at which they developed from one descendant group to another. In the previous chapter I introduced my own interpretation of ground ape presence and persistence in Ethiopia. The details of these apes' biology remain to be studied and their bipedal skills remain to be elucidated, but the discovery of *Ardipithecus*

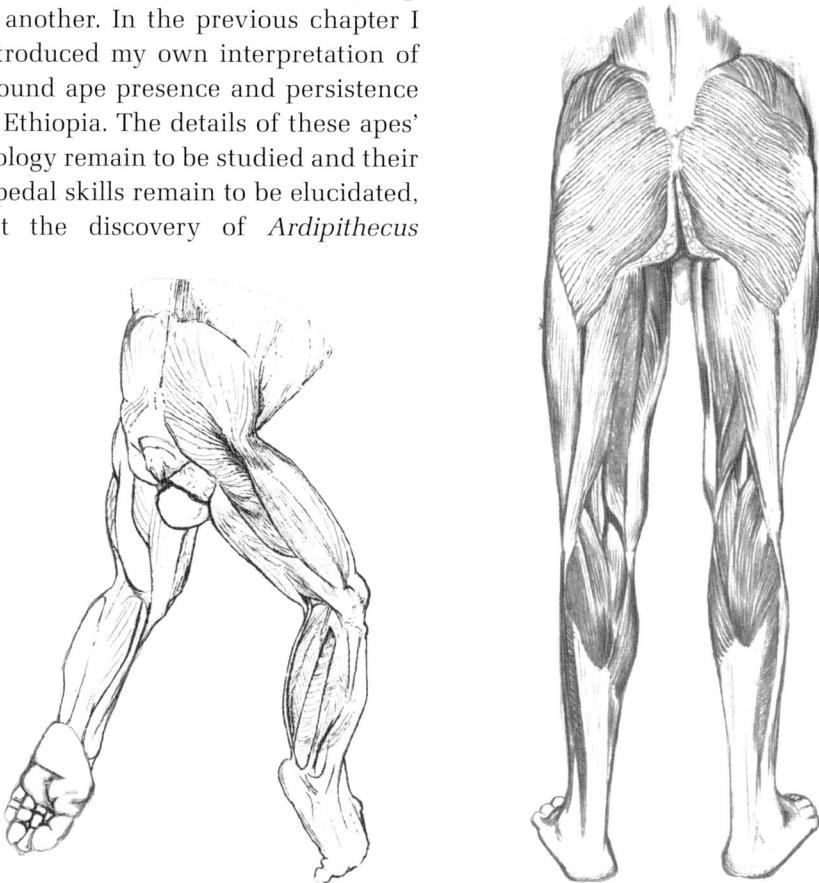

Back views of the lower body of flayed Common Chimpanzee and human, compared.

fossils has gifted us crucial insights into that no man's land in between being straight-backed, grounded and sedentary versus becoming bipedal, mobile and the pioneer of a new ecological role, reached via ten sensitive fingers in league with an observant, teachable brain.

Bipedalism required subtle but significant anatomical adjustments to convert habitually bent legs into habitually straight, weight-bearing ones. The most obvious of these changes would have concerned the creation of stable balance at the hip–upper leg junction, through enlargement and reorientation of the buttock or *gluteus* muscles. Formerly curved articular surfaces at the knee would have needed flattening and broadening to stabilise their newly vertical weight-bearing role. Feet and ankles would have had to be resculpted to absorb the new pressures of an upright body's weight bearing down on the many closely bound bones of the foot. Fossils have begun to emerge that illustrate the exact progress of reshaping bent ape joints into vertically aligned hominin ones.

During the 1960s and 1970s it was my good fortune to meet several pioneer Japanese evolutionary biologists in East Africa, and to be invited to join them in Kyoto University and at their primate research station at Inuyama. Led by Jun'ichiro Itani and Shiro Kondo, Hidemi Ishida was one of the first scientists to combine field studies with rigorous laboratory-based techniques to explore the evolutionary origins and mechanics of bipedalism in hominin ancestors. Hidemi, outgoing, quick-witted and hugely energetic, concentrated his studies on how the kinetics of walking changed over the course of evolution, and how substantial savings in energy accrued from a fully upright body posture. He was

Shiro Kondo, friend and Director of Kyoto University's Inuyama Primate Research Institute.

the first scientist to insist that standing upright must have preceded actual walking, a highly original conclusion because something must have straightened that back before it could stand. My own primary interest throughout has been to follow up on this most important insight, but brains, hands and all their creative potentials have claimed my loyalties before legs and football.

I have looked to the ecological settings that induced such an unprecedented departure from primate norms and behaviour, but the whole-hearted enthusiasm and sense of fellowship that I found among fellow primatologists in Japan, led today by Juichi Yamagiwa, has remained a source of inspiration. Thanks to the initiative of these and many other pioneers, Japanese science has continued to lead collaborative studies in evolutionary primatology worldwide.

There are important consequences for the recognition that erect balancing and, even more so, styles and speeds of walking, are distinct evolutionary increments that *derive from* erect squat-feeding. There is evidence that different populations of ground apes differed in speed, style and sustainability during their development of walking. Likewise, assuming that both hands became relatively independent foraging organs, small regional differences in dexterity, reach, handedness and, perhaps, the development of specific tool-assisted foraging techniques, could have become factors in the differentiation of regional populations. Even variations in adoption or in styles of habitual walking could have been slowed or speeded up by just such factors.

Expansions inland from the east coast took place at different times, and subsequent adaptations developed at different rates. We know, for example, that something close to a ground ape (*Orrorin tugela*) was already present in central Kenya around 5.7–6 million years ago (about a million years after the date for hominin emergence as calculated by geneticists, and close on 5 million years after dryopithecine arrival in Africa).

Fossil hominins from South Africa are a lot younger, so it is interesting to examine four inland basins off the east coast with hints of sequential emergences in mind. I began with boreal Ethiopia in the previous chapter, and now equatorial East Africa, Zambezia and South Africa invite scrutiny.

Equatorial East Africa immediately subdivides into a 'shallower', more southerly section, which I call the 'Mozambique Mini-basin'. Here, a series of tightly packed, relatively short tropical rivers, notably the Rufigi, Rovuma, and some Mozambique rivers, drain mountains north and east of Lake Malawi, all coastward of the Kingdon Line. Physically adjacent to the coast, these short, interdigitating basins would not have significantly inhibited movement between delta and delta, nor between coastal and montane forests. Conservative 'ground-apism' might have been more easily conserved in gene pools that maintained contact throughout very sizeable areas. It will be exciting to find hominin fossils from this region (however long it takes), but what tales of their own will they tell?

The 'deeper', more equatorial section is more problematic, as a corridor of drought (effectively the Kingdon Line) is interposed between the coast and the Eastern Rift Dome. Drained mostly by the Nile, I call this the 'Great Lakes Basin'. The volumes of waters draining eastwards off these relatively dry uplands (via the Tana, Galana and Pangani rivers and fluctuating by climatic epoch) might have been less significant conduits than larger, longer, more consistent waterways.

The previous chapter touched upon the role of Ethiopia as the only basin (actually a dome or massif) north of the equator. For hominins, the shores of Lake Turkana offered periodic connections between Ethiopia and the Great Lakes Basins – always subject to vagaries of climate. This corridor region has yielded a large and complex community of hominin fossils, to which I will return shortly.

The largest and most beckoning of all potential expansion areas for coastal ground apes is what I call the 'Zambezia Basin'. It consists of an extensive inland plateau, not without hills and mountain ranges but a region particularly well suited to an evolving ground ape. Yet not one early hominin fossil is known from this plateau. Much of it is no higher than 1,000 m, spanning the tropic of Capricorn between 8° and 20° south, a position that is influenced by winds and weather blowing in from both Indian and Atlantic Oceans. Here, riverine and shallow floodplain habitats have endured throughout the vicissitudes of the Plio-Pleistocene. Today, temperatures range between about 10 and 30°C, and rains fall between November and April. Waters are conserved by slow, impeded drainage which favourably moderates local microclimates. Thus, groundwater sustains forests and woodlands along the margins of rivers, swamps and lakes. Much of the Zambezia Basin drains southeastern Angola (plus neighbouring stretches of the Congo) and this area is laced by an extensive web of relatively narrow forested watercourses winding through grasslands. Some have come

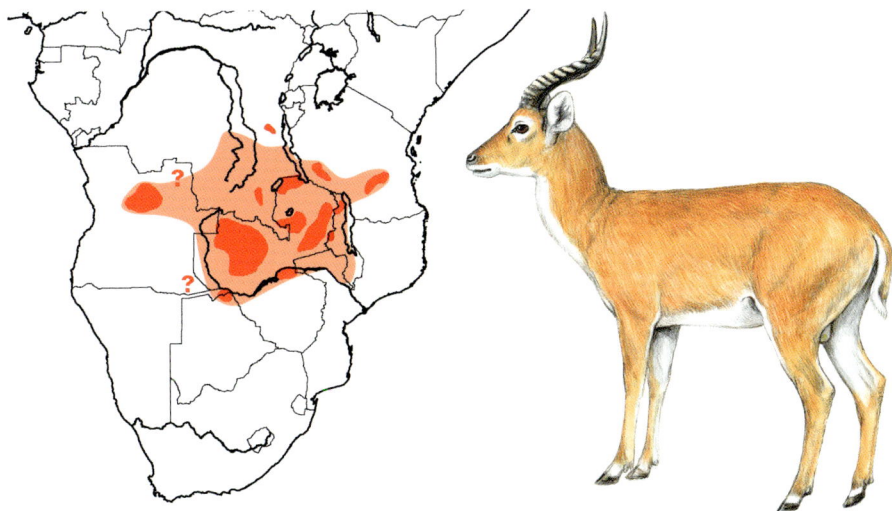

*Distribution of the Puku antelope (*Kobus vardonii*).*

and gone, all dependent on harvests of rain and water from a much more extensive area. That this is a discrete ecological and biogeographic unit is indicated by the distribution and biology of the Puku, a once widespread species of *Kobus*.

Up-river colonisation of this region by east-coast ground apes would have been made easy along the banks of the Zambezi, but once a sizeable population had become established deep within this interior, ecological differences must have reinforced and speeded up genetic differentiation between a minuscule parental source, still essentially 'coastal', and its expansive inland progeny.

A significant characteristic of floodplains is that resources tend to be most concentrated close to the edges of flood-waters. These pulse in and out on either side of their riverine or lacustrine spine on an annual cycle. Animals dependent on all the fast-changing activity and opportunities along such moving 'shore-lines' have to make lateral movements in and out on either side of their own stretches of water-course or lake-side. Certain to have been habitable for hominins, this extensive area of Africa is interposed between the two main sources of fossils – East and South Africa.

It is inconceivable that the Zambezi region was an empty corridor between fossil sites, so I am, once again, raising the possibility that some of the best-known fossil species, all encountered elsewhere, are just as likely to have *originated* in Zambezia as in any currently known fossil provenance.

At times, the Ugalla, Malagarasi, the Victoria basin and South Sudan floodplains could have formed a nearly contiguous strip ecologically similar to Zambezia. However, any one of this string of smaller regions is less likely to have spawned its own hominin population.

Nonetheless, early and fully mobile bipeds may well have intruded, even come to flourish there, albeit precariously. What remains incontestable is that Zambezia must have been a major theatre of hominin activity, of active proliferation and passage, over millions of years.

For what follows, my mentor is Aesop, with his fable of the Tortoise and the Hare. Our human ancestors were handy, front-end tortoises, while *Praeanthropus anamensis*, the 'Kanapoi hominin' (soon plagiarised as 'Kana Boy') and its immediate descendant, the much vaunted 'Lucy', or *Praeanthropus afarensis*, were leggy hind-end hominin hares. (Some taxonomists have sought, in my view mistakenly, to lump the two together as a single species.) I take Aesop further and locate the hares' likeliest homeland in the fossil-less Zambezia Basin, even though they turn up later, as fossils, in Tanzania, Kenya and as far afield as Ethiopia and the Sahel.

'Kana Boy' and 'Lucy' are my hares. Surprisingly numerous and scattered as fossils, they effectively pioneered the colonisation of areas outside their region of origin, and the sheer spread of their range testifies to their success as new, pioneering forms of African biped. To all appearances a walking ground ape, *anamensis* was first described and positioned (from mere fragments) by the marvellously perspicacious eyes of Meave Leakey. Both Kana Boy and Lucy seem to have had an affinity for floodplain, *dambo* or *mbuga* habitats, and that association points to their originating in a region where such habitats were dominant over an extensive area, and over great stretches of time.

Kana Boy has earned its status as distinct from *Ardipithecus* ground apes by virtue of its legs and thick enamelled teeth, not its brain or hands. Were we to restrict ourselves to cranium and forelimbs, Kana Boy was a rather primitive ground ape. Instead, I envisage this decisively mobile hominin as the first known colonist of regions as distant as Kenya and Ethiopia, even as early as 4.5 million years ago.

In spite of contradiction in the name (in my view *not* directly pre-human), I use the name *Praeanthropus* for both these early walkers, because the first of their kind was found close to Lake Eyassi in the late 1930s by a German expedition. Accompanied by our friend Hans Overdyke, its leader visited my father's office in Tarime at that time, his team having based itself in a Nazi-dominated settlement in nearby Oldeani/Karatu. Not described and named until 1950, well after World War II, this type specimen escaped the notice of francophone and anglophone scientists, mainly through the obscurity and inaccessibility of its belatedly published description in German.

Given how names depend on whatever specimens already exist in known scientific literature and museums, it was understandable enough that when the near-complete skeleton of a female *Praeanthropus* was found in Ethiopia the priority of that moniker was not known to her discoverers. Her scientific birth name became '*Australopithecus*' *afarensis* and she became the world-famous 'Lucy' almost overnight.

The fossil hominids of eastern and southern Africa have been examined from many perspectives and their variety and complexity continue to pose innumerable puzzles that have exercised many minds. Throughout, there has been unspoken pressure to try to shoehorn each new discovery into a storyline that leads on to you, me and our shared ancestors. Simultaneously, whatever dig it came from is endowed with special genealogical significance.

I have already confessed to being one more origins housekeeper, prone to a lot of guesswork and a bit of tidying up. Ethiopia, rich in a fine mix of fossils, once suffered some extreme neatnik shoehorning whereby the country's fossil sites were presented as host to a continuum of human evolution. That Ethiopia has offered habitats particularly attractive to whatever hominins were walking or grubbing around at any given time is a sound enough conclusion. To envisage Ethiopia's rift valley as a stage on which each ancestral species melded into another effectively shrinks human evolution to the dimensions of puppet theatre. Was this possible, in one very small locality in one small part of a very large continent?

The denouement of this theatre came with the thrilling discovery of an almost intact ape-faced *P. anamensis* skull (dated at 3.8 million years, with the catalogue code of MRD). Incidentally, this MRD Kana Boy was shown to have overlapped with Lucy's kind for more than 100,000 years. This torpedoed the theory

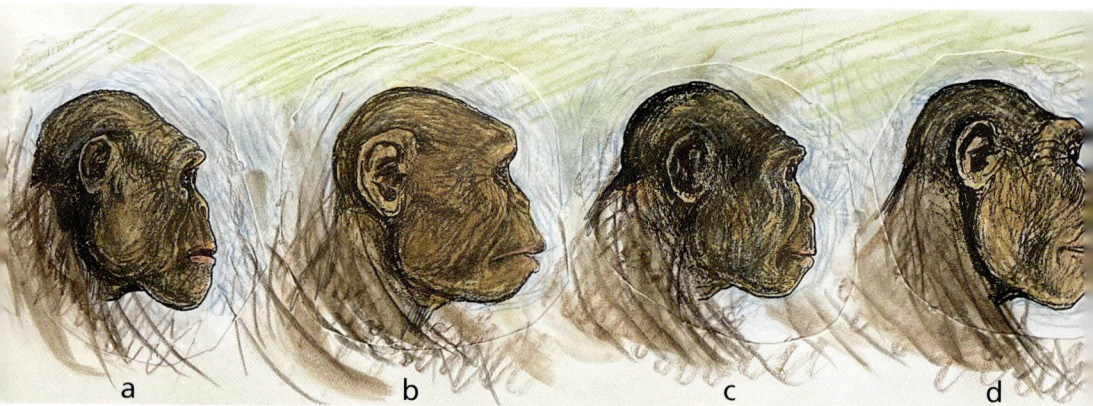

ABOVE: *Four profiles:*
a) Praeanthropus
afarensis;
b) Paranthropus
aethiops;
c) Paranthropus
robustus;
d) Paranthropus boisei.

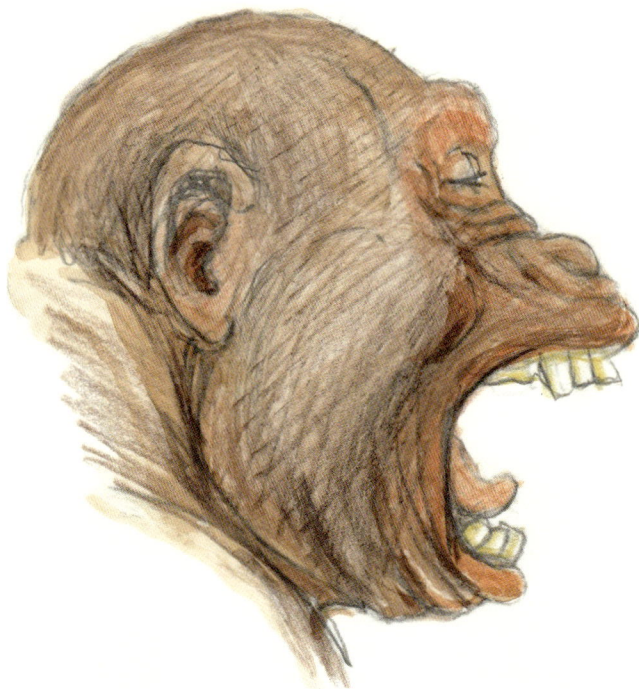

Kana Boy,
'Praeanthropus (?)
anamensis'

of continuous evolution in Ethiopia's rift valley, but raised new questions about where, how and when Lucy could have evolved from some distant Kana Boy stock.

Supposing that Kana Boy had already trailblazed vast areas more than half a million years before Lucy arrived in Ethiopia, I turn to other biogeographic precedents. Many mammals have speciated within shorter periods of time and over lesser spreads in range. Waterbucks provide a well-sited and contemporaneous model for the Kana Boy/Lucy divergence. About 4 million years ago, probably within Zambezia, a population of large, water-dependent *Kobus* antelopes split. Around the shores of Lakes Mweru, Upemba and southern Lake Tanganyika, marginal flats are relatively narrow with hills and mountains nearby. This could have been one of several localities where ancestral Waterbucks might have become heftier, while retaining a generalised *Kobus* body plan.

TOP: *Waterbuck* (Kobus ellipsiprymnus*)*. BOTTOM: *Nile Lechwe* (Kobus megaceros*)*.

By contrast, in southwestern Zambezia, dominated by the broad, frequently flooded Cuando Cubango (Okavango) flats, *Kobus* became ever more special-ised for true swamp dwelling, giving rise to long-hooved, ancestral Lechwe that, during very wet periods, were temporarily able to reach as far as 10° north of the equator on the Nile. Their more generalised (but still water-dependent) Waterbuck cousins spread much more widely across much of non-forested riv-erine sub-Saharan Africa.

This preamble invites some speculation as to how Kana Boy and Lucy might have split in the same region about 4 million years ago. Lucy took the more generalised route while Kana Boy probably drifted down a few blind alleys before going extinct.

My bets are on *Praeanthropus afarensis*, far and away the most successful and widespread of hominins (even known from the southern Sahel, so possibly reached Atlantic shores). Lucy may well have invaded Ethiopia from as far away as Zambezia, not nearby Afar.

Lucy had developed true 'walking feet' and longer legs but she retained a primi-tive, ape-like wrist and, as more fossils were unearthed, it turned out she differed from her male counterparts in size. In this she anticipated much more substan-tial sexual dimorphism in several descendant species of the same broad lineage: the 'Nutcracker hominins', 'Robusts' or *Paranthropus*. These stout, big-headed

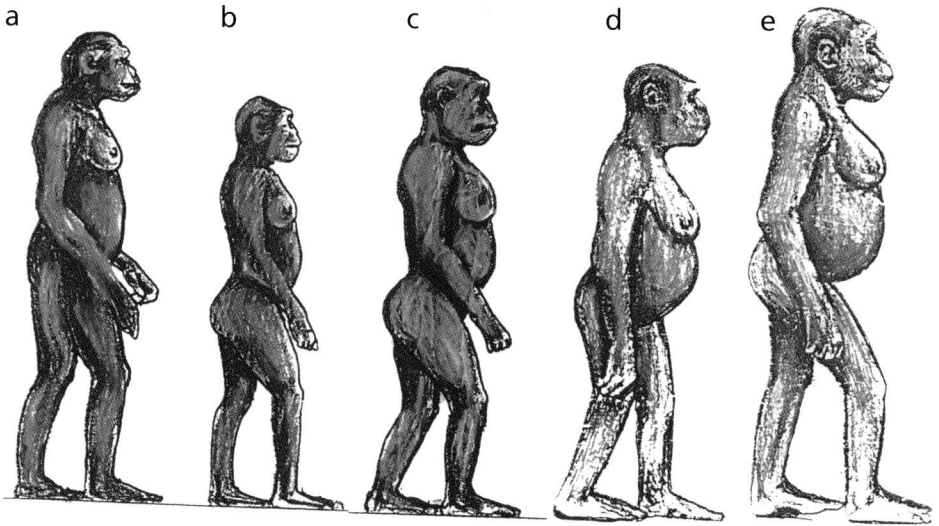

An evolutionary procession: (a) Kana Girl; (b) Lucy; (c), (d) and (e) 'Nutcracker hominins'.

hominins flourished between 3 and 1 million years ago, and clearly shared a common Lucy-like (and ultimately Kana Boyish) ancestry. They also maintained a similar preference for floodplain *dambos* or *mbugas*. Overall, the social groups for Kana Boy, Lucys and Robusts seem likely to have been made up of 2–12 individuals, dominated by one big male. Such groups probably remained relatively permanent, strongly territorial residents of as long a stretch of river and purlieus as served the group's needs. A given stretch of river would have defined a territory with defendable boundaries, including seasonal excursions away from each bank. Trees likely served as nest sites and night roosts and this would have had its own implications. Females were probably as hairy as chimpanzees, if only to provide babies with something to cling to at night (but perhaps to keep them warm as well). There is some evidence for a protracted shift towards more roots and herbs in the diet. For example, patterns on the surfaces of fossil teeth suggest that Robusts chewed many hard, scratch-inducing items in their diet. Sedges, grasses and their seeds, corms, tubers, roots and diverse fruits have been suggested.

'Lucy', one-time claimant for human ancestry, is now looking more like my fabulous hare. From a modern human perspective she and Kana Boy were the pioneering 'all-purpose hominins' but many of her descendant progeny ultimately found themselves in cul-de-sacs because they were eventually out-competed by *Homo sapiens*. Modern humans possess a toolkit that continues to thieve the livelihoods of any and every competitor (otherwise Robusts would still be out there, grubbing about on the *mbugas* or admiring each other's nest-making skills – oh! how I wish).

A direct genetic ancestor for the entire human race may never be found. Assuming that likelihood, every possible descendant of as-yet-unknown numbers of ex-east coast, ex-ground ape bipeds is of absorbing interest because every one of them represents the arrival, on Earth, of an entirely new extension of evolution by natural selection, via tools conceived by brains and operated by hands (which are themselves biological tools) all at the behest of brains. The fact is that early hominin brains were no bigger than living ape brains, and chimps' willingness to try to manipulate tools with maladapted hands implies that still earlier ape minds were in some sense 'prepared'. Perhaps their clumsy hands 'let them down', like a would-be pianist with an uncontrollable tremor.

We now know that however many ex-ground ape descendant lineages there were, only one led to us. To explore that, the most southerly, most isolated of east-coast basins beckons. There, we can reasonably suppose, was a long, well-wooded shore once inhabited by ground apes.

In a very real sense, South Africa is the most complex region in all of Africa. Some measure of Chicxulub's aftermath, even 66 million years later, is that South Africa, the southernmost tip of a pan-African, southeastern realm, has a flora of about 20,500 plant species, while 'Guinea', which occupies a large part of the

A selection of South African Protea *flowers, descendants of Chicxulub survivors.*

centre–west realm, has about 9,000. Similar disproportions appear in many other organisms. A particularly significant aspect of this contrast is that the south–east realm has many ecological elders – plants and animals that retain pre-Chicxulub adaptive features and occur nowhere else. They include 'primitive' cycads, proteas and magnolias among plants, and velvet worms, other invertebrates, frogs, reptiles and golden moles among animals. Most inhabitants of this treasure-house

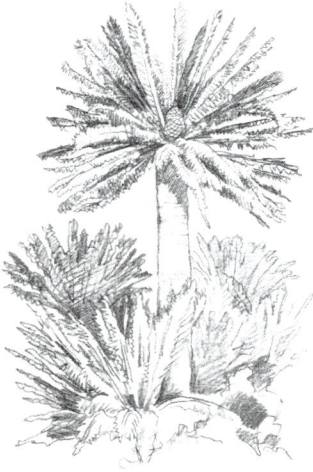

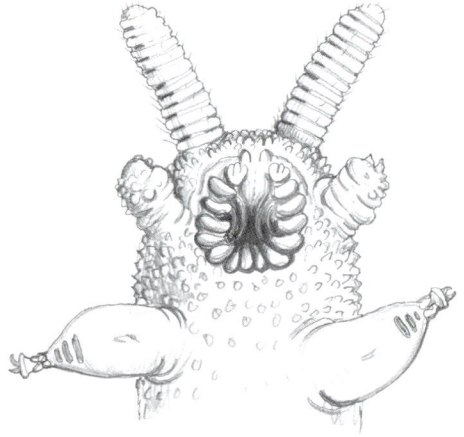

ABOVE LEFT: *Modjadi cycads (Cycadales).*

ABOVE RIGHT: *head and mouth of* Peripatopsis capensis *(velvet worm).*

Moreas exclusive to the cape.

of ancient species remain to be studied in the context of deep time and a deeper understanding of the evolutionary process (a subject for our final chapter).

When did our ancestral ground apes get here? During the driest, coldest centuries or millennia, the southernmost stretches of coast might have become inhospitable to any cold-sensitive species. If my model of ground apes inhabiting a coastal forest strip humidified by the Indian Ocean is correct, their arrival in the far south probably post-dates 5.3 million years ago, currently thought to represent a global trough in low temperatures and a likely deterrent for ground apes heading south. Between 5 and 4.5 million years ago, when conditions were at their warmest and wettest, southward penetration would have been at its easiest.

Compared to other 'basins', the Limpopo valley and delta mark the northern border of a radically different complex of South African rivers, all draining the 3,000 m high Drakensberg mountain chain. I call this multiple-river aggregation of watersheds the 'South African Basin'. More temperate latitudes and higher altitudes mean lower temperatures overall, more seasonality, and, on average, poorer resources for primates. This is confirmed, even today, by an impoverished primate fauna – both living and fossil. The ecological and climatic differences between a narrow Indian Ocean littoral and the interior are very substantial, but they are not physically distant. For an isolated population of ground apes, movement inland must have been severely inhibited by a combination of low temperatures and sparse resources during the southern winter. However, warm rains would have transformed these uplands every summer, and these montane habitats were close to the coastal plain, in some instances only a few kilometres upstream (albeit *steeply* upstream). These contrasts would have been exaggerated when overall climates were either colder and drier or warmer and wetter. Several contemporary species of birds are known to move back and forth within this region. Most famously the Blue Crane, South Africa's magnificent National Bird, bred in the uplands (1,300 to 2,000 m) between October and March before descending to the lowlands for winter. Industrial agriculture and plantation forestry are currently poisoning and destroying habitats and seem set to exterminate this magnificent bird, along with countless other organisms.

Other seasonal migrants up and down the Drakensberg are the Bush Blackcap, a uniquely local endemic, two nectar-dependent sunbirds, an olive bush-shrike and a wagtail, as well as fruit bats, all trying to join whatever vestiges of indigenous life can survive the depredations of modern industrial agriculture.

There are many short, mostly perennial rivers to connect high and low areas. When these steeper South African rivers are compared with those of Zambezia or more tropical regions, the pulsing of resources is related to altitude and temperature rather than to seasonally lateral movements of water over relatively flat surfaces. For hominins or ground apes, the contrast between seasonal movements up and down rivers, rather than in and out from a home stretch of flatland river, would have influenced a fundamental divergence in social behaviour between hominins in the two basins.

The southeastern littoral and Drakensberg therefore present an unprecedented model of the environment's influence on hominin behaviour and on the course of human evolution. Crucially, there is also a rich but rather late endemic fossil record well within the drainage basin of the Limpopo and other rivers of the deep south.

Take the earliest 'southern man-ape' from the far south – the skeleton of an *Australopithecus africanus* nicknamed 'Little Foot', which was released, feet-first, from its encasement in dense, hard limestone. Dated at 3.65 million years, 'Little Foot' consists of a nearly complete skeleton standing about 1.3 m high and retrieved from the bottom of what must have been a 30 m sinkhole. This and later South African man-apes cannot be derived from Lucys, and clearly belonged to a separate lineage.

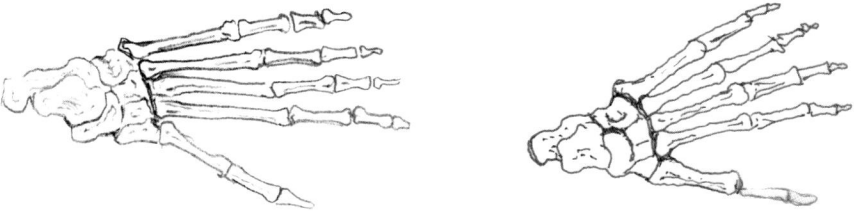

*Feet of (left) 'Little Foot'; (right) ground ape (*Ardipithecus*).*

Little Foot's limb proportions and platform feet are more like those of ground apes and *less* like those of modern humans, with big toes still capable of deflection from the other toes. This means that the transfer of weight onto the big toe was less complete and plantar arches scarcely existed, unlike modern humans (even unlike Kana Boy and Lucy). What is more, wherever southern man-ape feet have been found they invariably betray the reality of a ground ape ancestry and just such a peculiar line of descent.

There is the strong implication that South African man-apes were slower or less accomplished walkers, and that they retained elements of ground ape 'platform' feet. They differ most from Lucy and Kana Boy in the profile of the

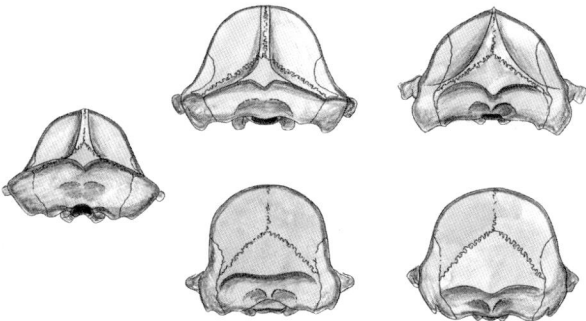

Rear view of hominid skulls: (left) Chimpanzee; (above left) Lucy; (above right) 'Nutcracker'; (below left) South African Man-ape; (below right) Handy Man.

LEFT: *Ground ape*
(Ardipithecus ramidus).
RIGHT: *'Little Foot'*
(Australopithecus
prometheus).

cranium immediately above its articulation with the neck. Both the thickness and the conformation of tooth enamel also differ between the two groups, as do variable patterns of tooth rooting in premolars. South African man-apes are consistently closer to humans in all these features.

Perhaps most significant of all, australopith hands are most like those of humans, with thumbs constructed to be opposable to the fingers in various subtle permutations of pluck, grip and clasp. Little Foot's fingerbones, although apishly curved, are of similar proportions to those of later humans whereas Kana Boy even retains the reinforced knuckles typical of quadrupedal apes. As for the wrist, a human-like flexibility contrasts dramatically with the tied-up, well-locked bones of Kana Boy and Lucy. In spite of ambulatory handicaps, australopiths *were* walkers, but there is a strong implication that they seldom had to hike any great distances. The articular planes in the knees of *Australopithecus africanus* fossils suggest a bow-legged gait that was emphatically ill-suited to fast movement.

However, fossils from the coastal lowlands have yet to be found so Little Foot, along with 'the Taung Child', 'Mrs Ples' and other stars of the southern fossil beds, are all palaeo-orphans, in that the tombs of their immediate Limpopo ground ape ancestors have yet to be disinterred.

Soon after ground apes moved into their restricted southern enclave, demographics would have created pressure not only to expand their range into the

South African man-ape,
imagined portrait and profile.

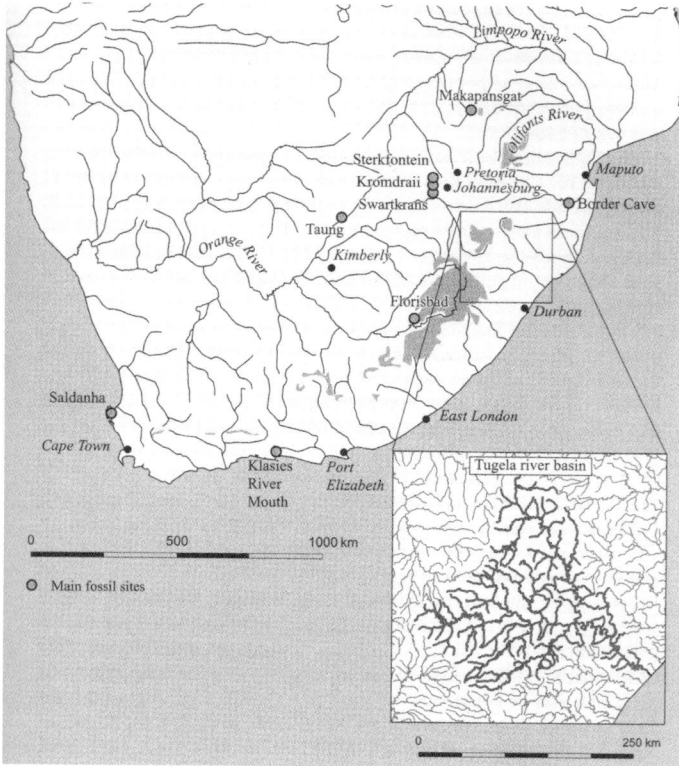

South Africa,
illustrating the
abundance of East
Coast rivers draining
the Drakensberg.
INSET: *Tributaries of
the Tugela river.*

uplands, but also to become less strictly sedentary. Local movement, driven by seasonal changes along short stretches of river, is a modest innovation. Indeed, in its very earliest manifestation little more than a localised shift from summers upstream to winters downstream would have been involved. Nonetheless, the longer-term implications of seasonal movements would have been immense.

To begin with, travel would have been eased by a majority of the rivers being short, falling along direct courses straight down to the seashore, or to the Limpopo River itself. Furthermore, the spacing apart of streams is and always has been rather even, averaging about 5 km, so single groups might, in many instances, have laid claim to single valleys. This arrangement would have been a departure from the ranging behaviour of river dwellers such as Kana Boy in flatter terrain, with more uniform environments.

If we are to try to bridge the gaps between ground apes, australopiths and *Homo*, I think the larger context of such transitions is likely to lie hidden in the detailed behaviour of slow, ill-equipped but fecund little primates, challenged by seasonal habitats and with strong inducements to expand out and away from steep and narrow gallery forests. For a hominin venturing this far south, the climate and (at that time) great diversity of habitats elicited unprecedented responses.

From a partly frugivorous ape's perspective, overall productivity in southeastern Africa is lower than in the tropics, with peaks and troughs corresponding to seasonal highs and lows, especially at higher elevations. Under these circumstances animal foods, already part of ground ape diets, could be expected to increase. Movement upriver would have faced no physical impediment, but environmental barriers, formidable in the dry cold winter, would have been trivial over much of the summer. How would ground apes respond to physical separation of strongly seasonal resources?

If the choice of foods was not to shrink as animals moved inland from the coastal plain, one mitigation could have been to intensify the strategy of ever more thorough probing of the habitat, and the development of new techniques for finding, gathering and processing food.

As with other apes, females would have been more mobile between groups than males. Overall, however, I suggest that ground apes, Kana Boys and Lucys were likely to have remained permanent, strongly territorial residents of as long a stretch of river as served the needs of a group that seldom numbered more than 12 individuals. Instead, the South African isolates, moving in response to seasonal and altitudinal changes along physically well-defined valleys, frequently encountering other dwellers in the same valley system, could have been under strong pressure to modify territorial behaviour. More specifically, foraging groups might have become less stable and territorial defence might have become less pronounced. In many instances, fellow valley-dwellers, numbering 100 or more, might all have been kin. But the environment alone would have encouraged social accommodation between larger numbers of more loosely connected individuals. Under these circumstances, narrowly territorial behaviour would have declined, and some sort of clan system, embracing a variety of associations within a larger area, could have developed. Social behaviour would have had to accommodate to larger numbers, to more complex relationships and judgements among many interacting individuals.

The distribution and nature of food would have determined what sort of accommodations were made between the individuals. Some could have encouraged periodic dispersion and inhibited permanent close relationships both between and within sexes, favouring greater social flexibility and closer body sizes between sexes. If ground ape small-item foraging was relinquished only very slowly, competitive confrontations over food might have been relatively infrequent. Tolerance between weakly acquainted individuals belonging to loosely connected, valley-by-valley clans could have been encouraged by resources that were consistently well dispersed or by foods that were occasionally super-abundant. The former would have promoted safe dispersion, and the latter would have encouraged temporary non-competitive convergences on, say, a grove of fruiting trees or a birth or hatch peak that suddenly rendered prey more catchable.

Food shortfalls could have elicited two very different outcomes: intensified conflict or further dispersal. The former would have enhanced territorial responses while the latter reduced them. Strong defence of territory, led by big males, implies well-defined boundaries and, for a social species, their protection by the whole group. Here, again, seasonality might have promoted cohesive territorial behaviour during periods when individuals were together, and relaxation when they were dispersed. Periodic confrontations could be expected between different clans on the borders of their ranges, but frequent dispersal of small units of fluctuating composition could have diminished the overall frequency and importance of intra-specific territorial behaviour within a clan.

Close relatives of *Australopithecus* turn up in much later deposits in equatorial East Africa (nearly 3,000 km north in Olduvai, the earliest currently dated to 1.8 million years). Louis Leakey insisted they were *Homo* but many palaeontologists doubted the wisdom of assigning two such similar species to separate genera.

For me their shared differences versus Lucys only serve to confirm that the human lineage derived directly from *Australopithecus* and not *Praeanthropus*. Assigning East African 'Handy Man' to *Homo habilis* signifies the emergence of what are clearly the progeny of South African man-apes outside their southern enclave. As ever more detailed studies imply, most of the social, intellectual and anatomical advances as well as the handicaps of *Australopithecus africanus* were shared by the bigger-brained *Homo habilis*, or Handy Man.

Enlarging the brain is as good an evolutionary threshold as any. Even so, all currently known fossils of *Australopithecus* and early *Homo* represent individuals living at too-late dates to be our direct ancestors. That near-certainty actually reflects our desperate dependence on so few fossils, so fortuitously localised in time and space. (Incidentally, it also reflects the low priorities assigned to human origins and prehistory by a culture that has submitted itself to a value system dictated by commercial interests.)

Yet, even if some of the presently known South African fossils turn out to be sideline buddings off our own lineage, their significance is scarcely diminished as guides to the time, place and nature of *Homo*'s beginnings. Given how

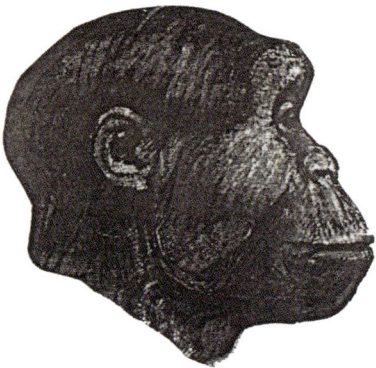

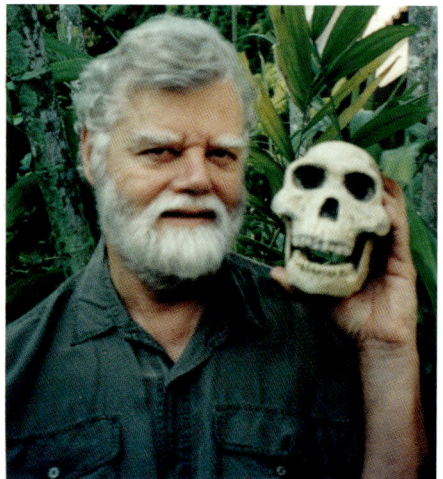

LEFT: Homo habilis, *the Handy Man, a reconstruction.*
RIGHT: *Jonathan Kingdon with reconstruction of* H. habilis *skull.*

dependent on fossils we are, much more effort is needed to find new fossil beds all along Africa's eastern littoral, as well as within all its inland basins.

As more sites and more fossils are found, it is becoming clearer that our lineage, like many others, once tolerated its ecological elders, allowing older, even more specialised members of the same lineage to persist, even co-exist, over many hundreds of thousands of years. Longer-established populations would have known better how to survive in more marginal regions or ecotypes. Minor specialisations in behavioural routines or in locally adapted tools and techniques could have helped blunt direct competition between cousins.

Multiple hominins from single fossil sites have presented difficulties for a few neat-nics but a visit to the vicinity of Mongiro hot springs at Bwamba, Uganda, would present them with a marvellous inter-layering of six species of *Cercopithecus* monkeys, three *Colobus* monkeys, and several *Galago* species within a community totalling 16 primate species. There have been times when the second-largest continent also had room enough to accommodate various sorts of people, even cousins belonging to different species, all co-existing (metaphorically 'side by side' but actually more often in occupation of separate sub-habitats).

Arguments about whether *habilis* should be assigned to *Australopithecus* or to *Homo* often boil down to any one scholar's choice of tesserae, as plucked from complex mosaics of multiple and variable adaptive features. Ambiguities between late *Australopithecus* and early *Homo* fossils repeat themselves with later *habilis* and early *erectus*, where the clearest discontinuities concern enlarging brains, the acquisition of sustainable walking/running gaits and larger but less sexually dimorphic body sizes. Even wider controversies have been triggered by the many fossils that blur supposedly specific features along a *habilis–erectus–sapiens* continuum. These digress too far from my avowed focus on Africa.

Although many equatorial species are unable to tolerate South Africa's cold, all need water. All primate fossils, no matter their date or provenance, have always existed beside water. This need has always been shared with other animals, especially big ones.

When rivers flow through drier hinterlands, banks and purlieus are often densely thicketed and webbed by well-trampled, mostly smooth pathways of large herbivores, from hippos and rhinoceroses to herds of elephants and antelopes, all coming and going to that necessary and shared resource – water. For relatively soft-footed hominins, there must have been little choice but to use these paths. They are as much components of the ecosystem as are tree trunks, river courses and termitaries. Early hominins would have followed animal paths as a matter of course (indeed, most travel would have depended on them). They would have noted trouble spots for the unwary, would have learned where and when each species preferred to travel and could have intercepted or diverted prey into prepared ambush or trap sites.

The users of such paths intrude on the domains of much larger group-living animals, and encounters become sufficiently routine to demand equally routine

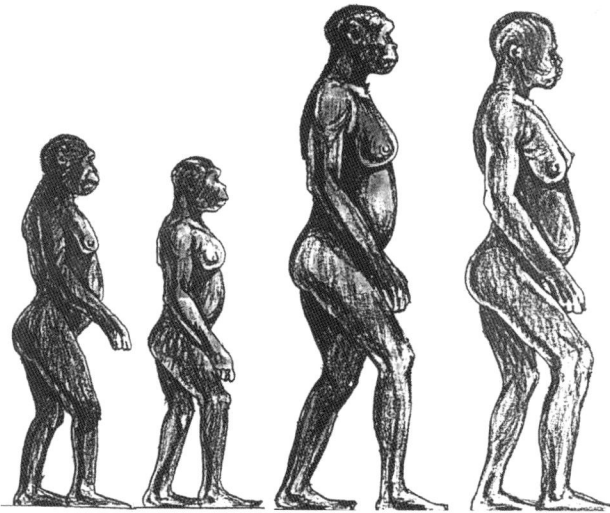

An evolutionary procession from left to right: Australopithecus africanus; Homo habilis; H. erectus ergaster; H. sapiens.

responses. If hominins, in their search for food, travelled any distance away from the riverine core of their range then they, too, would have had to develop routine responses towards such encounters, the most frequent of which may well have been with elephants. Improved gaits on the ground need not have excluded a continued facility for climbing trees, but scrambling up the nearest tree need not have been the only response to meeting mammals, especially big ones, on paths made by gregarious *them*, rather than by feeble little *us*.

As coastal ground apes extended their range into the uplands, a prime challenge would have been to get access to resources that were contested or pre-empted by a variety of other species, mostly other mammals. It is possible that one significant divergence between australopiths and Lucys living further north could have centred on more co-ordinated social responses to predators and aggressive competitors rather than pell-mell scampers up into the trees – each ape to itself.

Achieving access to contested resources could have called for the development of complex, strategic behaviour. A rather small, slow-moving, semi-terrestrial australopith might have diverged most significantly from Kana Boys and Lucys in the strategic intelligence and behaviour it brought to such routine confrontations.

One moonlit night, as a youth, I and three benighted friends walked unknowingly into the midst of a herd of elephants. Clustering together beside the trunk of a tree, we roared, we bellowed and hollered as loud as our lungs and throats could bear. Within a minute or two the elephants were gone and we resumed our path back to camp.

Australopiths were not the first to resort to such ploys, indeed, in a neat reversal of roles, I have been the one to hastily make way for a large troop of hooting, branch-bashing, ground-travelling chimps. I suspect the survival of highly vulnerable australopiths must have depended upon many subtle, systematic, even rehearsed responses to very frequent encounters with bigger, more numerous or hungry meat-eating mammals. As hands freed themselves from clutching every branch or substrate, just to move a body around, they became true servants of the brain. Fingers that had grasped a swinging vine came to hurl a hand-sharpened spear. A fist that stunned a cornered monkey into becoming a meal became the socket for a stone hammer cracking shells or nuts. Fingernails that once helped a ground ape excavate a morsel or scratch a rival's face came to curl around nail-knives, nail-chisels and nail-adzes. Neural channels between hominin brains and hands must have increased and diversified as hand-made artefacts forced once resistant nature (or recalcitrant individuals) to submit to the wielder of new tools or weapons.

What has had scant study, because it leaves scant imprint, is the confidence with which our socialised australopiths might have faced the outside world. We are familiar with the braggadocio of pubescent males in contemporary cultures and with their exploitation by warlords, politicians, the military or local gangs. The psychological and strategic role of young males in living mammal societies, including elephants' and our own, is clear enough. During the First World War, my father was a schoolboy being trained to be a soldier. The prospect of being shot, blown up or being gassed in distant trenches was integral to the bombast and vainglory implicit in that military education. Warriors, buoyed by crude manual skills, and weapons of the age have always been the most expendable class. Male hormones probably played as integral a part in the emergence of australopith and early human societies as did the quality of their weapons and other tools.

My colleague and friend, Richard Dawkins, describes termitaries, nests and countless other animal constructions, including human tools, toys and weapons, as 'extended phenotypes'. His brilliant analysis goes further – organisms at every level, especially parasites, 'manipulate' other organisms, even their behaviour, to maximise their own chances of survival. Richard's pioneering insights follow from his positioning of natural selection at the level of genes, the irreducible unit of life, in addition to the more obvious level of their hosts – whole, complex, compound organisms.

Bones, buried by a variety of accidents, have proved the existence of many australopiths and humans in South Africa. The burial site of *Homo naledi* is different; a substantial accumulation of their bones (and their bones alone) lies inside a deep, well-hidden cave, suggesting that dead people of all ages were interred in 'Rising Star Cave' over a protracted period of time. This might reflect little more than local dumping of the dead, but given that infected or rotting bodies reduce a community's survivorship, the designation of a nearby cave or sinkhole as a disposal site for bodies was effectively an extension of *H. naledi*'s anti-infection and anti-scavenger strategy.

So far, no trace of gnawing or cracking of bones has been found so it would appear that *H. naledis* were successfully frustrating hyenas and vultures as well as deterring disease by transporting their dead to a secure cave. The fact that this practice was sustained over time implies a long-term cultural tradition of

ABOVE: *Skull drawing.* BELOW: *Sculpture derived from drawing above.*

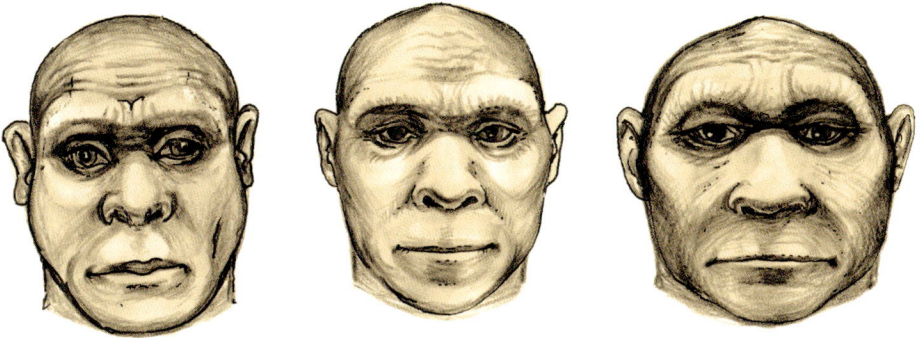

Three species (subspecies?) of Homo erectus *type: (left) Work Man (*H. ergaster*); (middle) Peking Man (*H. pekinensis*); (right) Erect Man (*H. erectus*).*

disposal/entombment, even if this meant little more than tipping bodies down a convenient hole. If tools and weapons have extended the human phenotype, Rising Star's cavernous bone-yard implies a less appetite-driven extension of the human phenotype, one that seeks self-preservation via thinking brains.

The bone-yards of South Africa document more than the deaths of early humans. They announce the birth of a primate with skills in thinking, making and acting that extended their presence and their phenotype wherever else they travelled. It may be an epic saga for humans, but for too many other forms of life the arrival of humans heralded eras of banditry that culminate in today's Anthropocene.

When did humans first leave Africa? What kind of humans were they? What are the implications of fossil finds, ancient DNA and other scientific innovations for human comings, goings and returns to the mother continent? What else do we know or not know? And when did all this coming and going take place?

Shockingly, there are very few certain answers to any of these questions, but let us return to the emergence and wider spread of australopith-derived humans, *Homo habilis*. Rather small (100–148 cm tall and weighing 20–35 kg), with a gait suited to walking and climbing but probably poorly adapted to sustained running, this 'first *Homo*' may seem an unlikely long-distance migrant outside Afro-Arabia. Nonetheless, there is a wide consensus that *H. habilis* gave rise to, even melded with, *H. erectus* (letting both terms embrace various closely related populations scattered through space and time).

Given speciation's need for geographic isolation in environments different from those of origin, I once posited Morocco and the Arabian peninsula as two possible regions for the transition from *H. habilis* to *H. erectus*. Nonetheless, a more distant, non-African transition is still conceivable.

Wherever *Homo erectus* evolved, their long legs had quickly marched these hugely successful humans to the Old World extremities of China and South Africa by about 2 million years ago. They were still around just over 100,000 years ago. On present showings, modern humans will be lucky to survive a fraction as long.

ABOVE: Through a Porthole – Frigatebirds.
OPPOSITE MARGIN: *Pterosaur (*Pterosus*)*.

NICHE-THIEVES

*In which the ultimate ecological role assumed by
Homo sapiens is defined.*

When my three-year-old self asked my mother, 'What are people for?' she repeated her mother's answer. For the equivalent, 'What are chickens for?' farmer's daughter Bibi Minnie offered, 'To provide eggs and roast rooster.' She would have laughed off a sportswoman's suggestion: 'To be best at all the things chickens do best', or a vegetarian's notion: 'To wake us up in the morning.'

Down on the coast we sometimes saw kink-winged, fork-tailed Great Frigatebirds, or 'man-o-wars' as Teddy insisted on calling them. They patrolled the shore in long, slow soars, with nary a beat of their more than 2 m wingspan. Had I asked what frigatebirds were for, one answer might have been to snatch sea-food off the waves, but a trawlerman (blithely blind to his own predatory profession) once came up with an indignant 'the bastards steal lunch-bags off seabird mums as they fly back ashore to feed their little chicka-bids'. I had been much impressed by that fast and ferocious harassing of other seabirds to disgorge their crop contents, but for that sailor there was moral out-rage. For many much earlier mariners, the birds' fast and furious pillage brought to mind warships, man-o-wars, frigates, pirate corsairs – hence their naval name.

To ask what anything is 'for' in nature needs to be rephrased in terms such as 'how has natural selection arrived at something as perfectly proportioned an air-borne forager as a frigatebird?' Yes, frigatebirds thieve the hard-won proven-der of other seabirds but, as products of natural selection, frigatebirds are per-fectly proportioned airborne snatchers that include piracy or thieving among their skills. Their skeletons are reminiscent of a scaled-down version of the pter-osaur *Quetzalcoatlus*, another airborne snatcher with the wingspan of a small aeroplane and the beak of a gigantic Marabou Stork.

Returning to that three-year-old's query about what people are for, paraphrase his question as: 'Do people have an ecological role?' Among many possible an-swers, 'niche-thief' is as close as my ecologically minded self can get. It can be said that every evolving organism tries to steal a march over some other organ-ism, quite often over close relatives. In a sense, all adaptive advances involve moving in on the niches of precursors.

We can also say that most organisms feed by 'instinct'. Likewise, any 'instinc-tive' feeder has to bypass any potential meal that is protected by an impervious shell, poisoned chalice, chemical screen or other off-putting device. Instinctive

feeders ignore such organisms in preference for well-established menus which their own species have become particularly adept at exploiting. Plants and animals protect themselves with such barrages of devices that most successful predators and their prey have to evolve equally specialised counter-devices, such as mollusc shells and the tough muscles that seal them. Take the open-bill storks, which have become such specialised snail-eaters that their muscle-cutting beak-tips can hardly manage any other diet.

The shells of some molluscs would challenge a jackhammer, so I like to imagine their occupants, contemporary or fossil, building the equivalent of bomb-proof shelters to foil shore-foraging dinosaurs or worse. But such constructions took millions of years to evolve.

What I am proposing in this chapter is a much faster type of takeover, in which tools and techniques, operated by strategic intelligence, play pivotal roles. I envisage humans beginning to steal niches from the moment hands began to invent tools to serve their brains' dictates, or their stomachs' imperious appetites. I am suggesting that hominins might have begun the process of becoming systematic 'niche-thieves' because they pursued technological and behavioural solutions to get access to foods that were previously only accessible to species that had evolved the techniques and physiologies to outwit their preys' defences.

These pages have largely by-passed *dambo*- and *mbuga*-loving Kana Boys, Lucys and Robusts, because I expect 'southern Man-apes', the original *Australopithecus*, to emerge as a separate lineage that adapted to habitats more complex, diverse and demanding than those faced by their Lucy cousins.

Isolated down in far southern Africa, australopiths adapted to an environment that was not only pre-Chicxulub in parts, but which embraced a greater spectrum of landscapes, altitudes and climate changes than can be found anywhere else in Africa. The range of habitats they inhabited was so exceptionally broad and tightly packed that it had already challenged the adaptive potential of many species before ground apes got there. Few primates persisted there for very long.

Australopiths might have differed from Lucys in extending the ancient ape tradition of brainy generalist. In their inventiveness, perhaps they became the brainy 'frigate-ape', the super-thief of niches. Lucys may have represented an early flush of hominin evolution, but in any later interactions with the *Homo* lineage they and their Robust descendants never left their continent or habitats of origin. Once their niche-thief cousins had emerged from their southern enclave, their progeny became the ultimate pioneering travellers (as well as likely agents of the Robusts' decline and eventual demise).

I follow Louis Leakey in accepting the systematic use of stone tools as convenient palaeontological markers for genus *Homo*. Louis went further and correlated brain size with tool use (he appreciated, of course, that tools alone have less implication for brain size than the ability to develop, share and impart knowledge of their manufacture and effective utility).

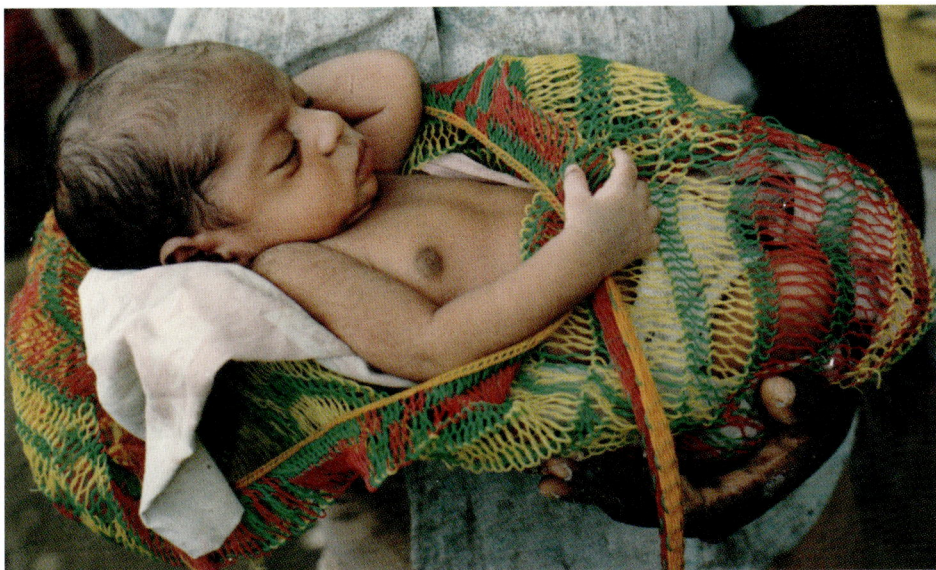

OPPOSITE MARGIN: *African Openbill stork (*Anastomus lamelligerus*).*
ABOVE: *Newborn baby in woven string 'bilum'.*

Inheritance as the pathway for a package of information and action has its mimic in chains of learning, but the survival of any technique within a human culture is both hazardous and prone to obsolescence. Genetically programmed, evolved skills, such as the fishing of kingfishers or the oyster-catching of oyster-catchers are superior to artificially invented technologies in that almost all the skill components, from identifying the correct context for the tool, to its actual use, are built-in and heritable. Before knowledge could be stored, human inventions easily died out unless practitioners could pass on their skills. So, look to mental quickness, adaptability and aptitude for learning (and teaching) new skills for the greatest survival value.

In a moment I will turn to string, carrier bags and other non-stone plant and animal materials that get lost to archaeology because they decompose. Just such ephemeral tools and structures, in the form of nests, ear-probes and termite rods, 'extend the phenotypes' of contemporary apes. This tells us that all hominins used some sorts of tools. Together with many others, I contend that the origins of *Homo* lie in South Africa.

What were the consequences for other animals and plants that shared *Homo* or australopith habitat? I contend that our evolved ecological province includes the thieving of niches, no less than frigatebirds burgling booty from boobies.

It is appropriate that a critique embedded in deep time should examine Africa, because our landscapes and our inhabitants have been assaulted by humans longer than on any other continent. The most significant dimension of this critique is that while humans are the single most interesting and most dangerous product of Africa, we are also evolved mammalian primates, ultimately subject to all the environmental forces that govern the evolution of any animal or plant, anywhere on Earth. Furthermore, an ecological perspective invites us to try to define where, within the huge complexity of nature, the human animal fits. Or mis-fits. I take the invention of tools and technology as central to such a quest.

The lighting of, control and use of fire is one of the earliest skills learned by our ancestors. It is a tool with extensive impact on the environment. Plants that are adapted to withstand (even exploit) lightning-lit fires long predated pyromaniac humans – such plants were pre-adapted to take advantage of hunters' subsistence techniques.

The setting of fires to 'clean the land' is a practice widely used by inhabitants of fire-climax vegetation such as tropical savannahs. Alteration of very extensive tracts of land to suit human hunters put many other animals at a disadvantage but certainly favoured others. For example, more than half of surviving antelopes are mostly relictual leaf- and herb-eaters while only a small minority are grazers, yet fossils confirm that grass-eaters increased very greatly in importance over the Pliocene and Pleistocene.

Humans are likely to have shaped savannah ecosystems much earlier than is generally thought. The present spectrum of large savannah animals is already a vestige – they have survived because they can cope with both fire and hunting. If any credit is to be allocated for their survival it lies with the animals' intrinsic vitality, rather than with their hunters' abilities, appetites, or even any contrarian impulse to conserve them.

In an extraordinary mimicry of natural adaptations, australopiths and their descendant *Homo* populations began to manipulate elements and use materials in ways that rapidly multiplied the number of ecological niches they could invade. Each new tool opened possibilities that were formerly the prerogative of very specialised animals. Where diggers had needed heavy claws, now there were hardwood, horn, bone, shell or stone picks; spears mimicked horns, porcupine quills or canine teeth, and so on. For the very first time, here was an animal multiplying its ownership of niches via technology. Here was a new competitor that would encroach on at least a part of many another species' former niche. In some cases, the overlap may have been so great (among, say, scavengers) that hominins took over and the competitor was out-scavenged.

The more skills or niches that humans have appropriated, the more versatility and flexibility has had to be learnt. This is in marked contrast to any other animal. All lower animals are rigidly programmed to perform their genetically predetermined roles.

Where do we most differ from other mammals? The paedomorphic or infantile trend that marks out human evolution hints at links between anatomy, the development of technology and co-operative behaviour. The swollen, infantile shape of a modern human cranium is, in a sense, the anatomical expression of multiple skills, in communication, manipulation, language, co-ordination and the pleasures of *play*.

Lone child yelling (photograph by J. Kamminga).

Furthermore, most juvenile animals are prime targets for natural selection. In this, the ultimate agency may be external diseases, predators or accidents, but a more immediate selective force is exerted by other members of a juvenile's own group or family. If, for example, there is a potential for parental neglect, then there are advantages for the young that can 'manage' their negligent parents best, as well as for those with responsible mothers or fathers.

To illustrate the control exerted by young animals over adults, consider how the newborn or newly hatched summon their parents with a cry. That wail or shriek is much more than a simple expression of need. It is the young animal's only means of forcing its elders to give it attention, protection or food. Its coercive effect probably has evolutionary roots in the danger, to family or group, of alerting predators, but the signal itself and compulsive parental responses have become automatic – hard-wired. As the growing youngster's range of needs widens, such cries are soon augmented, developed or differentiated. Both crying and begging behaviour may change continuously, and eventually it may even get incorporated into adult situations where fellows have to be appeased, cajoled, seduced, courted or outwitted.

Parental status can influence the subsequent fortunes of their offspring, but babes grow up and any novel behaviour that has favoured the survival of a social species may well depend on ranks first acquired in youth. It is here where social *Homo sapiens* might learn most from other group-living animals.

In striking contrast to antisocial Striped Hyenas, matriarchal Spotted Hyenas sometimes form large groups or clans in which cubs, left in nursery dens, develop ranks and alliances that have been shown to reflect their own mothers' ranking. The alliances that squabbling cubs form in the den can be integrated

Spotted Hyena (Crocuta crocuta) *sketches.*

into durable adult ranks. During 27 years of very detailed observations in Masai Mara, East Africa, the brilliant Kay Holekamp and her many collaborators have confirmed that cubs can be as bright or even brighter than adults and that rank gets passed on from mothers to daughters. They have shown that genes as well as life-sustaining behaviours infiltrate much wider networks in Spotted Hyena society via rank. The complexities of such societies help us understand the enduring success of royal (even nouveau-riche or recent 'revolutionary') dynasties as a mode of governance and source of ritual in all too many modern human societies. In no sense do I seek to disparage or belittle centuries-old human institutions but I do assert, with conviction, that we have a great deal to learn from hyenas, chimps, whales and many other social animals. After all, we too *are* both social animals and kids with a lot to learn.

An infantile behaviour that is common to all higher primates is the tantrum, but when the contexts and character of chimpanzee and human tantrums are compared, there are significant differences. Where the young chimp's pique is a response to another chimp's failure to allow the infant access to food or fun, the human child goes further than 'attend to me or I'll scream and stamp'. The human shouts, gestures or pulls to assert a wider range of messages. 'Look at me, see my work, I'm smart.' Here the borders between learning, play-acting and play itself become blurred.

Young apes like to maul adults and to strut or see-saw around, stick flowers or leaves on their heads and generally show off to one another and to adults. But the most conspicuous difference between the high spirits of ape and human children is the latter's assertion of 'self' in a variety of roles. The progressive development of self-consciousness, the awareness of role-playing and the ability to express that awareness, is much more likely to be the product of competition and selection among juveniles than it could be at the adult stage. Smart kids, able to manipulate others in their own favour, had the best chance of surviving.

Forcing attention out of adults is linked to acquiring technical skill, so children compete among themselves for their elders' and their peers' attention.

Who can hit the ball? *Infantile competition.*

Project this back to early *Homo*, and children's hunger to excel at adult skills might prove to have been a major force shaping human evolution. Links between skills and rank remain a fertile area for future research in many areas of natural history research, especially human natural history.

For juvenile tool-users, demonstrations of skill in the operation of artefacts become major opportunities to express prowess. Herein lie the seeds of almost continuous innovation (also, in some societies, reinforcement of social stratification). Indeed, it may be in the biological origins of our earliest learning techniques that humans can most truly be said to be 'self-made'. A prime incentive for displays of imitative skill seems to be the praise of peers or parents, something that rewards both.

Because *Homo* is an animal that has slowed down growth, the presence of children permeates human society as no other. Childhood is a longer proportion of a lifetime, and children's needs have to be accommodated into many if not most adult activities. Modern societies that seek to prolong learning, especially among young females, will always be at an advantage over those that diminish the potential of half their population. Prolonged involvement of *Homo* grownups with their children not only allows knowledge to flow from parent to child, it eases and has always eased the diffusion of childish innovation into adult practice. Smart kids grow up and *their* kids are often at some sort of social advantage.

Exceptionally helpless babies made every member of a human or pre-human group more vulnerable than its ape equivalent. Unlike the highly dimorphic Lucys, all classes of australopiths and *Homo* were to some degree juvenilised. Even if fierce adult males could enhance the security of relatively helpless females and children, all shared the vulnerabilities of the weakest. Behaviour that put the whole group at an advantage would have been tested and selected for. Behaviour that diminished the chances for survival among women and their children would have been selected against.

Like other organisms, early human groups probably engaged with any aspect of the environment that provided food. To this end the evolved defences of many

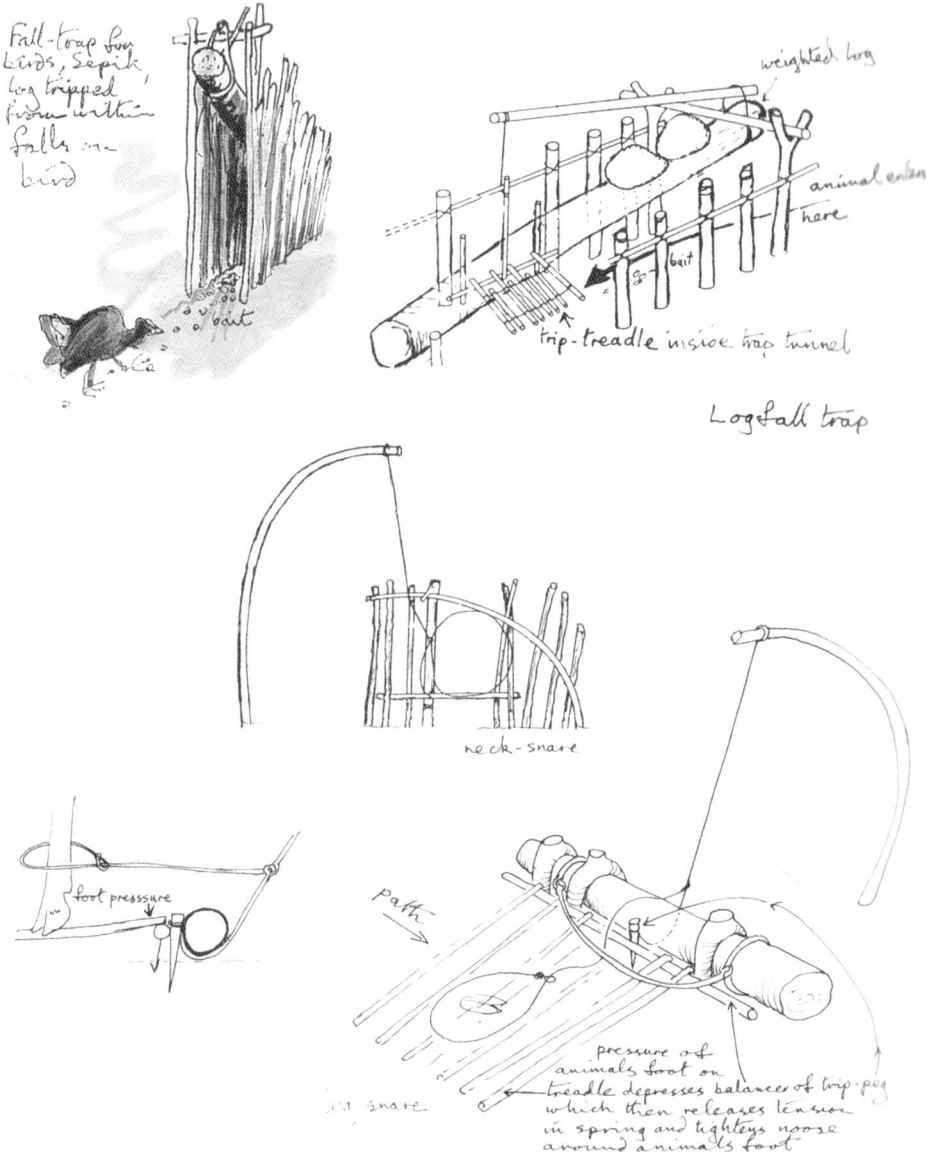

The handwritten labels in the illustration read:

Fall-trap for birds, Sepik, log tripped from within falls on bird

bait

weighted log

animal enters here

bait

trip-treadle inside trap tunnel

Logfall trap

neck-snare

foot pressure

Path

snare

pressure of animals foot on treadle depresses balancer of trip-peg which then releases tension in spring and tightens noose around animals foot

Snare and log-fall traps.

plants and animals had to be overcome and obstacles had to be removed. Some of this centred on group trapping, snaring and hunting, with cords, staves, sticks, stone cutters, crushers, scrapers and diggers as the agents of such appropriation.

A self-driven compulsion towards social participation implies a need for effective communication. This must have provided powerful incentives to elaborate or improve language. Improving communication would have had obvious benefits for an economy that no longer relied on innate foraging behaviour, because every successful harvesting technique had first to be invented and then learned and passed on, as just one of many components in an extensive strategy for subsistence. Intelligible codes of communication may have helped small groups to join up with or even negotiate with larger units, in which exchanges of goods and ideas became possible.

In Tierra del Fuego, Darwin was horrified to witness a father kill his supposed and still small son for dropping and breaking a basket of eggs. Conspecific assassin males are just one of the many hazards faced by young animals of many species – humans are evidently no exception.

If high juvenile mortality was the norm, survival into adulthood, in prehistoric societies, may have been highly dependent on sustained demonstration and performance of aptitude, especially efficient communication during childhood. Children were not just sponges for lessons from the adults and observant learners from the environment. In effect, they may have been specialists in innovation, a role that could have been crucial in relations between neighbouring groups. Whether neighbourly relations were friendly or hostile, children or child-like adults, like nerve-endings, may have been effective interfaces between groups and thus a major agency in exchange.

In any competition between different populations of closely related humans, the ability to impart technological knowledge and adapt to novel situations probably had more implications for brain size than any direct facility in the use of tools. Fast-developing technologies served by language and observant eye-brains were the main advantages possessed by modern humans. Great selective advantage would have lain in any social or demographic features (such as language-linked extended families or the invention of media) that enhanced the spread of technology and its social contexts.

In all this, older, experienced animals are critical for the maintenance and diffusion of traditions in many species, but it seems unlikely that they were significant sources of innovation among humans, as is often assumed. It is more likely that inventiveness would have come from the younger and less experienced, in playful, experimental, but nonetheless ecologically relevant contexts. If the main characteristic of our lineage has been to extend childhood, then analysis of early stages in the development of an individual should be a real guide to our prehistory as well as to our intrinsic nature.

The great majority of prehistoric artefacts must have been constructed from degradable animal or plant material. Since these have not survived, how can we retrieve any understanding of that immense lost industry? How can we begin to catalogue their diversity and appreciate their fitness in terms of functional form? How can we tease out their possible origins?

Perhaps a start can be made by recognising that the materials for artefacts and the main commodities of prehistoric existence spring suddenly into use through the very act of foraging, in the decision to collect or pick up an organism or to trap or kill prey. Take what we call string, thread, cord or thong, an article that can have plant or animal origins. The incentives and skills that invented and implemented

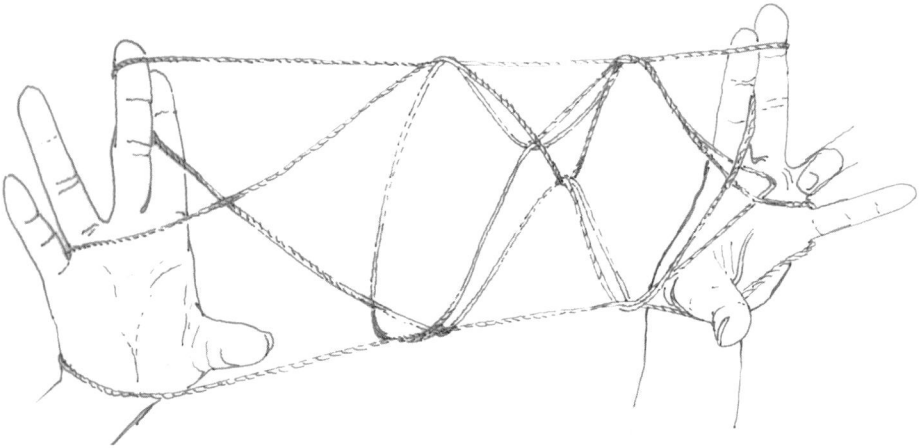

Cat's cradle string games, especially popular among foraging societies.

uses for string might have predated the emergence of *Homo*, because natural cords and vines sometimes ensnare animals without any external agency. In addition, children's curiosity (rather than adult competence) was more likely to observe varieties of trapping, luring and wrapping techniques as, say, exemplified by spider webs. Sometimes observant and imaginative children grow up to become innovators and inventors. I contend it was such innovators that invented string, way back in prehistory. String, cord and vines can help with the porterage or binding of food, tools, branches and even stones, back to camp, so it is no great stretch of the imagination to find uses for cordage catching on. A strong feeling for string, cordage and basketry is evident in all gathering, hunting and fishing communities, right up to the present. Men and women spend long hours processing, repairing and tying fibres into shelters and nets, or weaving ingenious traps. The breaking points of various fibres are not only hand-tested over and over, discussed and well known, but also tested in tug-of-war games. There are also string games played with threads over multiple fingers, games that are almost certainly among the earliest ever invented. Some of these traditions have a history of more than a million years. In my view, traps, especially cord snares, probably harvested more food in foraging societies throughout most periods of prehistory than did active hunting.

To a significant extent, the use of organic materials and artefacts is subsidiary to plain appetite. At the moment when a prey animal is killed, its independent otherness departs and it becomes a possession, a mass of diverse materials which must, in the literal sense, be rapidly sorted out by an owner or owners before they spoil. Typically, the animal's outer membrane, its skin, is transformed from an essential biological entity into a material that its new owner can either use or discard as debris.

One potential for skin, and its most immediate usefulness, is as the carrier bag in which the whole or parts of its content will be transported from the kill site to the eating, cooking or sharing site(s). Thereafter it may be processed further and find different uses, or it may be discarded. But that brief utility of the skin, as a carrier bag, illustrates some significant characteristics of the earliest artefacts.

For a start the material is right there, on the spot, just where there is a need for it. Anyone who's forgotten their bag or haversack will know how awkward it is to carry more than one or two small items, let alone the various and large masses that prehistoric people would have had to carry. The other demand is for material economy, in which weight and bulk must be at a minimum when there is a long walk back to the group or to a base. In this the anatomy and employment of human shoulders and their harnessing with shoulder-straps became important adjuncts to foraging.

Other considerations are multiple use and economy of effort. Both the pliant accommodation of fresh skin to its load and the ease with which

Noosing the head of a crossbow-shot monkey through a slit in its tail converts the prey into a ready-made carrier bag. Babu, a Baka from Cameroon (photograph by Lisa Silcock).

it can be separated from tissues commend it – a ready-made carrier bag. If an animal skin is found to be of further use, say as mats, body-covering or thongs, choice of that material over alternatives may even be the prime incentive for hunting a particular species in the first place.

Part of this choice involves specific qualities. Many subsistence hunters concentrate on certain prey species for the quality of byproducts. For example, Impala hides can be worked into soft leather that is superior to most other animal skins in combining ease of processing, strength, texture, light weight and durability. Ornamented with beads, pigment or shells, such materials can acquire aesthetic or ritual virtues.

The warm fur of an otherwise inferior prey may make it more desirable, but the utility of any tissue, animal or vegetable, depends on the forager having a sharp blade. Any material will be useless if the means of transforming it for specific purposes is not at hand. A single precondition such as a blade may be attached to still further preconditions. For example, the earliest hunting hominins probably opened up large mammals with slivers of stone, bone, tooth, shell or plants – improvising a tool from materials in the immediate vicinity.

However, in swamps or plains where few or none of these materials were at hand, hunting would have been severely constrained unless both blades and

weapons were carried in preparation for a successful hunt. Since hands must be free for most routine activities, including the hunt, taking along a blade would have required not only a tie or thong, but the blade itself would have required some sort of envelope or sheath to protect the carrier. This shows that the most elementary processing of a primary product, such as the hide of an animal, not only implies but essentially *demands* the use or development of further technology, involving further working of materials.

This dynamic helps explain how technologies become elaborated. Speeded up and mass-produced, the innovations of contemporary technology take us into a dystopia where small-scale, local niche-thieves become voracious, global monsters hiding behind bland faces. Automatic weapons in the hands of boys and repressive regimes may be a contemporary manifestation but warlord-led Ruga-ruga children once spread terror over many parts of slave-era Africa. I know, from personal experience, how young boys, with unbroken voices, can behave when given AK47s and a military role at a road-block.

All aspects of human welfare and all domesticated plants ultimately depend upon our need to understand *their* and *our* evolutionary and social history. The realisation is growing that all plants and animals, including ourselves and our domesticates, have spent significant periods of evolutionary time subject to precise ecological parameters, and confinement to quite precise geographic regions.

There are innumerable potentialities for human exploitation in the adaptations of highly specialised biota. There are also many foods and condiments that were once part of human resource use in Africa, which are now being displaced by imported or introduced, mass-produced plants and animals.

The workings of the evolutionary process tie our time and place to past and future, and tie our survival to all the animals and plants on which we depend – fruit, oil palm, rice, wheat, maize, livestock, fish and many more, every one of which owes its existence to the same overarching process of natural selection. Each species' virtues as well as its vulnerabilities derive from that evolved history.

Many outmoded and originally non-African agricultural and pastoral processes, particularly monocultures, will eventually have to be scaled down or even abandoned. In the meantime, maintaining viable mammal communities is integral to the long-term objective of ensuring that Africa develops locally relevant systems of resource use, not primitive or exotic methods as imported from or demanded by powers from other continents.

The ability to co-operate in larger groups and to diffuse knowledge is generally thought to have contributed to the long-term triumph of modern humans over other hominins. It is a process in full swing today.

From the beginning of the middle Stone Age onward, more versatile tools suggest greater flexibility and adaptability in exploiting a range of resources greater than that used by any previous species. One corollary of modern humans taking

OPPOSITE: *Clad in Impala leather, a Barabaig woman, as portrayed in 1957.*

an ever-larger variety of both traditional and new food types (as well as expanding territories) was that other types of humans, no less than other animal species, had a new intruder and had to compete for a similar range of resources.

During droughts or famines, the balance would tip still further against the survival of competitors, because 'niche-stealing' (or outright predation) effectively deprived less competitive species of their living.

A perennial resource for competition was water. So long as large herbivores were primary items in the diet of people in arid areas, killing sufficient antelopes as they came to shared water was a prudent, even sustainable, harvesting technique.

Once bovids had been domesticated, the water needs of cattle, sheep and goats became extensions of human needs, leading to the exclusion of wild animals. Thus the Sahara, a desert larger in area than the entire United States, has effectively lost all its indigenous, desert-adapted fauna, mainly because every single, scarce source of water has been appropriated for livestock and their owners. Mounted soldiers in search of target practice or fun have gunned down any residual remainders of indigenous herbivores.

I was once taken to a site not far from Nairobi that I had often passed by, where an entire herd of tame and much-loved giraffes was encircled and machine-gunned by young soldiers mounted on cross-country jeeps. The giraffes' tail hairs were taken to make bracelets – the rest rotted. Skulls within a circle of skidding tyre-tracks marked out the scene of an all-too-often repeated crime against defenceless nature.

As everywhere in Africa, it is way past time to call out the irresponsible and aggressive livestock industry, in most of its many guises. Principal agency in laying waste to our landscapes and exterminating our natural resources, a disciplined evaluation of a passing need for meat, hides and milk must take its measure against less damaging alternatives – alternatives that are already being devised or synthesised.

The moderate appetites of small family-like groups of frugal subsistence farmers, pastoralists and foragers have transmuted into the horrifying gluttony of

Giraffe
Necropolis.

their mechanised, market-oriented contemporaries. Prairie grasses on former forest lands are grazed by millions of two-legged meat-eaters, via surrogates, the chemical-doused bovids on which they gorge.

In the last forests, corporations eat trees as if they were chainsaw-voiced death-watch beetles, and it is not only on land. Trawlers, long-liners and their fishing fleets hoover up seas and lakes, blitzing their floors, like rampaging aquatic dragons.

In the immediate aftermath of World War II, while martial and quintessentially male belligerence was still the spirit of the age, too many young African leaders took their lead from Mao Zedong who, influenced by Field Marshal Stalin before him, famously announced, 'The People must conquer Nature.' Subsequently, 'development' has all too frequently envisaged and promulgated Mao's essentially military concept of War against Nature.

Early *Homo* and pre-*Homo* societies centred on concern for their youngest, weakest and least capable. Are there lessons there for today? Are there any real evolutionary, historical, even anatomical and innately behavioural lessons to be drawn from the mutable and opportunistic behaviour of niche-thieves towards fellow group members, both human and not-so?

In spite of human history being one long tale of brutality, hypocrisy and ecocide, there have been moments when more benign, even altruistic, facets of human nature, intelligence and ecological sensitivity have found expression. Among all my contemporaries and their immediate forebears I like to think of Rachel Carson and Wangari Muta Maathai as transcendent exemplars of this potential within the human heart and brain. It is no coincidence that both were female, neither ever wore uniform, and both were disparaged and censured by whatever surrounding powers-that-were. Some such powers have since reached plague proportions.

Meanwhile, none of us can escape seeing the face of a niche-thief staring back at us from the mirror.

Therein lies the rub.

Shell of a rhino, killed for its horns (bronze and chalk).

Dust jacket of Self-made Man.

19 OUT OF AFRICA AND BACK AGAIN – THE BANDA STRANDLOPERS

In which the adventures of voyagers are recounted and island foragers darken their complexions.

I once got caught in a rip current that I remember for its inversion of appearances. Suddenly the shore, but a few score metres away, was streaking past me – not vice versa. After a thrashing struggle across the mighty force of the current I broke past an invisible boundary between deep, hurtling water and still shallows. Just short of the point where the current would have taken me out to an open sea of sharks, I got to shore and lay on oh-so-beloved sand, panting like a race-horse. Born many hundreds of kilometres from the sea, always ill at ease with the ocean and its fearsome depths, I got up, grateful to be on solid ground and intensely aware of being a true land-lubber.

My fright was a reminder that primates and people are intrinsically terrestrial, not aquatic animals. Yet if we go far enough back, things went the other way. Primitive aquatic vertebrates evolved into the amphibian ancestors of reptiles, birds and mammals.

That this great divide was crossable both ways is proved by a more recently evolved abundance of sea snakes, penguins, sea-cows, whales and seals. All of these animals have had to evolve enormous modifications to their previous land-lubbing anatomy and physiology. It was access to immense riches out beyond the land/sea boundary that was their evolutionary prize.

I have seen monkeys plunge into shallow waters, in the case of Crab-eating Macaques diving down among Bornean mangroves where the bait was sunken fruit, molluscs and, yes, crabs.

Why have primates not re-crossed that dangerous divide between land and sea?

This rhetorical question is interesting because natural selection can be at its strongest along margins or boundaries between major habitats. Any animal pushed up against a big boundary is tested by perils external to the animal, like a scorching sun or that awesome current.

Our answer, as humans, has been rafts and boats. For an animal with continental forest origins, hand-made water craft are the pathway to new lands and whole archipelagos of islands.

For this very pedestrian primate, that rip current marked the outermost boundary of his range, but just as significant were the internal limitations of a very terrestrial mammal's anatomy and physiology – I am no seal!

At about the time of my encounter with that terrifying current, my personal physiological limits got branded onto my skin. I had joined a gaggle of children (and a mother or two) to forage through the shallows of a cove in New Guinea. Neap tide was at midday so I got sun-scorched, notably beneath my chin, brows and under-arms. These usually sheltered bits of my anatomy were hit by radiation reflected upwards off the sea, sand and coral *below* me. My Papuan companions were totally inured to much longer and repeated exposure, regardless of whatever time of the day or night the tide was out. Low tide was the only time when abundant sea foods could be easily reached. Apart from their hands, sharp, steel-tipped fish-spears (cadged off tourist divers) were the children's main foraging implement. In their heedless, day-to-day good cheer the children expressed an unremarked conquest of what, for me, was an invisible but dangerous foe. Was the blithe spirit of my companions the legacy of long lines of ancestors subjected to and selected by solar super-stress? As the tide turned, we waded back ashore and I was equally impressed by how much fodder bulged the hand-woven dilly bags that hung from youthful shoulders.

Then I met Betty Meehan and Steve Davis, anthropologists who had documented the annual and daily cycles of major foods as foraged by local people along the shores of the Arafura Sea. They found that molluscs were a year-round staple and an average harvest of 11.5 kg of clam shells per forager yielded 2.4 kg of highly nutritious flesh – and more than three times as much when the women waged a competitive challenge! Betty emphasised that the tropical shorelines of Southeast Asia and Australasia offer one of the richest and most productive habitats ever exploited by humans. She raised the distinct possibility that the tropical shores of the Pacific and Indian Oceans sustained what might have been, for a while, the largest and most extensive human population and economy in the world.

Fishermen along the reefed coasts of East Africa were as welcoming to me as a small boy as Papuans were to me as an adult, but there was one very significant difference. Papuans belong to a discrete population of seafaring Melanesians who inhabit large parts of an immense Pacific Ocean realm of islands – literally, millions of kilometres of shorelines, all out beyond the edges of mainly non-Melanesian Asia. The far fewer fishermen of eastern Africa pursue a minor occupation, restricted to the narrow margins of a proletariat that inhabits an entire continent.

Where the fishermen of Papua and East Africa converged was in their inherited dark skins and hair with flat shafts (and therefore spiralling, 'curly' and a better insulator than the round-shafted, more primitive, chimp-like hair of other peoples). In both instances the local kids' careless indifference to what I had learned to call UVR (or ultraviolet radiation) made a compelling contrast with my own maladaptation to those piercing rays.

Long ago, medics discovered that both too much and too little UVR induces disorders of metabolism, enough to kill off kids and/or mothers, in the past or present. They showed that thinning or thickening melanin layers in the skin correlates with vitamin D metabolism. Too little UVR starves the body of vitamin D, leading to rickets, weakened bones and muscles, so depigmentation was adaptive for prehistoric people trying to survive sunless northern winters. More melanin shields essential metabolic processes from being burned up by ultraviolet radiation, and nothing exposes people to the sun more than tropical beach-combing. The contrast between my own immediate and discomfiting sunburn and my companions' *sang-froid* served to emphasise where their advantage lay.

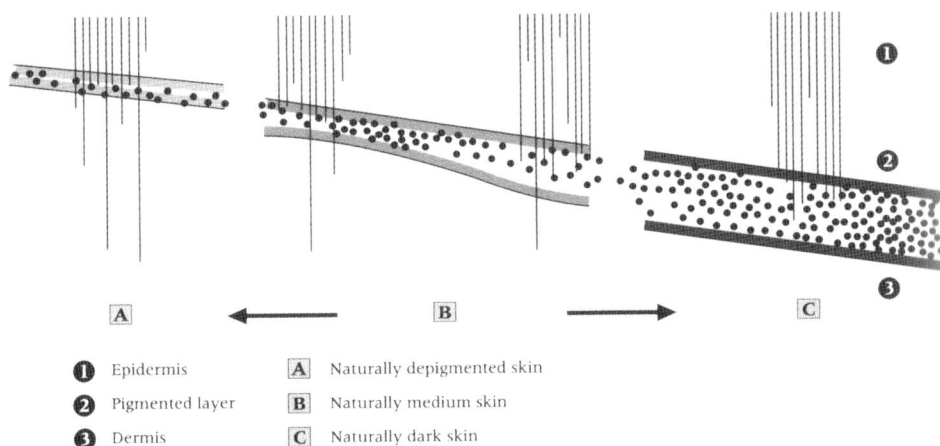

❶ Epidermis	Ⓐ Naturally depigmented skin
❷ Pigmented layer	Ⓑ Naturally medium skin
❸ Dermis	Ⓒ Naturally dark skin

The skin and ultraviolet radiation. Both the epidermis and the pigmented layer may be thick or thin. Some ultraviolet rays may fail to penetrate the pigmented layer.

The contrast between my skin tint and that of my Papuan companions was but one of several less obvious markers, but consider the physiology involved.

Human skins vary in thickness and pigmentation (mine is thinning as I age) and vary in how bits of melanin are packaged between epidermis and dermis – excuse the clinical language. The result is that most people are somewhat pigmented ('mid-mels') and darken in summer and turn paler in winter, while very sizeable minorities are super-pigmented ('super-mels') or, at the opposite extreme, have de-pigmented skin, hair and eyes ('de-mels', like me). There are even three-person, three-mel families and I can think of one particularly eminent mid-mel whose super-mel father was a fellow East African, his mother a de-mel Anglo. In any event, every wedding of super-mel and de-mel augments the founding mid-mels, who will always be the numerical majority.

In common with many of my generation I was taught that differences in skin colour simply reflected a relatively recent, rapidly acquired and generalised evolutionary response to ultraviolet radiation. Quite apart from the hopeful logic of that correlation, there are political attractions in supposing a vaguely egalitarian adaptation to sunshine.

However, when Papuan and African genes were compared, research has been unambiguous in linking Africans *by descent* with Melanesians and Australian

Aborigines. The easiest explanation has always envisaged early Africans migrating eastwards around the coasts of Asia.

I first questioned this supposition and a simple short-term sun-skin-colour correlation in a book entitled *Self-made Man* (1993), pointing out that the conditions for acquiring super-melanic skins *as a response to UVR super-stress* are obvious in the southern Pacific and eastern Indian Ocean – equivalent conditions cannot be found anywhere in mainland Africa. Determining which way prehistoric people flowed between Melanesia and Africa (and the timing) has important implications, among others for the history, knowledge and control of diseases. This and its logic have helped maintain my interest in the topic.

The African landmass offers no archipelagos, no potential for the isolation of a distinct fisher-folk gene pool – in this respect our continent is emphatically unlike the Pacific islands.

Flying foxes (Pteropus *spp.) – tropical seaside bats.*

The situation reminded me of flying foxes. Before mass-market tropical fruit-growers, nets and shotguns converged to usurp their habitats, these giant seaside fruit bats once existed in great 'camps' all over southeastern Asia and along tropical Pacific shorelines. Of some 60 species, three or four have reached offshore East Africa, where suitable conditions are confined to islands such as Pemba, Mafia, the Comoros and the Seychelles. Flying foxes have never been seen on the mainland (which bursts with other species of fruit bats) and flying foxes could only have evolved in southeastern Asia and its offshore islands, never in Africa.

An interesting detail of their behaviour is the habit of 'dipping'. Flying down to the sea's surface, they scoop up a part-patagium of water, and douse themselves with sea-spray. The benefits to the bats of this dipping await further study.

This is not just a fancy digression. A little-known and clinical aspect of my diversion into bat biology is that flying foxes not only host their own malarial parasites. In deep evolutionary time they are thought to have been among the very first mammals to host malaria after its earlier parasitising of birds and reptiles, but the bats soon passed on malaria to rodents, then primates and eventually humans. This story is only just unfolding, as a detail of global human travels and the researches of contemporary epidemiology.

In resolving the tropical travels of early modern humans, we can hope to learn something of their susceptibility to diseases and parasites (even observe the taste, among some, for flying fox stew).

It is not just a matter of tracing whether darkening skins travelled westwards or eastwards. Wherever our equatorial travellers set off from, their genetic baggage must have included susceptibilities and resistances that are as relevant for epidemiology today as they were for those now ancient travellers. The African/Melanesian/Aboriginal complex has evolved several adaptations to malaria, as have flying foxes. We don't yet know where, when or how these assets were acquired.

Self-made Man began with the Darwinian premise that all animals adapt to circumstances and circumstances change: 'For humans, by definition tool-making animals, "circumstances" have become more and more self-made.'

Knowing that I owed my life to quinine, sulfa drugs, mosquito nets and lots of loving care, it was easy to accept the idea that my human family and countless tool-making predecessors had created the environments within which their descendants, including me, had to adapt or contrive to survive.

With memories of Papuan reefs, the origins of skin pigments were among the questions I raised. Where did natural selection, in the shape of UVR, join diseases as a significant culler? Where did the extremes of super-pigmentation and drastic de-pigmentation arise? What sort of circumstances underlay such adaptations?

These questions are important because for ever afterwards, survivors would have retained genetic advantages over less resistant lineages. Throughout history, the superficialities of appearance, so-called 'racial' physiognomies, have also marked out opposing warriors, from Ancient Egyptian armies to urban gangs today.

In the deeper perspectives of *Self-made man*, I proposed the propagation and diffusion of ever-better resistance to numerous lowland, humid-adapted diseases or parasites, even whole suites of genetic advantage.

Did a sun-blasted, seaside subsistence economy, such as that of my Papuan companions, explain super-melanic skins? Did deeply black skins, flat-shafted, spiral hair and tropical disease resistance all get selected for under the super-stress of being *forced* (by tides) to forage under intense solar radiation? How could this oceanic distribution of super-melanic skins be explained when the inhabitants of mainlands nearest to the islands of Melanesia have pale tawny skins capped by round-shafted straight hair? These skin and hair types are common among primates, and are the most likely combination among the earliest modern humans.

Self-made Man marshalled the slender threads of evidence available during the 1980s for a chronicle that traced an overland spread out of Africa and across

Asia by newly distinct 'anatomically modern humans' reaching as far as the Pacific. This was in the context of a still earlier diaspora. The early humans I nickname 'erects' had populated Eurasia for about 2 million years, a range so vast that identifiably distinct regional populations evolved.

Even so, back in the 1980s few of us imagined tracing our more obviously human adaptations much further back than 'anatomically modern humans'. We fancied our novice race as a species of such unprecedented significance in the history of life on Earth that pre-moderns and other ancestral lineages got relegated to lesser limbos of interest or consciousness. In *Self-made Man* I effectively ignored the very numerous, pre-modern but wholly human populations that preceded us. Yet they had occupied the length and breadth of Eurasia many times longer than 'moderns' have been around. These 'archaics' (the earliest, longest lasting and most frequently fossilised being 'erects') tamed fire and almost certainly invented sophisticated technologies and tools such as axes, spears, nets, traps (and possibly rafts and/or boats). Considering the fact that Eurasia had hosted *Homo* for some 2 million years, the Afro-centric negligence of *Self-made Man* was more than remiss.

We now know that, along the way, these pioneers encountered well-established and widespread pre-modern *but emphatically human* precursors.

During the several decades since that book was published we have been forced to acknowledge that our particular branch of 'anatomically modern humans', *Homo sapiens* or 'moderns' have much deeper, more diverse and thoroughly mixed genealogies than we could imagine in the 1980s – we are all fertile mules.

The prime mover in this seismic shift has been the brilliant founder of ancient DNA science, Svante Paabo, and his teams. They have extracted DNA and analysed the genomes of several archaic populations, known collectively as Neandersovans, a widely scattered profusion of archaic Eurasians who might have separated from the Africa-based ancestors they once shared with Mods as much as a million years ago.

The Old World had been inhabited and explored by humans that long ago *before* 'moderns' began to filter out of Africa some 200,000 years ago with remains of 'modern people' turning up as far east as China (Fuyan, Luna and Zhirendong) and as long ago as 70,000 to 120,000 years. As these Africans invaded the territories of diverse and scattered Neandersovans, these mutually fertile peoples coupled often enough for natural selection to operate on most of their progeny, sometimes (perhaps often?) favouring mixed over unmixed strains. All originally foragers, the very numerous 'archaics' *that already inhabited* the far east mainland were too similar to the incoming 'moderns' in their physiques and their economies NOT to have interacted and interbred with them. Paabo's teams have revealed that 'archaic genes' hide within the genomes of most, if not all, of our contemporaries. Why? Because those early, so-called 'hybrid' kids got a share in the legacy of hundreds of thousands of years of adaptation to uniquely Eurasian conditions.

A nice example occurs in Tibet, where it seems that specifically Neandersovan genes help protect Tibetans from the mountain sickness that so afflicts or kills foreign visitors and, much more significantly, can tragically compromise pregnant women. The 'Roof of the World' or its vicinity was likely occupied by 'archaics' long before 'moderns' appeared.

Further south, two Denisovan genes (IL4 and CDH13) confer some resistance to malaria (an advantage similar to that shared by many birds and bats). That should induce some grateful respect for such pre-modern forebears.

Before pursuing human travels in the tropics, return to Paabo's earliest discovery of Neanderthal genes in virtually all contemporary western Eurasians. Once this was revealed the world was needlessly shocked to learn that archaic genes survived in living people – after all, we already knew that much of our genome is shared with chimps and shrimps (even one or two alleles with bananas).

Paabo heard from several women who condescended to inform him that *they had always known* that their husbands were Neanderthals. (I even heard one lady likening her hubby to an ape while another called her spouse 'one big banana', so some strange sorts of evolutionary message seemed to be getting through.)

If the genomes of other 'archaics' are ever recovered, there is the distinct possibility that we will find that contemporary, living humans walk about our streets with still more archaic genes to account for, including 'erect' and 'handy man' ones.

With the excavation of long-lost ancestors and extraction of their genes we glimpse some of the ways in which natural selection has gifted us our very existence. But temper this with the thought that natural selection, under whatever manifestation of chance, can just as well snuff us all out.

Paabo and his teams have reconstructed the entire genome of the big-bodied, big-brained, big-mouthed, talkative people we call Neanderthals. In being shown that Neander genes persist in virtually all western Eurasians, we are forced to change the way we think about ourselves. Once regarded as late upstarts out of Africa, northern Eurasians suddenly find some of their regional roots going deeper than was ever expected.

Over some 300,000 years Neanders survived three Ice Ages in western Eurasia. They witnessed the comings and goings of vast glaciers around the Baltic region and, like bears, acquired genes and physiques backed up by techniques that rendered them capable of surviving such terrible ordeals by climate.

Many of the Neander genes inherited by Eurasians can now be shown to improve the ability of these 'hybrids' to weather the cold and other afflictions of northern winters. Genes, we are also told, render their descendants prone to diabetes and psychological depression – so winter blues may have started as Neander blues?

Paabo's revelation that all western Eurasians are 'Modern'/Neanderthal hybrids was followed by his teams' discovery that the Neanderthals' far-eastern cousins, Denisovans, occupied mainly warmer climes where they also interbred

with incoming 'Moderns'. The team went on to determine that male Neandersovans were the main fathers while mothers were predominantly 'moderns',
even reckoning that female offspring enjoyed some hybrid vigour whereas their
hybrid brothers were *less* viable than their sisters!

None of this genetic prehistory was known in 1992. Today, the highest measures of Denisovan genes in any modern peoples occur among Melanesians (up to
6 per cent) followed by Australian Aborigines and some of the super-pigmented
Asian peoples once lumped as 'negritos'.

The patchy survival of Neandersovan genes implies an equally patchy pattern
of comings and goings in which many, perhaps most, Asians descend from later
overland expansions. All the earliest peoples were foragers but the extent and
prevalence of shoreline foraging before 'moderns' arrived remains unknown.

LEFT: *Andaman Islanders.* RIGHT: *Semang, Batek and Aeta, the original southeast Asians.*

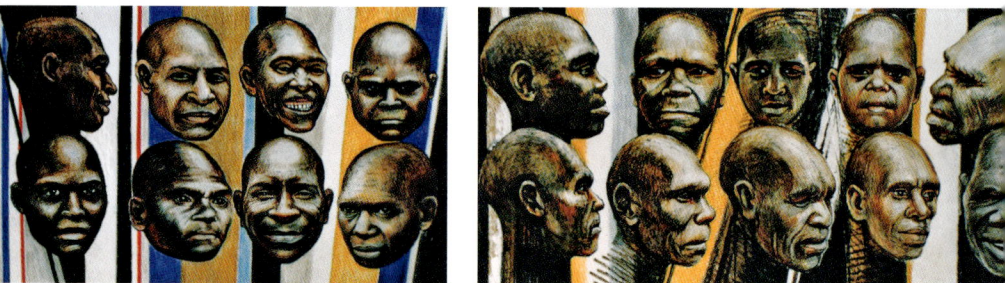

LEFT: *Papuans.* RIGHT: *Aboriginal Australians.*

From an evolutionary perspective the crucial artefacts were canoes or rafts
because the isolation such craft engendered would have allowed island colonists to develop some island-specific adaptations. Water craft would also permit
them to spread over whole archipelagos of islands.

At present we cannot know whether pre-modern genes in Melanesians included adaptations to shoreline living. Genes for both darkening and lightening
skin colour are common to Neandersovans and Mods so the timing for changing

Fishing raft, Coramandel coast, India (photograph by Belinda Breeden).

skin colour also remains unknown. All we can be certain about is that the outer islands of south eastern Asia and Australia were reached by raft or boat at the very latest by some 65,000 years ago. We also know the obvious – that stretches of sea separate super-pigmented Melanesians and Australians from originally mainland-restricted, less pigmented Asians.

In searching for a name to describe prehistoric shoreline foragers I adopted the Dutch term 'strandloper', adding the name of the sea at the vortex of Australia's bruising northward glide past island Indonesia – the Banda Sea – an oceanic whirlpool that is surrounded by and peppered with islands as well as churning them. These islands, I surmised, were originally colonised by newly mobile shoreline foragers – 'Banda Strandlopers'.

With the invention and development of water craft a huge extension of habitat for humans opened up.

Something of Banda material culture was explored in *Self-made Man* – they developed an economy that used a wide range of small items intensively. Where continental mega-hunters only existed at moderate densities in the best game country (in part because they and their prey were frequently nomadic), Banda economies could sustain larger communities at appreciably higher densities, with the full participation of the entire community in foraging, all subject to much less movement than their continental cousins. This detail would have favoured the exploitation of settlement 'weeds', particularly the germination of spat-out or excreted seeds on nearby middens, giving birth thereby to agriculture.

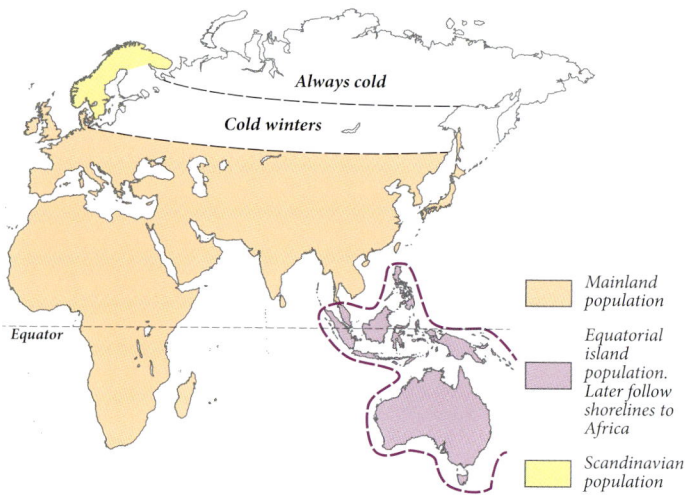

Always cold

Cold winters

Equator

Mainland
population

Equatorial
island
population.
Later follow
shorelines to
Africa

Scandinavian
population

The isolated 'extremities' of Homo sapiens' *earliest dispersions northwest and southeast.*

For people who had developed rafts and boats, there were no physical barriers to their settlement of a plenitude of offshore islands and also of virtually the entire seaboard of continental Asia and, eventually, the Indian Ocean shores of Africa.

Self-made Man outlined these modest beginnings among Banda Strandlopers, exploring how combined genetic and technological adaptations could have opened up one of the most extensive areas of settlement to be found anywhere in the prehistoric world. Because that realm was so extensive and fragmented, the Banda would have become ever more diverse but their isolation would have been somewhat offset by the mobility that water craft gave them.

To get to Africa the Banda had to skirt India, so I searched for archaeological data in the subcontinent. In the absence of any other clues, stone tools are often the only indication of human presence in south Asia's past. This is particularly

Fisher boy among mangroves.

true of India, where human fossils are extraordinarily rare but stone tools are very abundant. In Chennai, southern India, substantial numbers of stone tools, previously thought to be the handiwork of modern *Homo sapiens*, are now dated to 250,000–385,000 years. In the absence of human remains we can only assume the makers were Denisovans. This can be confusing because the founders of archaeology made scant allowance for time-warp convergences (and even abandonments) in tool-making techniques. Perhaps a few creative and skilled tool-makers in prehistory might have played havoc with those contemporary classifications that are based upon over-tidy progressions?

As for anthropological sources, there are two marginalised peoples, the Vedda and the Andaman islanders, who hint at ancient marine connections between India, Australia and Africa. The Vedda closely resemble some Australian Aborigines while Andaman islanders hint at more distant connections with both Asia and Africa, but both supposed associations remain just that – hints.

A central theme and calendar of *Self-made Man* was immediately challenged by a school of scientists that were persuaded that the colonisation of Eurasia and Oceania by 'moderns' was as recent as 60,000 years ago. A reason for their revision of dates is clear enough.

Some 73,500 years ago a volcano, Toba, erupted in Sumatra, devastating very extensive areas of the far-eastern mainland, extensively decimating human populations. It now seems likely that late excursions out of northeast Africa or its vicinity found broad tracts of Eurasia to be *terra nulla*, open to colonisation. So the Toba eruption may turn out to be a central event in determining the ever-changing dispersion and distribution of contemporary human populations.

For other animals, the most likely outcome of Toba was a complex patchwork of annihilations and survivals. Over time this scatter of nuclear populations was further fragmented by fluctuations in sea level that transformed former

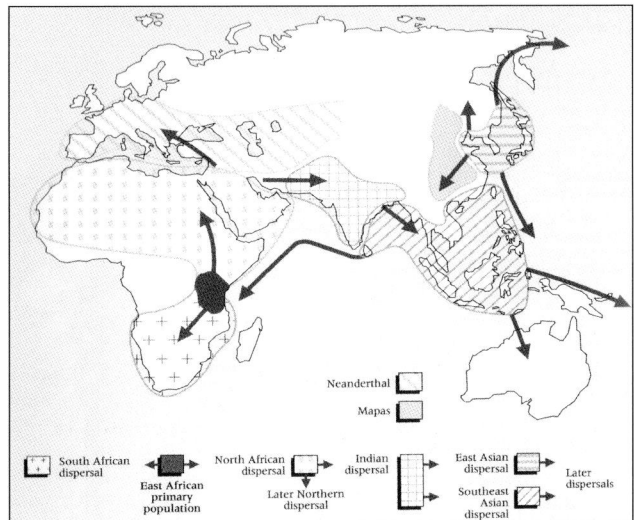

Primary and secondary dispersals by modern humans.

islanders into mainlanders and vice versa. In the aftermath of exploding volcanoes, through ups and downs of sea level, beneath big swings in climates and in our ancestors' newfound mobility over land and water, we must extract the deeply puzzling story of our contemporary existence and the distribution of distinctive peoples, especially all over the Indian and Pacific oceanic regions.

Consider now the diverse nature of the coasts that those hypothetical Bandas would have encountered, especially after Toba.

Wherever there were hinterland hills or mountains sufficiently high to catch rain, mist or dew from the Indian Ocean, springs, streams or rivers were generated, all potential foci for coastal settlement. Even barren coasts held temptations; for example both the South Arabian and Somali coasts are still major turtle hatcheries and the beaches abound with molluscs, crabs and various seasonal gluts of marine small fry. Offshore are some of the richest oceanic fishing grounds in the region.

The resources of the African coast are finite. Supposing that the habitable foreshore had become settled by Banda Strandlopers along its entire length, the natural course of population expansion would have been up the larger rivers. Here, fish and reptiles would have been the main aquatic staples while waterside animals and plants provided the rest of the diet.

Following the obvious attractions of harvesting sessile molluscs and crustaceans by the earliest Strandlopers, their next challenge was to devise the means of hunting fish. Apart from trapping or building fish-weirs, any active hunting of water animals poses quite peculiar problems, starting with aquatic prey being sought by terrestrial foragers where the water is at their feet, while for fish the surface is their ceiling. This is a problem that was faced and solved by penguins, herons, kingfishers and other piscivorous birds from the very start of their many and separate evolutionary trajectories. It would seem a small step to imagine humans becoming waterside hunters and inventing beak-like projectiles to spear fish. It is also quite likely that human communities developed such skills and weapons at both inland and seaside sites and at different times.

This is a scenario that poses problems for some archaeologists because they envisage the hafting of harpoons, spears, arrow heads and axes to their respective shafts or handles as signifying an advance in cognition – humans suddenly getting cleverer. Instead, I contend that something more like mimicry was involved. Large piscivorous birds such as herons spear their prey by uncoiling their long, arm-like necks to hurl the lance of their beak at alert, fast, bone- and scale-protected prey. The birds' eyes and brains also have to compensate for visual refraction which bends the angle of the stab as it passes through air, then water.

Today most Banda living sites must be deep under water, but it is more than likely that hafted spears and harpoons were near essential parts of their equipment. If so, hafting looks less like some momentous cognitive emanation and more like an ecologically determined innovation in which human harpooning got to out-heron herons.

Whatever professional perspectives we bring to topics such as hafting projectiles or stabbing spears, their origins could have one or more provenances, possibly scattered through time but always ecological in origin.

How the hypothetical Banda might have interacted with pre-existent 'moderns' is unknown (as is their very existence!) but we can be sure that genetic mixing would have been frequent. On the coast, plants were likely to have played a subsidiary role in diets. Upriver, where the aquatic habitat narrowed and biological diversity increased, emphasis would have shifted so that plants, especially those growing close to riverbanks, could have been eaten more frequently.

If my saga of people flowing out of Africa and then finding their way back again is to be taken seriously, that Asiatic excursion should find some sort of genetic confirmation.

In pioneer surveys of the Y (male) chromosome the great geneticist and humanist, Mike Hammer, and his colleagues discovered that many contemporary African men have polymorphisms that seem to derive from a far eastern ancestry. Did these alleles hitch-hike a ride on those seashore-foraging Banda Strandlopers? If so, here is further support for a westward migration from a Pacific centre, or centres, of origin.

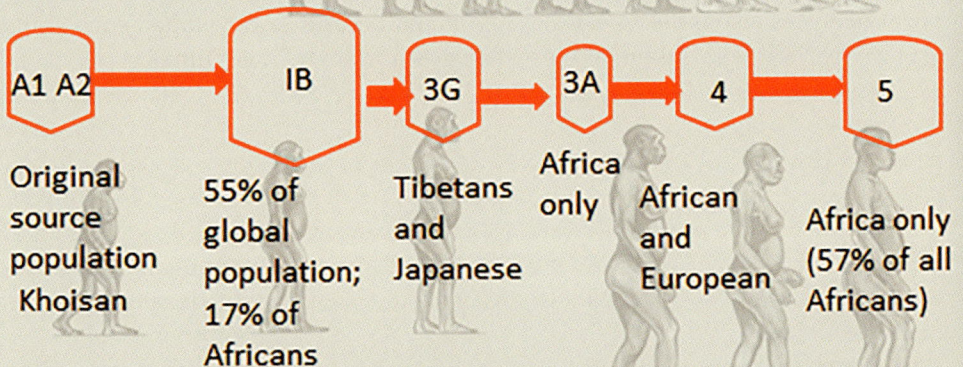

Genetic Evidence for the immigration of Banda Strandlopers to Africa

Polymorphisms on the human Y chromosome suggest African origins (ie. A1 and A2) followed by excursions to Eurasia, followed by a return to Africa (3A, 4, 5).

A1 A2 → IB → 3G → 3A → 4 → 5

A1 A2 — Original source population Khoisan

IB — 55% of global population; 17% of Africans

3G — Tibetans and Japanese

3A — Africa only

4 — African and European

5 — Africa only (57% of all Africans)

M. Hammer et al. 1998 Out of Africa and Back Again: Nested cladistics analysis of human Y chromosome variation. Mol. Biol. & Evol. 15(4):427-441

Polymorphism on the human Y chromosome.

It is already clear that part of the overall diversity derives from the passage of time – genetic drift. Other sources of genetic diversity in Africa include infusions from returning diaspora.

My suggestion also requires, among whatever other proofs emerge, archaeological evidence for both a time and a land-fall for such an arrival – so consider the options.

The most inviting and largest of all east-coast rivers is the Zambezi. To its immediate south is the Limpopo. Between them are the remains of a culture that was still called 'Charaman' in the 1980s. Here archaeologists had recorded a sudden change in the range and diversity of tools. With whatever dating techniques available at the time they tentatively dated the Charaman to between 50,000 and 40,000 years ago – just about the time I hypothesised that Banda Strandlopers might have arrived.

Among the primary tools used by Banda Strandlopers would have been fishing spears and barbed harpoons. To pierce all sorts of scaly, shelled, tough-skinned, often muscular prey, very hard spear points were required, and they needed to be made of material harder than wood, such as shell, bone, ivory or stone. This carefully prepared point could be 'hafted' or bound into its prepared shaft by means of a fine super-strong cord bedded in tar or resin.

Precisely such tools, together with beads, barbed bone points, chunks of ochre pigment and beautiful bi-faced arrow and spear-like tools, mark an abrupt change in Charaman sites in the vicinity of the Zambezi and Limpopo deltas.

Some authorities regard the terms in use when I was writing in 1990 as obsolete. Nonetheless, the localised Charaman and much more extensive and likely later 'Lupemban' represent real changes in stone tools that I first learned about from Sonia Cole who, with Merrick Posnanski, had excavated at a rock shelter-cum-stone-tool-factory at Nsongezi, on the Uganda/Tanganyika border in 1964. There they found, with rare clarity, advanced Lupemban quartz tools unambiguously layered above much older, cruder Sangoan workings.

Just such a dispersion of contrasting Middle Stone Age cultures has been found across very large areas of Africa. The older, more arid-adapted (called 'Stillbay' or 'Sangoan' in the 1980s) with cruder, mostly heavier tools, stretched from the Horn of Africa and Ethiopia to the Cape, mostly over the healthy uplands (vestiges of this culture and its tools were still viable among the Khoisan when Europeans first invaded the Cape).

Archaeology is made horribly difficult in Africa by too many churners and dispersers of our soils. Burrowing animals, from swarming termites to Aardvarks, powerful, writhing roots of figs and other trees, landslides and wash-aways and untidy penetrations get to mess up prime living sites. Because tropical life seldom has much need for winter retreats few tools originate from caves with clearly layered floors.

Refinements in stone technology first manifest with the Charaman (which I posit had Banda Strandloper origins) seem to have spread up the Zambezi and on into south-central Africa and the Congo basin where similar 'bi-face' flaking and chipping techniques were refined to a very high degree. New composite technologies developed small quartz adzes, chisels, even stone planes and scrapers. One Lupemban technique used a punch to flake off long sharp blades of stone which were then retouched with further very fine and controlled chipping. Their long, pointed stone spearheads are among the finest stone tools ever made – anywhere.

Although the most conspicuous features of these innovative techniques involved smaller more precisely shaped quartz tools, double-ended stone picks, stone-headed clubs and heavier pointed picks suggest sedentary living, more frequent digging and more shaping of wood. Although dates for the 'Lupemban' remain uncertain, this was predominantly a tropical, riverine, forest or forest-edge culture that can be imagined insinuating its way along most river valleys and lakeshores.

It is not just that descendants of the Banda imported their seaside techniques deep into that evolutionary whirlpool we call the Congo basin. I contend that they may well have faced the many hazards of tropical lowland diseases with new levels of resistance, even immunity. Being skilled harpooners was just one of many attributes of these 'returning Africans'.

Areas west and south of the Limpopo remained essentially Khoisan territory long after the Banda are supposed to have arrived and there are indications that much of upland eastern Africa was also inhabited by Khoisan-like peoples long after the hypothetical Banda had arrived and spread inland. Among my Sandawe acquaintances I once met an elderly lady with the most beautiful, most logical and most symmetrical pattern of facial wrinkling I ever saw. I imagined her resembling some Pleistocene ancestor from my own lineage – she even had me visualising her as a model for some sort of universal grandmother.

Newer, much more diverse, industries eventually came to dominate all the moister regions of southeastern, central and westerly zones, where higher rainfall generated an abundance of rivers and more heavily wooded or forested country.

Implicit in this equatorial dispersion is the likelihood that such people were resistant to at least some of tropical lowland Africa's numerous diseases and maladies.

There are enough deadly diseases, such as Ebola and a long list of cryptic viruses, bacteria and other microbes, to have periodically depopulated entire equatorial forest regions. Nonetheless, if Banda genes came to dominate extensive areas of the equatorial forest belt it is almost inconceivable that some localities were not already occupied by precursors, genetically modern or archaic. In any event, if the Banda prove to have been a real presence in Africa, with the advantages I have suggested, their descendants would have become sufficiently numerous to have interacted and interbred with most of the peoples they came into touch with. After all, their cousinship was not all that much removed.

Nonetheless some, such as the southern Khoisan (Bushmen), were probably more insubordinate than others. As the primary source for the most ancient and

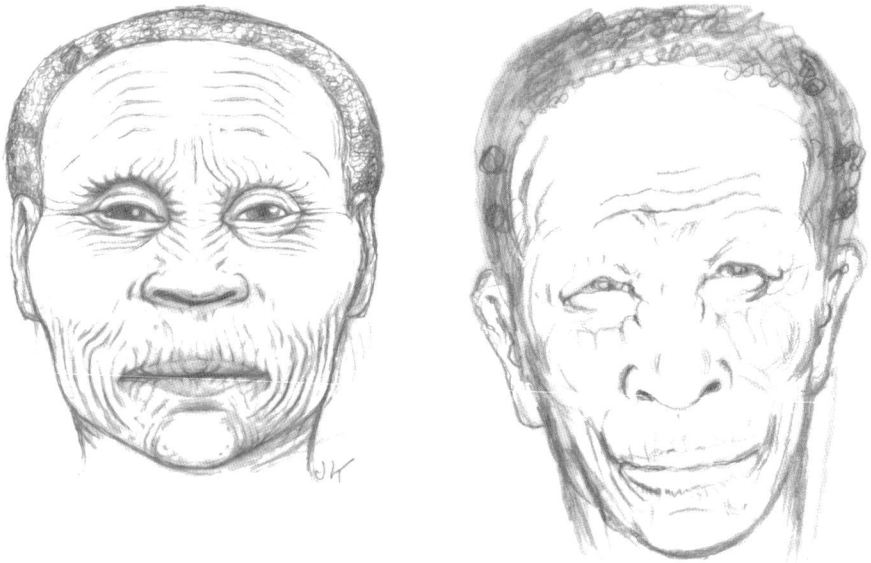

Elderly Sandawe woman from East Africa, and elderly San woman from South Africa.

most durable of genes, they have been consistent in differing from most equatorial Africans in remaining hunter-gatherers that repelled agricultural invaders and conserved more of the earliest *Homo sapiens* genome and habitats right into recent history. Some eventually became shepherds, some even got to nanny the kids of invading farmers. Their direct descendants, such as pan-Africanist and musician Miriam Makeba, President Madiba Mandela, Bishop Desmond Tutu and many others, will continue to assert the presence of our most ancient ancestry on local as well as world stages.

Today the dominant language group in sub-Saharan Africa consists of Bantu speakers. The emergence and subsequent spread of these clever, iron- and steel-forging, agricultural cultures from their Bight of Benin heartland was as recent as 5,000 years ago and their dominance in eastern and southern Africa could be as recent as 1,000 years. Bantu speakers may well have large measures of Banda and Lupemban genes in their ancestry but, as with most Africans, here are the reproductive outcomes of men and women crossing ethnic and cultural boundaries over all those thousands of years. Today, Africans are the most varied and the most anciently diverse people on Earth, and attempts to unravel the genetic complexities of our people will probably occupy generations of geneticists to come.

It is, perhaps, a futile task to try to untangle such origins but some useful, especially medical insights may well emerge in the continent that has been both the cradle of our race and an ever-renewed melting pot of genes.

East Africa was probably the world's earliest such melting pot, as wave after wave of mobile people came in over some 50,000 years or more, mostly to stay on in this unusually hospitable region of the world.

Today, Americans are undisputed stirrers of the human genetic pot and it was an early visit to the United States that precipitated my embarking on a series of 14 life-size sculptures which explored the external, easily perceptible outcomes of churning modern human genes – I sought to portray *jamaa ya binadamu*, the Human Family, as the subject of my group portrait. Pretending that a single gene could code for a single property, like shape and size of nose, ear or forehead (in reality single genes cannot be that expansive), I plotted the conjugations arising from just four sets of genetic variations to create my wholly synthetic family.

Turning to Greek mythology for names, Gaia (or Rhea) and Kronos became the founding pair while Oceana and Themis brought in two sets of ten different genes to mix with those of Gaia and Kronos via their two sons, Victor and Hippios.

Different skin colours were among just four sets of variables. Gaia was 'mid-mel', a shade of beige. Kronos was a 'red-skin' while Oceana was deeply pigmented or 'super-mel'. By contrast, Themis was de-pigmented or 'de-mel'.

Eight offspring (all given Greek names) were the outcome of all this conjugation of facial and other features.

Understanding the more conspicuous signs of diversity and difference still represents a major challenge because ignorant non-Africans have so grievously misrepresented and mistreated Africans – peoples so much more complex and often more creative than their too-often thieving/deceiving detractors.

I began this chapter with my visits to Australia and Papua New Guinea. Among the many friends I made were people from the remotest of islands within the region called Melanesia. The very name reveals the currencies and vocabularies of those newly mobile, rapacious pirate sailors who webbed the world with big-ship sea-lanes. For the most part ignorant, often contemptuous of other ways of living, their points of reference were so few that entire regions acquired names derived

Jonathan and some of his 'family'.

Gaia and Kronos and their progeny.

from the sailors' preoccupation with the pigmentation of local peoples' skins – Niger, Nigeria, Zanzibar ('the Negro Coast' in Persian), Asmara ('brown-skin-place' in Arabic) and Ethiopia ('where faces look burned-brown' in Greek).

Lamenting history is all very well but finding ourselves newly linked up into a *jamaa*, a family, that shares a small planet calls for our curiosity to be guided by the best of science accompanied by a lot more sibling empathy for apparent strangers.

New possibilities open with the implication that measures of melanin in our skins might reflect little more than what profession our more distant prehistoric

East Africans.

ancestors were born into. De-mels had hunting forebears, among them Nean-ders, that helped drive woolly mammoths and rhinos to extinction along the shores of the Baltic, while super-mels had seashore Banda Strandlopers or Den-isovans among their ancestors. One answer to Louis Armstrong's plaintive song, 'Why did I have to be so black and blue?' might rhyme with ''cos granddaddy caught a fish or two'.

I suspect Saidi might have found a light-hearted Swahili way of going molec-ular with a garbled 'hey, didn't I say we super-mel Nyamwezi used to be fish-ermen?' or 'yes, yes! Me, I'm a super-mel! But what's a zebra? A super-mel? A super-dee? Or a de-super-de-super-de-super-er-er-er – what?'

Subsistence in the Sudd.

Fovea – ceramic shards as metaphor for visual neurons (shattered dish created with Marino Moretti, Maestro di Ceramica).

TRANSLATING NATURE

*In which the felicities and perils of representation are explored.
VNOs and sex. Senses as channels, with a personal emphasis on
vision. The lives and appearances of the beholden, shaped by the
eyes of their beholders.*

The stallion was cantering round and round a cluster of six or seven zebras as
they watched his noisy, braying pursuit of a frightened-looking mare. Every now
and then he threw his head up in the air, curling back his upper lip in a gesture
that exposed all his protruding teeth. Teddy, who had grown up among horses,
said, 'She's in heat and he's in ecstasy!'

Ecstasy. *A zebra performing flehmen.*

Was he in ecstasy over the scent of the young mare? If so, his nostrils should
have been flaring. Instead, his snarling gesture effectively sealed off his nostrils.
No, the mare's vaginal secretions were being assayed by a hidden sense organ,
the VNO (vomeronasal organ). For that zebra stallion this could be said to be
his sixth, sex-sense. Not smelling, not tasting, not something in between but a
separate channel in which a bunch of neurons clustered just above the zebra's
palate and bedded in mucous membrane were super-sensitive to ions, steroids
and secret sex cues.

If the VNO could be said to be a minor sex organ for zebras, it is THE prime
sense channel for snakes and for a familiar icon of my childhood. One particular rafter over the verandah of our farmhouse had become a convenient refuge

for a half-grown Rock Monitor lizard, who lay stretched out as if he were some undulating flake of bark.

Called *leguaan* by the South African farm manager, our house-guest soon assumed that nickname. 'Leguaan' had a pair of VNOs large enough to influence the shape of his head. Each of his attentive eyes looked out from its own capsule, just as each nostril swelled its own bit of snout. His VNOs occupied the space in between, swelling an imperious nose while their servant, a blue and pink forked tongue, only came into play when Leguaan started to move. He went nowhere without exploring his way forward with fast, strangely liquid flicks of that forked, snake-like organ. It shot in and out from an almost invisible crevice between his horny 'lips' at the foremost extremity of his blunt, purportedly prehistoric face. The lizard's choice of what paths to follow, his mental maps, were determined by the information embedded in floating molecules, possibly single ions, gathered by every extrusion of his tongue and analysed by every quick-fire tongue-pass against his palatal VNO. Leguaan lived in a universe of ions. I have problems even trying to get my mind around an ion, let alone my tongue.

'Leguaan', a Rock Monitor (Varanus albigularis).

Lizards and snakes define their place in nature in terms of endless tongue-work, and the sorts of information they need to know are gathered and analysed at electronic speeds. Here is an animal in possession of a set of senses broadly comparable to our own. Yet its prime gatherers of knowledge seem way beyond anything a mere human could imagine. Why? Because we have no such organ with which to sense, and our tongues cannot tell an ion from an onion.

Watch Leguaan's jaws, hinged like a nutcracker, crush a land snail, a dung beetle's thorax or a tortoise hatchling's carapace. Closed, the lizard's mouth shows no more than a thin line, almost invisible within a face formed and freckled like granite. Above and below this hinge, jaw muscles bulge out each cheek. Behind these protrusions, vertical trenches separate face from neck while opening into hidden and rudimentary hearing pits or 'ears'. Leguaan is a living masterpiece of both camouflage and of bionic architecture but his most humbling quality is his possession of a sense now lost among humans. He could pick up, perhaps in lizard ecstasy, signals that lie beyond human imagination. Sharing a brain with Leguaan, but not his sixth sense, invites reflection on what connects senses, brains and their translation into actions and behaviours, as this chapter attempts to explore.

From the beginnings of life on Earth, countless perennial contrasts – day/night, hot/cold, dry/wet and so on – have governed animal existence. Primitive animals developed sensors, which of necessity differentiated into 'eyes' that registered light waves, 'ears' tuned to sonic vibrations, VNOs fired up by ions.

The most primitive of animals needed to evolve responses to signals – at their simplest, away or towards the stimulus. This in turn demanded some means of 'translating' those signals into appropriate actions. Natural selection chose electrical pulses, passing along biochemical threads, to allow those earliest animals to both sense and react to stimuli, signal by signal, particle by particle. Brains have evolved as processing centres for the collation of signals into what we can call 'meaning'. Brains, the human super-speciality, have evolved as command centres that respond to, ignore or censor incoming signals, and then send out appropriate responses.

The Pelican in her Piety: (left) bronze in St Francis non-denominational chapel, Makerere University, Kampala, Uganda; (middle) drawing; (right) vertical version.

My childhood mentor and family doctor, Fairfax, might have put it this way: 'Seeing, hearing and making-sense-of-things are all separate functions, channels, of an electrically structured, centrally placed and single brain. Eyes, ears, tongue, fingertips – all just foragers, like the searching limbs of a spider, obedient to your brain, presiding at your centre.' (Fairfax did not then know that each spider leg has its own barest rudiment of a brain and can perform its various roles in concert with other legs, but independently of the legs' owner.)

Even so, I still like Fairfax's image of a brain hanging in a web, with busy sensors pulling data in from all the space around it. That spiderish image could even be imagined hidden within myself, but with my brain, not my limbs, acting on what all my sensors can gather.

My preoccupation with signalling – especially visual signals and my translation of personal experience into hand-made artefacts – has invited a curiosity about imagery and its origins in plain animal sensation.

The most elemental of all sensations is pain. I was once offered an unusual opportunity to explore an expression of pain, in the form of an invitation to design a lectern with the specific theme of 'The Pelican in her Piety'.

Mediaeval clerics appropriated a belief that pelicans pierce their own breasts as a symbol of Christ shedding blood for his brood. Because pelicans often have russet stains on their breast feathers, and because they can be seen to regurgitate fishy cargo deep into their nestlings' gullets, the story seemed plausible enough to inobservant Neolithic peasants. Besides, the symbolism had its own appeal.

Long, lance-like pelican beaks also manifest the act of spearing or piercing. In ancient myth, pelican breasts absorb that penetration, but not without pain. I therefore tried to invest my lectern with the visceral, universal sensation of flesh being lanced, tissues being torn, the body being violated.

I once saw a python seize a pelican, way out in a swamp where I was too late to rescue it alive. Anxious to draw the dead bird, I had to fend off a very angry snake all the way back to shore, but recollections of that pelican's last flaps infused my early models with pain rather than piety. Having modelled an

ABOVE: *Preliminary sketch of pelican (charcoal).*
LEFT: Self-inflicted Pain *(bronze).*

What the Eye is Hungry to See – an early exploration of visual cortex properties; a precursor of Fovea (as shown at the start of this chapter).

acceptable piece of pulpit paraphernalia, I went on to seek a less cluttered expression of pain as one of the many vicissitudes of being a sentient animal.

To return to Fairfax's word, 'channel', to describe another sphere of sensory traffic, this found its perfect illustration when I caught a Mountain Fruit Bat and released it in a large darkroom, lit by a single electric bulb. While the light was on, the bat flew around the room silently, a silly-looking half-grin on its face and its doggy eyes monitoring my every move. The instant I switched the light off, the bat started to snap its tongue in clicks that were audible to me, and opened up the echo-locating channel that most bats use to fly by. Over and over, the movement of my finger on the light control could literally switch the bat from one channel to another, from vision to sound, back and forth!

*Mountain Fruit Bat (*Lissonycteris lanosus*).*

Every living thing has evolved sensors and channels appropriate to surviving in the world it finds itself in. Frogs, heroes of my earliest chapters, have ears and brains tuned to respond only to the wave-lengths evolved and used by their own species – all else is cackle. The world in its entirety is much bigger, more complex and multi-layered than any brain, human or otherwise, could ever accommodate. Primates mostly use the seeing and hearing channels – plenty of other beings (most notably some ants) get by without eyes or ears. Optical neurons are no different from the receptors of other senses in translating input into electrical impulses, a medium quite other than that in which 'the seen' actually exists. My primate eyes and brain select the useful, the pleasurable or what can be learned from individual experience.

Most of what any eye picks up is trashed because there's too much stuff out there – too much for *any* animal to sort out. Nature, culture and 'common sense' blind the see-er to anything that is not useful to it. People are pre-programmed but also *raised* to edit what they see. Deep in the recesses of brains, both nature and culture delete what is not selected. As the great essayist, Anais Nin, put it, 'We do not see things as they are, we see things as *we* are.'

Nature and culture have long converged in the work of artists, who have been copying nature from the first moment human hands smeared marks on rocks or made some other material assume shape at the behest of a brain served by hands which have evolved to become the vassals of this primate brain. If organic shapes are the outcome of selection (natural or as artifice), an important distinction arises when the artist comprehends that the shapes she enjoys copying are the outcome of a comprehensible process – the process of natural selection. The imagery that emerges from such enlightenment expresses or merely hints at the thrills of engaging with the very process whereby we exist, whereby we are.

When our emergence from nature is acknowledged, our origins among so many other sentient beings immediately forces acceptance that our animal senses remain the vehicles for our transmission of signals, always in code.

Of all those incoming signals, the visual channel is the most immediately connected to the brain and the most crowded with signals, especially for sighted humans. The visual sensitivities that have to be sorted by neurons can be listed, item by item. Edges between light and dark, edges signifying movement and mass, horizontals and verticals (signifying gravity and substrate in their many dimensions), 'focal' spots or circles (typified by eyes as perceived by other eyes) and, among primates, birds, butterflies and fish), natural or contrived colour contrasts.

We owe one of the best (certainly earliest) demonstrations of an animal's sensitivity to colour to the punctuality of Holland's postal service, along the canal-sides of Leiden. There it was that my colleague and friend, Niko Tinbergen, had a couple of aquaria which held sticklebacks (small, freshwater fish in which sexually mature red-bellied males build and defend their own little masterpieces, tunnels or sleeves of torn water-weeds). Every weekday, at 9am,

TOP: *The patterns of fishes and birds compared.*
BOTTOM: *Bishop birds (*Euplectes *spp.) moulting over five grassland seasons.*

Nicko's male sticklebacks hurled themselves against the far sides of their glass ponds in short spasms of piscine fury. Puzzled by the predictability, but also the brevity, of these exhibitions of near self-harm, it was only when Nicko had to rush to catch the post one day that he realised that the postman always stopped his bright red van right across the canal from his aquaria – at 9am.

Nicko went on to show that certain eye-catching combinations of tone or colour-contrast, such as bold red and black pattern splats on both goby fish and bishop weaver-birds, work just as well underwater as out in the open air because their selection has been 'in the eye of the beholder' for a visual channel that has its own susceptibilities, regardless of setting or medium. He demonstrated that a mere tuft of red feathers was enough to enrage a European Robin, and a tiny disc of blood-red on a parent gull's beak was what he called the 'releaser' for a gull chick's beak-tap begging and its parents' response of regurgitating fish. 'Releasers' can be seen, heard, smelt, felt, tasted, even vomited and 'vomerised', but it is the visual and auditory channels that release the most interest in me.

Serengeti's grasslands are often aflutter with seed-eating finch-like birds, all in streaky camouflages that match the earthy ochres of the dry season. At that time ornithologists grumble that distinguishing one 'little brown job' from another is almost impossible. Come the rains – the great greening – and every adult male bishop, weaver, widowbird, whydah and quelea sheds his camos and grows some unique combination of reds, yellows and blacks that maximise the contrast against all that greenery. Then, as green drains out of the landscape, testes shrink and males moult back to match their 'little brown job' mates. When males are exposed to a plenitude of pouncing predators – hawks, snakes and small carnivores – their exhibitionism becomes a liability.

*Long-toed Plover (*Vanellus crassirostris*).*

You cannot imagine my delight as an art student when I learned that human brains do NOT process images as cameras do. While drawing, my brain finds edges and builds constructions that are at least partially based on my awareness of the countless moving masses that clutter my three-dimensional, fast-lane life. Even the quickest of my drawings are *not* snapshots. This implies that even an outline sketch that bears little relationship to the so-called objectivity of a photograph can actually transmit information to another human being more selectively, even more usefully (sometimes more comically), than a photograph.

Few people can imagine the lifestyles of people from the great floodplains of the Sudd in South Sudan. Living among them, my sketchbooks filled with pencil marks that sought to translate my experience of these magnificent people. Back in camp I tried to amplify those quick finger-squiggles into colourful symbolic panels in the hope that they might help elicit empathy from outsiders, especially unsympathetic outsiders. These panels became the dust-jackets for our Ecological Impact official reports (*see* p. 385).

Singing in praise of his 'Song-Bull', a young man in a Nyany encampment likened the beauty of his bullock to stars in an open sky, clear of those starless rain-clouds that brought floods and disease. His was a mental geometry that invested flickering pinpoints of light in the sky with private preoccupations.

Whole constellations can be reassembled in human minds into bullocks, diagrammatic sheep, scorpions, fish, bears and Lions. Neolithic imagination imposed its diagrams as if they were tame beasts pursued by wild ones across a boundless sky.

People of the Sudd.

ABOVE AND RIGHT: *Iconography of the cattle camp.*

Dinka cattle camps, South Sudan.

Young see-ers inherit just such superimpositions upon nature. Imagery and symbols on walls, in caves, books, temples, within gilded frames, comic strips and billowing flags at vast youth rallies – none escape the impress of the cultures that generated them. Encoded within each artefact are cross-links to earlier knowledge, to 'ideas-about-objects'.

In the Preface, I reminisced over one of the many expeditions I shared with sculptor Gregory Maloba, trips in which we invited students to come along with us. On one of these trips, our objective was to see if we could translate disparate and arbitrary observations into three-dimensional sculptures. At the time I had acquired a broken-winged owl. Free-living owls can simulate a tree stump or gnarled, truncated branch. By opening their feline eyes and raising paired forehead feathers that resemble a cat's ears, an owl can flash the mask of a threatening wildcat at any disturber of its daytime sleep. Tracts of feathering can lift, shift, fluff out or flatten to suggest all manner of outlines, forms and mouldings, all as malleable as modelling clay, paint or a wandering pencil point.

The best-known, certainly the longest-lasting of African civilisations developed along the banks of the River Nile. There, Ancient Egyptians developed a system of hieroglyphics (deciphered from the Rosetta Stone by Jean-François Champollion and Ippolito Rosellini, and further studied by my cousin, Stephen Glanville, an expert on Egyptian papyrus scrolls). Champollion revealed that a particularly elegant and unambiguous representation of an owl stood for that owl-like humming sound we make with lips closed (Romans geometricised this into the letter M).

BELOW AND RIGHT: *The mutability of owls.*

The mutability of owls and source of an Egyptian glyph.

Inspired by the whole idea of mouth-made sounds and their associations being rendered as graphic (even three-dimensional) ideograms, Gregory and I explored with our students how arresting imagery can be conjured from arbitrary, even contradictory, associations.

If the sound 'oummm' can be borrowed from an owl, its source turned into a visual glyph, then get linked up with a succession of other glyphs – windblown Fff, keening Uuu, sleepy Lll or squealing Eee, to signify such arbitrary words as MULE, ELM, FLUME, ELF, even FEM – then why not let some arbitrary memory or supposed attribute of mind, voice or shape reformulate itself? A rap, a rhyme, rumba, romance or representation in shallow relief or boldly carved granite – something entirely new snatched from an imagination as wayward as permitted by any artists' particular era, culture and faith.

Choosing the glyph for M, I borrowed the feather-transforms of my invalid owl to bulk out the glyph's flat architecture, I hung my sculpture between a diagonal back and vertical front. While the rear view was as abstract and linear as architecture, I rendered left and right sides to suggest the mutability of owls, even into glyphs and all their diverse associations with human imagination.

Our students' riffs on owls, Ancient Egypt and experiment are now dispersed beyond recall, but I still have my own terracotta artefacts to illustrate something of our little exercise in translation.

Gregory and I sometimes persuaded our students to try out 'exercises in play', and I remember a particular visit to the Game Department's animal orphanage. On the way, we stopped at a brickworks where students filled their satchels with coloured clays. On arrival in Entebbe, each of us chose a subject that had some personal association and then spent the next few hours manipulating clay. Belonging to a fish clan was enough for Gregory, who took shelter in the coolth of an aquarium to make a very piscine portrait of a large, lazy Nile Perch. Meanwhile, a young colobus monkey, hot, lonely and phlegmatic, allowed me to try to conjure up his image in coloured Kajansi clay. Gregory described us as 'sketching' with clay rather than chalks. We let the animals excite us in unpremeditated, even arbitrary or whimsical ways while we made clay obey our hands and take the impress of our eyes, minds and memories.

ABOVE: Nile Perch *(terracotta by Gregory Maloba).*

RIGHT: Sleepy Colobus *(terracotta by Jonathan Kingdon).*

Many years later, some of us found ourselves remembering those days with affection and nostalgia. We shared a newfound dignity and worth as artists during the era of hope and purpose that accompanied the early days of Uhuru (political independence). We were free and momentarily independent.

With Uhuru, anything and everything seemed possible, and playful creativity found its most brilliant and fecund expression among Makonde sculptors in Tanzania, many of them refugees from Portuguese oppression in Mozambique. Expressing an extraordinary explosion of creativity, among leading masters were Samaki, Sangwani, Mpagwa, Dastan Nyedi and Maestro Chanuo.

All East Africans are familiar with rocky outcrops with suggestive names – Lion Rock in Dodoma, Camel Rock in Tabora, Zebra Rocks in Serengeti, so

Rhino Rocks.

another of our Makerere projects was to mould and assemble lumps of clay which human imagination could translate into the image of some animal. My choice was a rhinoceros – sadly that playful jumble has now come to suggest something more like Golgotha or a cemetery.

LEFT: *'After-image', perhaps 'search-image' as might occur as an etched silhouette in a prehistoric hunter's brain.* MIDDLE: *A rock surface (derived from a photograph) taken in Oaum Regaya, Libya.* RIGHT: *A Grass Rhino outlined on the same rock surface 8,000–9,000 years ago.*

Chyulu, Rhino Cemetery.

Scholars of rock art in Africa have shown that the random irregularities of cave walls often provided the starting point for small charcoal or ochre markings that suddenly assumed the shape of a recognisable animal. Only the hungry eyes of hunters endlessly scanning for prey or predator could extract mental images of animals from the multiformity of nature. During long hours of leisure in the security of cave or rock shelter, those same eyes and brains, though often sleepy, were still active. Now and then, hands at the command of those reminiscing eyes and brains could smear and daub a mental imagery that translated the private into the public.

My fellow artist, Elias Jengo, and I once reiterated our shared Makerere and Mkomazi experiences on the walls of a rock shelter – for that we earned our Swahili title '*Wazee wawili wa Mpango*', two old cavemen!

Kingdon and Jengo at Mkomazi shelter site.

More recently I devised a very different metaphor for the eye-brain activity of seeing, by assembling a colourful pottery roundel entitled 'Fovea' (*see* p. 386). I then shattered this giant dish into hundreds of ceramic shards. My apparent vandalism (a photo of which introduced this chapter) stood in for the way in which every fragment of incoming light and colour gets shattered by the retina and allocated, like my shards, to a separate neuron. Within a trillionth of a second the brain, NOT the eye, decides what to extract or 'see', and what to reject. Each species (and every human cultural tradition) retains whatever is relevant but rejects whatever is inconsequential for its way of life. Sometimes, what we call 'styles' emerge. Sometimes nomad minds find a contrary consolation in the barrens of deserts, or in obdurate obedience to blind ascetics.

Learning that the eyes' receptor cells separate, sort and decode the many properties and patterns of light falling on retinas proved to be the key to exploring a set of questions that bothered me from early childhood. How do we explain a zebra? What agency created such a flamboyant masterpiece of nature? Why and how did black-and-white edge patterns evolve to decorate part or all of the hides of the wildest and most beautiful of all horses?

Common Zebra sketches.

LEFT: *Mountain Zebra foal.* MIDDLE AND RIGHT: *Mountain Zebra (bronze sculptures).*

Start by imagining the impact upon a newborn zebra's retinas as it staggers, giddily, to upright itself on still-soft hooves, inside a tight little herd of stripes. I spent many hours over many weeks, stretching to months, then decades, ogling zebras before articulating an answer that began 'in the eye of the beholder' – in this case the eyes of zebra foals.

Stripe patterns are a product, a consequence of and a response to the ways eyes and brains work. However, for any living zebra its primary, day-to-day response to bold stripes is perhaps no more complicated than, 'Aha! Another zebra – I'm safer.' Perhaps even, 'I'm attracted.'

It was Darwin who first suggested that zebroid stripes evolved on long-extinct ancestral equids, and I am convinced that he was not only right but that the origins of those stripes lie deep in the secret saga of equine evolutionary history.

Taking up Darwin's suggestion that striping evolved in equine ancestors, I went through thousands of photographs of equids (and, incidentally, many hundreds taken from Palaeolithic paintings of prehistoric horses). These precipitated a plate devoted to equine polymorphism (now in the library of the Natural

TOP: *A suggested progression in the evolution of stripes in equines. Hypothetical ancestor (top left) with Onager/Grevy's lineages above.* MIDDLE SEXTET: *Ass/Mountain Zebra/Quagga lineages.*
BOTTOM: *Rock art from Mashona, posited to be 8,500 years old. An extraordinary rendition of zebra stripes, as if the artist became entranced, 'carried away' by the optical sensation of striping. In this, that ancient artist may have had more insight into the meaning of stripes than some contemporary biologists.*

*Three columns of equids: (left) Common Zebra (*Equus quagga*); (middle) Quagga (*Equus quagga quagga*); (right) Palaeolithic horses as represented in cave paintings.*

History Museum in London). Then I turned to known fossils. More than 50 million years ago, little horses, now fossils known as *Hyracotherium*, lived in forests, where they are thought to have spaced themselves out, like any contemporary woodlander, as defensive territory-holders. Animals conditioned by millions of years of territorial behaviour are not easily socialised, so how did small territorial forest browsers evolve into large nomadic grazers, prone to travel in herds?

The first clues appear in a decline among maturing zebra foals of mutual nibbling, their primary bonding mechanism. Growing youngsters detach themselves from partners but still champ their teeth and nod their heads in a faintly comical charade that a horsey friend calls 'empty grooming'. To me this suggests that a social function has flipped its sensory channel, as if by the flick of a switch, from tactile to visual. Perhaps each such foal is reiterating an ancient switch in which the one-on-one tactile bonding of once-secluded animals became an all-inclusive visual bonding whereby *any* zebra, no matter how bad-tempered and bad-mannered, however close or distant, becomes attractive by virtue of its stripes. Eventually I concluded that, as primitive horses moved out into more open habitats, striping provided the mechanism that forced these famously curmudgeonly animals to override embedded anti-social tendencies and find one another irresistible after all.

How?

Equines of all species most frequently nibble one another on the shoulders or rump, where flexible skin forms stripe-like creases that are easily enhanced by the evolution of real black stripes (as on donkey withers). My hypothesis is that all-over striping expanded from just such grooming targets in some early ancestor. 'Empty grooming' in juvenile zebras betrays a switch from touch to vision in the life of a single zebra. In my view it also reiterates a similar switch over evolutionary time.

Zebras actually use their raucous voices to tell us how dependent they are on vision and stripes. Grazing zebras are generally quite quiet during the day, but if you camp around Mara-Serengeti, Rukwa, Mkomazi, Laikipia or anywhere zebras are still allowed to congregate, then listen on overcast nights when the sky is moonless, starless too, when even the silhouettes of trees and trunks are lost in the blackness. Zebroid decibels can become deafening as every individual, seemingly from oldest stallion to youngest foal, brays, barks and catches its breath in wheezing squeals.

Why the cacophony? The most likely explanation is that zebras are so reliant on vision for social cohesion that calling out tries to compensate for night-blindness. Each animal signals its position, its existence and, perhaps, its discomfort at being night-blind.

In paint I once tried to conjure up an indistinct cluster of stripes out of darkness – perhaps the rallying focus for a lost and lonely foal? My painting was not entirely preposterous.

Stripe patterns are a product, a consequence of and a response to the ways eyes and brains work. Stripes have been selected in the eye-brains of many animals. There are zebroid duikers, marsupials, wasps, numerous birds and countless zebra fish. We must examine the susceptibilities of eyes if we are to explain all these convergences in black-and-white patterning.

I grew up with zebra hides as floor-mats – grim trophies for the lover of horses but daily reminders of how easily the eye is seduced by lines in black and white.

I belong to a very long tradition in which skills in drawing have been enlisted in the service of and as an expression of science. The act of drawing is, quite literally, an act of 'figuring', and looking at drawings can be an active retracing of that translation process.

How does this relate to the evolution of animal patterns?

The same principles apply to the questions that arise when we try to understand the evolutionary origins of colours and patterns. In this regard birds pose many an exciting challenge and their behaviours sometimes offer clues. Why are some groups exceptionally colourful while others are consistently drab?

Bulbuls and greenbuls exemplify the latter; their loud and very varied musical repertoires betray song as offsetting obsessively cautious behaviour and the dullest of green-brown colouring.

Yet other songbirds have patches of brilliant colour or white on their throats, often outlined into 'bibs' or facial masks of contrasting colours. These serve to draw attention to pulsing beaks and distended throats – sometimes the mouth's lining (especially those of nestlings or courting males) is of a startlingly brilliant

Red Queen. *Coat-of-arms for a biologist (by courtesy of Matt Ridley).*

*Northern Carmine (and one White-throated) Bee-eaters (*Merops nubicus *and* M. albicollis*) at their colony, with many holes occupied.*

colour. Facing front, then left, then right, song and songster reach out to their intended audience, from some high perch or cloud-backed sky, even from within the depths of some tangle of thorns.

The most consistently 'bibbed' birds in my experience are *Merops* (I have already asserted a personal preference for the genus name over the English 'bee-eaters'). Not one of our African homes or encampments lacked for *Merops*. The 'Little Merops' is a 14 g aerial acrobat that foraged over our flower-beds on short wings that flashed apricot orange underneath but bright green above, its bib the brightest of lemon yellow. By contrast, the 50 g 'Northern Carmine Merops' arrived in Acholi on falcon-like wings to hawk above moving herds of Uganda Kob, elephants and Tiang, sometimes sprinkling the broad backs of buffalo as if they were animated colourful confetti. While their heads, rumps and vents were kingfisher blue, their torsos, tails and wings were of a crimson only matched in brilliance by the red of their irises, glaring out from what could be called black masks but are better seen as visual elongations of their hard, coal-black bills.

Like some of their closest relatives, kingfishers, most *Merops* sport such beak extensions and all species like to dunk themselves in water on hot afternoons; very occasionally the larger species may even catch fish. Like kingfishers and woodpeckers, *Merops* have big heads, reinforced skulls and iron-hard beaks that they sometimes thwack against branches as if they were boxers practising on a punch-bag.

Those slender, slightly curved implements can also be used with a surgeon's skill to detox noxious shieldbugs, hornets, wasps, flying ants and, yes, bees, spilling away their venoms before the arthropods' battered remains get swallowed or proffered to a mate.

Those pick-axe bills also dig burrows that can be a metre or more long. Here, bills are served by funny little pedalling toes that shunt loose soil backwards like super-fast motorised rotary spinners. During this furious action the feet are relieved of their body-supporting role by wings that are adapted to become props throughout the miners' feverish excavations. The burrows can angle horizontally into a bank, or go down at a steep, near-vertical angle on sandbanks or dunes.

Like other self-made hole dwellers, *Merops* then have to defend their holes and nestlings from marauding snakes, lizards, dormice and those arch-extractors of nestlings, Gymnogenes (also called African Harrier-hawks). Breeding pairs must also face off parasitic honeyguides, as well as other *Merops* unable or unwilling to dig a hole or rear a nestling of their own.

Having to defend such laboriously dug retreats from such a host of hostile intruders or would-be overtakers has a bearing on the evolution of those Zorro masks and brilliantly coloured bibs. The most conspicuous of these are piebald masks sported by 'White-throated Merops', which arrive all over equatorial Africa after breeding amid dunes in the south Sahara during the short boreal wet season. Once they are back on the equator, they often join their carmine cousins to hawk over bush fires.

This apparently perverse annual cycle of living it up all along broad equatorial expanses only to retreat into Saharan wastes to breed is clearly driven by the many vulnerabilities of any would-be *Merops* breeder. At the mouth of its precious burrow that bold facial mask, facing out from the interior or perched beside it, signals: 'This hole is ours – keep off.'

That visual message is augmented by a less savoury, olfactory message emanating from all *Merops* burrows, lined, as they always are, with insect debris, accumulated faeces and the rotting corpses of nestling losers, frequent victims of notoriously murderous fellow nestlings. Any hungry cat or cobra must also pass a ferocious masked pecker before it can reach the dubious dinner of a thorny nestling. Yes, the feathers of these chicks sprout within sharp, chitinous sheaths as if, for a few vulnerable days, they were baby porcupines.

A *Merops* with kingfisher-like colouring came into my hands after I had marvelled at the aerobatics of 'Black Merops' flying high above rainforest canopies in western Uganda. A few weeks after stating my intention to, somehow, render an impression of those flashing colours, that ambition met with a challenge. My companion, Tony Archer, clearly expecting a field-guide, specimen-based portrait, thrust a carefully labelled skin into my hands, grunting, 'Very well, paint this one.'

Murderous bee-eater nestling.

RIGHT: *Conjugating the colours of a Black Bee-eater.*

The bird's bib was of purest scarlet, its brow blue, while its underparts and rump were also azure blue. As with many *Merops*, the undersides of its wings were rufous, the rest was black. Tony was nonplussed by my painting, which attempted to explore how a bird can flash conjugations of colours that remain in a human brain long past the moment of their witness. I have illustrated and written field guides, but a single living bird has a thousand more lessons to teach us than its ID mugshot image in a book or on a screen.

The prime designated viewers of that bird's colouring must have been its mate and conspecifics, but if its bib resembled those of other *Merops* in deterring burrow marauders, then why that startling red? Did evolution select that colour for its visual contrast (as I chose to stress in my painting)? Or did it have a more symbolic role, such as resembling the blood-red spot on a gull's beak? If *Merops* are provoked to defend mate, nestling or burrow, their stiletto beaks can certainly draw blood. Could there be some unknown association with blood?

Hoopoes, fellow-foulers of stinking nests, likely possessors of fetid flesh (and related to *Merops*), have wings and crests that flip open and shut like Chinese fans, signalling 'keep off – I stink'.

Why on earth should a whole phalanx of species ornament the back, not the front, of their heads? Take the males of those woodpecker species that seek out hollow branches to drum out their presence to potential mates or rivals. Bright red caps or napes, ornamenting the birds' only moving parts, head and neck, help locate a bird while some measure of the drummer's vigour emerges in the persistence, the rhythm and resonance of its insistent tapping.

Rereading some previous paragraphs reminds me of Leonardo da Vinci's critique of logorrhoea and his defence of drawing and picture-making – 'a

Mweya euphorbias perched upon by Marabou storks.

Woodpecker array, all advertising their power as peckers.

type of knowledge, which is impossible for writers, ancient and modern, to convey without going into infinitely tedious, confused and long-winded detail'. Leonardo's critique inspired a chart of patterns mapped onto eight woodpecker silhouettes in which I found a delightful visual logic in the exact patterning of those jack-hammer heads angled against the feathered outlines of their bony, sinewy necks – the driving force behind each drumming blow.

Recording and interpreting things that have their own evolved logic not only involves different media of cognition, but the translations that emerge are often a lot faster and less clumsy than their verbal equivalents. In my case, I set out to deploy simple media skills, first taught to me by Dorothy, as the means to explore evolution as I met it face to face on my own home ground.

I judge evolutionary biology to be the most exciting, truthful and intellectually challenging topic of our time. Any involvement in biological science creates fellowship across oceans and some confirmation of the best universals in human culture, at the same time as fierce rebuttal of the most perverse, irresponsible and barbaric era in all of human history.

In an environment dominated by markets, the only publishable form my earliest studies could take was a handbook of our greatest natural resource – the mammals of Africa, a roll-call that includes *Homo sapiens*. Drawing has long been an adjunct of anatomy and illustration but, in 1964, I wanted my *Atlas of Evolution in Africa* to take the practice further. Wordless 'questioning of form' can document feeding, copulating, excreting, fighting, even mini-signalling of information among conspecifics. 'Sketches' of animals behaving were both the solution to a problem and a personal pleasure in their making. The process of watching and drawing such activities generates an understanding of shapes and shaping that I once called 'translating Darwin-speak'.

It is one of Darwinian evolution's most profound insights that morphology has emerged, incrementally, as the outcome of individually varying animals probing all the possibilities of their lineages' existence, under the limitations of short lives lived out in specific localities. Over evolutionary time, and to a significant extent, natural selection of small details of individual behaviour seems to have driven and shaped shape. Those individuals with behaviours appropriate to their setting and time have been the most likely to survive. Riding on the coat-tails of such behaviours are all sorts of small, physical and physiological differences that further enhance the effectiveness of behaviours. Incrementally, such physical changes can reorder colours, size, the shape of beaks, ears, limbs, even entire body proportions. Along the way, new species emerge.

I'm repeatedly disconcerted by a mismatch between my ability to collect, measure and interpret the physical, sometimes dead, products of evolution in a detached, scientific spirit, and my personal appreciation that mammals are as alive as I am and as full of 'moments' in the pursuit of their lives as I am in mine. Nowhere is this more apparent than in my struggle to marry the fleeting moments of ethology or behaviour to the more lasting solids of morphology (the subject of an earlier chapter). For me, drawing is an essential medium and adjunct for trying to bridge that divide between vivid, vivifying life and the material world that living things fleetingly inhabit and eventually leave behind, as bones, fossils, images, buildings and relics.

Drawings have illustrated all my books because the very process of making them was integral to my discovery of previously unnoticed facets of the animals' biology and evolution. When my work was remodelled into field and pocket guides, such details helped me lift out those features that differentiated one species from another.

From the start, I had to ask myself how viewers of those hand-drawn images as well as readers of my texts might respond to such an alliance with evolutionary biology. For sheer contrast two responses will do.

Illiterate Twa pygmies identified the paintings of small mammals immediately, but it had to be explained that images of larger animals were dwarfed in my field guide because they were 'not very close' (or 'shrunk' to fit the page). Once the hunters had made the necessary mental and visual adjustments, they enthused over paintings of buffaloes, pangolins and monkeys, remembering where they last saw the animal in question. The Twa conferred on me the title 'Doctor of Work', an honour of which I am proud.

Invited to review my *Atlas of Evolution* for a local journal, a super-literate laboratory student of kangaroo cadavers delivered an opposite reaction. He began by lambasting a lack of verbal keys, while their alternative, drawings, 'served no useful purpose' and the work was summarily trashed for inadequate deference to august authorities. My critic was a visual illiterate, a total verbalist for whom his national dialect and associated conventions were almost his only link with fellows. Somehow, somewhere, his links to the external world had shrunk to verbal translation, with words his only medium and precedent his imprimatur of authority.

Among the drawings that served no useful purpose for my roo-phobic critic were sheets of sketches of Caracal cats. These cats exemplify an inconspicuous animal that uses small movements of its head and black-tufted ears to flag information

*Tufted ear tips as flags – Caracal (*Caracal caracal*).*

to other Caracals. The slightest twitch of an ear-tip can signal mood, status and intentions in a manner analogous to flags on sailing ship masts. The iconography of Caracal ear or head flagging is intricately crafted, and fingers on a pencil can scarcely keep up with the rapidity of their flickering movements. Nonetheless, I believe such drawings can be a clearer medium for exploring such visual morse codes than laborious written accounts or quantified records of filmed frequencies. Future students will, no doubt, use new photographic sources for further analysis, but when I did my sketches, the economies of my time made it enough to observe and advertise the existence of such interesting behaviour.

The recent discovery, by molecular scientists, that the Caracal is not a close relative of the northern lynxes (which have independently evolved similar ear tufts) gives an added interest to ear-tip signalling. Tuft-eared primates, squirrels and some antelope species converge with cats to suggest that the elevated ear-tip devices used by all these animals sometimes resemble sailors' hoisting of mast-top flags to communicate. Any visually oriented animal, needing to discriminate among their own species between friend, foe, rival or mate, must judge, select, be selected, signal or slip away on the basis of 'appearances', some subtle, some gross.

Artists and scientifically minded humans are not the only animals that seek to lift out significant or informative 'form' from the chaos of nature. To survive, every visual predator, whether tiger or tigerfish, must repeatedly 'see through'

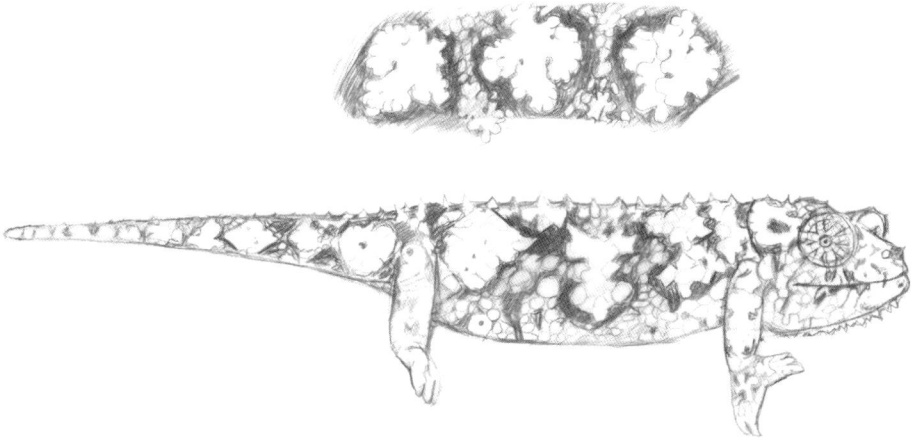

TOP: *Growth pattern of 'crottle' or 'stone-flower',* Parmelia *sp., a fungus host to carbohydrate-generating algae, that is, a lichen.*
ABOVE: *Spiny-flanked Chameleon (*Trioceros laterispinis). *In 1930, Teddy described the discoverer of this reptile, Arthur Loveridge, as 'as prickly as his new-found lizard'.*

the disguises used by their prey. Because visual predators tend to select the most easily seen individuals of their prey, the evolutionary explanation for progressively better and better camouflage in prey animals is obvious; it lies in the processions of survivors and offspring of survivors that were somewhat less easily seen and caught. The visual acuity of predators has therefore been a prime agency in the prey's external appearances. Thus, concealing coloration is as much a manifestation of predator acuity as a reflection of the behaviour of their prey.

We call the outcome 'camouflage' but it can just as well be seen as a striking manifestation of appearances being translated into another medium. In effect, selection *by* predators causes some aspect of the landscape to be drawn or painted onto the bodies of surviving prey animals. Hair, feather or chitin become the medium for miniature landscape paintings. Some apparently crazy excrescences and structures on prey animals can only be explained in the context of their predators, and the natural backgrounds against which predation takes place.

Studying how camouflage hides an animal is the precise opposite to lifting its profile out of its setting. For the analytical viewer, camouflage is as much a subject for drawing as any other manifestation of evolution. Perhaps even more so, since drawing is little more than tonal scratches onto surfaces, and the externals of animals are made up of innumerable scratchy bits of variably toned skin or chitin. Camouflage is found on both predators and prey and at least two distinct classes of camouflage emerge. One, usually very small-scale and commonest in arthropods and marine organisms (and exemplified by the underwing, lichen-mottle of cosmopolitan butterflies), consists of patterns that match the animal or plant's immediate substrate exactly. I remember marvelling over an East African chameleon that was body-moulded and colour-coded to exactly match the sculptured lichens of its montane forest habitat (*see* opposite).

Other, commonly larger-scale, camouflages consist of abstract patterns in which the disposition of light falling on broken ground, or the chaos of plant growth, is mimicked by patterns that average out the relative proportion, disposition and shape of a limited number of tones. Leopards provide a classic example on a predator, while their mammalian prey, from fawns to giraffes, offer another.

Research has shown that at least 70 per cent of giraffe calves are killed in their first year of life, mainly by Lions and Leopards. Like deer fawns, the survivors probably owe a lot to markings which are very close in their size and distribution to light and shade in their dry woodland habitats. Unlike most deer, there is no pressure on giraffes to change the coat with age. So the marks enlarge as the animal grows, with the end result that adult patterns become out of scale with their surroundings. A close examination of the ragged-edged blotches on a Masai Giraffe suggests remnant outlines of pale, fawn-like spots that have coalesced to isolated dark, rosette-like islands.

With paint and brush, and taking the analogy of a rosette literally, I once plucked two or three blossoms out of the intricate trellis of a giraffe's hide to isolate its abstraction. The result resembled formalised stone flowers on a capital or architectural frieze, yet the image was snatched from the actual pattern of a passing giraffe.

My giraffe-hide bloom was translated, yet again, into the multi-medium of a television programme. The latter was a TV translation of what I had rendered *in paint* of abstract landscape elements that had already been translated *in fur*. For the oxpeckers that once swarmed over giraffes, their hosts' hides *were* the landscape.

Earlier on, a hieroglyph became the starting point for an exercise in translation with Makerere fellow-sculptors. In equivalent classes with painters, I

Giraffe-hide rosette.

Kitahurira Rapids.

invited my companions to retrieve some arbitrary yet memory-provoking detail of a distinctive landscape, then re-think and re-envision the intricacies of that detail, treating just such visual intricacies as the medium for their paintings. I rendered my own memories of the rushing, forest-girt Kitahurira Rapids of Ishasha in the medium of butterfly wings, more specifically the patterns of the lycaenids that swarmed all along the river banks. Like the convolutions within an evolved human brain, butterfly wings reconcile order and pattern with the many felicities of chance. Ishasha's waters, frothing over black rocks, throw the imperatives of gravity over a chaos of broken-up lava.

Translations are neurophysiological processes within animal brains. Their cultural equivalents or 'translation memes' were among my major preoccupations when I was invited to submit designs for a mural in the hallways and arcade of a large public building in Nairobi's city centre. Left naked of ornament, this skyscraper certainly promised to look as bleak as most of its neighbours – overcoming that anonymity was the challenge. Walking back and forth in the concrete cavern, I felt that here was a single species of skyscraper among many. I was charged with evolving a pattern for it that would render it distinct from its neighbours. This was an opportunity to link up my interest in the evolution of visual signals in nature with the architect's brief. He wanted me to endow his cavernous halls with a 'Sense of Place'.

Mezzanine mural in ICEA building, Nairobi.

People enjoy picnics beside waterfalls, seashores or piles of monumental boulders because they are landmarks, focii for refreshment, even rapture. The memorability of such places can be reduced to their dominant visual characteristics – slashes of white water cascading over cliffs or rocks, looming masses that suggest giant presences, or a marine blue horizontal. I decided to vivify the hall and arcades by borrowing my methodology from nature, my chosen material ceramic tiles in bold, flat colours. Taking the simplest elements in pattern – vertical strips, like patches of zebra hide, with variations in shade, tone and breadth – I set them moving in irregular oscillations, and built images out of the interactions where widths changed or edges fractured or slid.

Butterfly wings, feathers and other animal patterns provided vital points of reference, but I was also aware of ancient traditions of decoration in which

Outer extremities of giant ceramic mural in ICEA building, Nairobi.

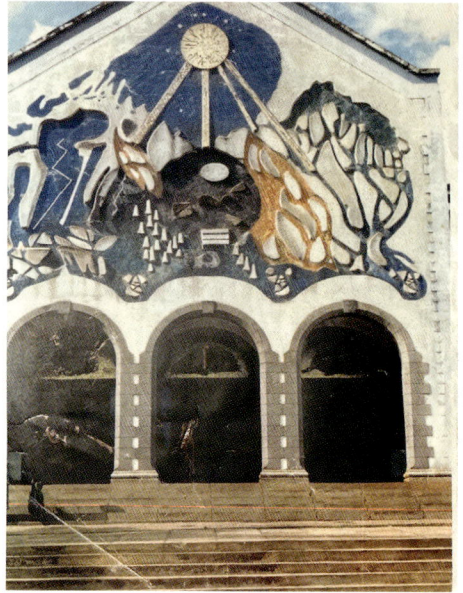

LEFT: Canticle to the Sun *seen through banana gardens (see also p.455).* ABOVE: *Bas-relief mosaic fronting St Francis non-denominational chapel at Makerere University, Kampala, East Africa.*

images emerged out of strips or stripes – Jomon whorled goblins, Scythian strapwork stags, Sung interlaced dragons, volutes around Celtic saints and Islamic floral arabesques, to mention but a few. The artists wove their dominant preoccupations into the ordinary fabric of their lives, in metals, felt, appliqué, painted velum and tin-glazed terracotta.

Surprise and delight are not easily communicated by the routines, methods and practicalities of science. Notwithstanding this, the biological world and evolution are a central inspiration and a continuous source of renewal for an ever-increasing number of thoughtful people, including artists and poets. In Nairobi, I was grateful for the opportunity to let such half-play, half-earnest 'translation' spill over, large-scale, onto the walls of a building.

Yet another large-scale mural involved multiple translations. My beloved university, Makerere, built a non-denominational chapel and dedicated it to Saint Francis of Assisi, 13th-century author of *Canticle to the Sun* (often cited as the first poem to bring fame to the vernacular Italian language). In his poem, Francis articulated a vision of humanity integrated into nature and the cosmos. The sun, 'who brings us the day', is worthy of praise as a brother, as is our sister, the moon, who is 'clear and lovely'. Praise, too, for sister death 'from whom no man escapeth'. I won an open competition for the design of a 'treatment' (effectively some form of 'translation') of the chapel's west front. My concept began with a

Evolution on the Wing, *mural in the Common Room of the Biology Department, Duke University, North Carolina.*

white cement bas-relief, into which I set strips of glass mosaic in my own praise-song for Francis's Canticle.

I was offered yet another opportunity to translate natural imagery into ornament (and intellectual provocation) when I spent a year as visiting professor at Duke University. One of my colleagues was Fred Nijhout, who was working on the genetics of pattern formation in butterflies. In homage to this brilliant biologist I devised a composition called 'Evolution on the Wing' that began with a butterfly variously called 'Cosmopolitan' or 'Painted Lady', *Vanessa cardui*, along with its congeners.

Here, in America, this was a familiar and much-loved being out of my African childhood (seen again on later travels in Europe, the Far East and Australia – a truly cosmopolitan insect). Its exquisite venation and pattern on the underside seem most closely to mimic lichen but, in this mural, accurate reproduction or representation of pattern was not my objective. At the time of painting I recorded my side of a conversation with Fred: 'The panels should evoke the dynamics of evolutionary transformation in the simple markings on a small set of butterfly wings. I want to follow your initiative in exploring how coding for very small alterations in separate details adds up to big differences in the appearance of a whole organism. In charting the wing patterns of *Vanessa cardui* on the right, of *Vanessa virginiensis* on the left, and bits of *Vanessa anabela* in the lower corners, I want to suggest an insect fluttering through evolutionary time, transforming itself as it flies. As if the physiological transformation of egg to larva to pupa to adult was not wonder enough, this small radiation of *Vanessa* seems to me to exemplify the wonders of evolution in an almost pure, distilled form.'

These are transformations that seek to translate vivid, vivifying life.

MIND AND MEMORY

21

A critique of locusts of the Anthropocene as they consume the fruits of the Neolithic. Discordant memories in the minds of a motherland's children.

The birthing of mutual love can be beyond memory. My gaze once met tender, affectionate eyes that returned my smiles while the caress of hands catered to my every need – that much I may know but I cannot say that I remember. Today, those beginnings can only be 'remembered' vicariously by watching other babes nestled contentedly in other arms, or by contemplating blurry, black-and-white family snaps of my early self being cuddled by 'my' adults. Whatever happened to me happened on the watch of the three familiars I began this tale with, Dorothy, Saidi and Teddy, the guardians of my very existence. Without them I would have been one big bag of maggot-food, as Saidi might so indelicately have put it as he squeezed out the squirming larvae of Tumbu Flies from the flesh of my buttocks.

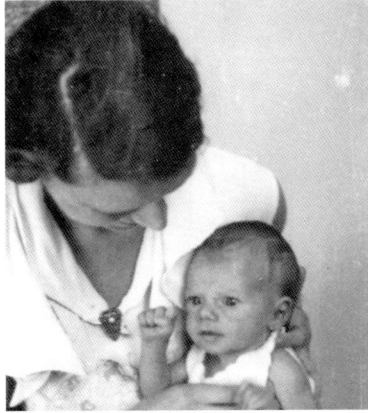

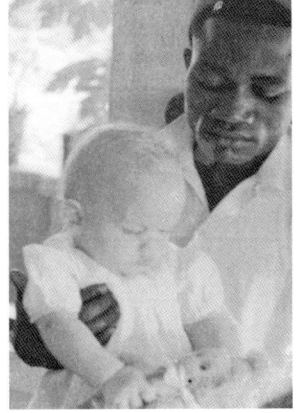

Three photos of newborn Jonathan with: (left) Teddy; (middle) Dorothy; (right) Saidi.

As for those black-and-white family snaps, it is curiosity about the past, and family empathy, that invest Dorothy's photographs with meaning and interest. Her camera simply made 'Leica light-charts' of dappled daylight falling upon surfaces, scenes and skins in 1930s East Africa. Teddy, who loved explosives, augmented her efforts by setting off 'Magnesium Flash Powder' to create a bang and a bit of a stink that momentarily lit up the night for Dorothy's lens.

OPPOSITE: Yaida Consumed by Cain's Crab-claws.

Dorothy's Rain on the Rift Escarpment – *here during an episode of volcanic eruptions nearby (Dorothy had a feeling for the ephemeral).*

Her prints may trigger momentary flashbacks, even nostalgia for evenings crouched over hypo solution in a cupboard darkroom, yet my memories are not clothed in black and white. Accidental moments caught on strips of acetate film are just that – accidents of timing, accidents of lighting during a pre-digital era.

Yet my mother's photo albums, her folios of paintings and her written notes all clutched at time as it hurtled past. These were the media for her experience of ephemeral moments in a contested region of Africa, between two world wars. She differed most from her contemporaries in keeping her senses and lenses open, not closed to the cascades of stimuli that so disoriented too many of her more domesticated and less talented contemporaries.

Dorothy relied on animal senses and highly selective sorting systems in her brain to translate her sense-models into mind-models. Like any other brain, hers tugged her into still more layers of translation – into language, music, dance, sculpture and her speciality – picture-making.

Dorothy translated day-to-day experiences into words, numbers, sketches, paintings and photographs. These conformed to the conventions and devices of her time, each serving to inform or simply please other humans (most especially in stamp-pasted sky-blue aerograms to her mother, Bibi Minnie). Dorothy learned the techniques and conventions of picture-making and word-spinning from mentors, peers, books and exhibitions. Her efforts were rewarded by having her work published in magazine articles and exhibits in museums. Some of those conventions were recent and highly technical – others went way back into prehistory where noble antecedents engraved an elephant's outline onto a Saharan or Karoo boulder, rendered the unmistakable silhouettes of ochre elands and agate giraffes stampeding across the wall of a Kondoa rock-shelter, or chiselled away in bas-reliefs of Egyptian adolescents obedient to the architects of their father's tomb.

If knowledge rests on conscientious translation from one medium to another, my earliest coaching in any such skills was unusually early. Because there was no school nearby and because my mother was a trained artist and art teacher, my first lessons from her were not in reading and writing but in drawing directly from life. I remember her sitting me down at the age of five, with pencil and paper, to draw an acacia tree in the yard while she busied herself with her own sketchbook. After a while she came over to see my efforts. 'Splendid! But haven't you noticed how the trunk narrows as it rises, and see how the branches flatten out sideways, not like that oleander over there, where they all go up at a steep angle. Now, don't rub that one out, just do another drawing to compare with the first one.'

My mother's injunctions to observe, accurately record and compare things were, I only discovered very much later, the essence of scientific inquiry. Yet they were experiences that slipped by with no more significance than play, small but enjoyable details in the fabric of day-to-day existence on the shores of Lake Victoria in the early 1940s.

Dorothy saved early paintings by Jonathan: (top left) Elephant, *1940; (bottom left)* Fire Finch, *1941; (right)* Judy the Donkey, *1941.*

There was one, perhaps unexpected, byproduct of the sequence in which my home-schooling took place. In my perception of the world around me, and in my repertory of communicative skills, visual imagery preceded literacy and its attendant, essentially linear, logic.

Of my three top mentors, it was Saidi who seemed the most knowledgeable to me and, for those first seven years, certainly the most influential, best able to translate every experience, every encounter, into something new, fun, funny and the trigger for a glorious repertoire of stories and dramas. Illiterate to begin with, acquiring literacy became but another game, another joke for Saidi.

Literacy is commonly promoted as the prime virtue of education, and most parents want their children to be literate. So do tax-men, employers, ruling-party commissars, priests and idealogues, every one of them seeking to link up their own vested interests with the promise, energy and credulity of newly literate youth.

Kilimanjaro Vortex, *designed for television, to shift, whirl and zoom.*

Childish curiosity about the 'meaning' of animal shapes was set off by Dorothy's tutorials in representation. If a small boy's fingers holding a pencil could put an outline around an image of Judy, his donkey, how come she is so different from, yet so alike to a zebra? How did animals acquire shape, presence and pattern? How could I acquire a personal understanding of whatever the shaping process was? I have already mentioned the serendipity of an eight-year-old being gifted an uncensored book in which its half-disguised message and meaning was the wonder of evolution by natural selection.

Even so, there was no book-learning behind my repeated efforts to communicate with Josephine the baboon and Chipper, the initially chained then feral Vervet Monkey, but our hard-working hands and swollen brain-cases implied that we were the same sort of animal – all of us were primates.

For me, my mother's ability to translate visual knowledge into line, colour, mass (and later, print) began as an adjunct to hunting with Saidi, an effort to convert rotting specimens into permanent trophies fit to be ensconced and enthroned in a museum cupboard. Such links to the kindergarten of science became as exciting as hunting, and eventually transmuted into a lifelong quest to explore and share something of the wealth and vivacity of nature in Africa. I pursued some of the ways in which I might take intellectual possession of the long and convoluted African history that I inherit with Chipper, Josephine and a chimpanzee named Lulu.

Lulu.

Is it possible for us, as a species, to embrace our cousinship with all the other primates? Is voyaging into unexplored regions of our animal minds possible, or even desirable? In answer I see turning our backs on natural history renders us foundling orphans, at the mercy of every passing political, economic, religious or social tyranny.

To acknowledge that we are born, by obviously biological means, into a matrix that teems with other forms of life, including diverse cousins, diseases, predators (and beings many orders of magnitude larger or smaller than ourselves) demands modica of humility, even traces of empathetic fellowship with other evolved beings, from microbes and mini-bats to whales.

Within self-spun cocoons, it is easy to believe that habitats are man-made and that the greater part of our makeup is exclusively human. Not so – by far the greatest part of a human is animal and much of what is left is anciently adapted for life in Africa.

The innumerable artefacts that we insert between ourselves and our environments either dull or stress our sensibility to the extraordinary vivacity, diversity and hidden complexity of *every* manifestation of nature, as I have tried to suggest throughout these pages. Along the way telescopes, microscopes, spacecraft and eyed drones have emerged, and not just as optical enlargers. They scope out new frontiers to enlarge our minds.

Over and over again, populations with better tools and better weapons have expanded their spheres of influence from one region to another, from one continent to another.

With every such expansion, especially since the dumbing-down of humanity by a triumphant Neolithic (and now the Anthropocene), those human societies with the most intimate and applicable knowledge of nature have, over and over again, declined, suffered and died. This is the foundational, symbolic story of Cain and Abel, but the genes of both brothers mingle in every one of us.

If my own childhood can be taken as any sort of a guide, every child is born with empathy for others but it is no contradiction that fellow-feeling can co-exist with insensitivity towards the life or death of a calf, a chicken and, in times of war, to the killing of fellow humans.

Part of this insensitivity can be traced to the ease with which another life can be taken – to the point where only a button, a trigger or a very sharp blade lies between life and death, sometimes for thousands. We were not always such ingenious and easy killers.

In an earlier chapter I described seeing the last major habitat of the Hadza being bulldozed into funeral pyres. Later I made a painting in which those machine servants of the Neolithic assumed the outlines of a phalanx of crabs, reduced to their claws, all intent upon consuming the living mosaic of *miombo* and grassy *mbugas* of Yaida. These were claws intent upon superimposing Cain's checkerboard upon Abel's fractal foraging grounds (*see* p. 418).

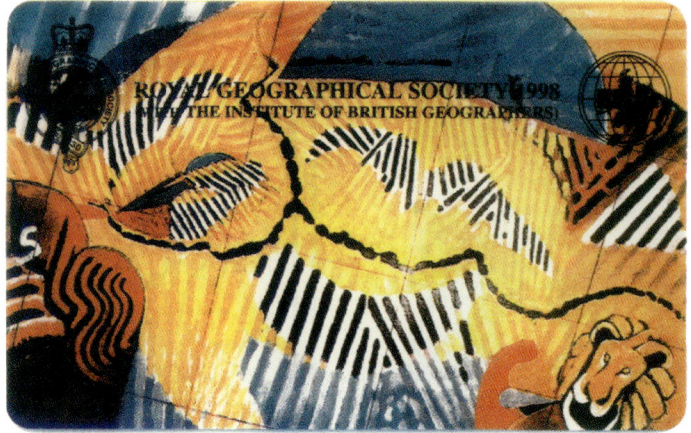

1998 Membership card for the Royal Geographical Society: Mbono Valley, Mkomazi, stage for a daily drama of life and death, zebras and Lions, the major players.

In the Cockaigne we inhabit, some passing primate, armed with crab-like machines, can operate on the illusion that he *owns* trees that were standing before his great grandfather was born, *owns* the disposal of every living being that has inhabited *their* landscape for millions of years. Now some such grotesque cartoon of an ape claims to *own* those woods, forests and all their inhabitants.

Some years after my vision of Yaida going up in flames, I was offered the opportunity to explore what a mind can 'make' of a place. What does it remember of its experience of that place? Can mind and memory be 'mapped'?

As one member of that university eco-exploratory expedition to Mkomazi, Tanzania, a locality I had known from childhood, I joined my colleague of more than 30 years, Elias Jengo. We agreed on the title 'Mind and Memory Maps' to chart our thoughts. For myself, I began by 'mapping' eight localities that I called 'The Transect Series'. Each 'map' began with literal outline maps which sampled each locality, at very different scales.

Each map of 'The Transect Series' had associations with a season and a time of day or night. The dominant colours linked with other associations, events, animals or plants. One 2 m panel maps the Vituwini Ridge and Gulela Hills. It commemorates watching the dawn come up from the top of a giant rock near the base of Ndea hill. Bushfires from the previous day still smolder in the semi-darkness. Suddenly, some small birds, Red-and-yellow Barbets, clambered up a tree beside the rock, and as the first rays of light broke the horizon, they sang a sweet and complex syncopated chorus. The male's disproportionately large head, its red plumage fluffed and pulsing, bobbed feverishly as it stomped and bucked about on its perch, wracked by the sheer energy of its performance. The precise music of the chorus, its extraordinary volume and the flamboyance of the display brought back memories of many other dawns in the Tanganyika of my youth. Other flocks answered from different directions – then the sun was up and the day took over. The colours of this map are those of the barbet, of fire and of dawn. It commemorates my memory of a very particular experience in a very precise place and it asserts that there are seasons when very small organisms can pervade the atmosphere of a place – like us, they have their moments.

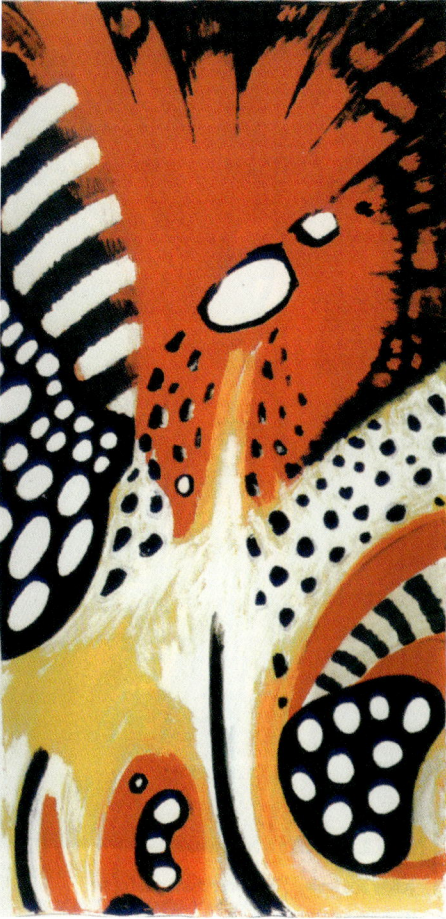

LEFT: Dawn Chorus – Sunrise in the Season of Fires, *painted in the plumage colours of the Red-and-yellow Barbet (*Trachyphonus erythrocephalus*).* RIGHT: Late Afternoon Before the Fire Season. Gazelles, Falcons and Grey Hornbills.

A month earlier, as the Kifikua Flats dried out, tiny African Pygmy Falcons hawked for dragonflies and locusts disturbed by gazelles as they waded through the wilted grass. African Grey Hornbills faced off atop whistling thorn trees, tock-tock-tocking their moments in animated heraldic symmetry.

A predominantly green map of the whole Pare mountain chain suggests that time, towards the end of the wet season, when trees and grass are full of caterpillars. The air pulses with the monotonous calls of the Emerald and Diederik cuckoos, which are particularly common at that time, their barred, iridescent plumage a good match for the colours of the vegetation during the caterpillar season. For me, the cuckoos quite literally colour my perception of the landscape.

We often describe animals as being 'well camouflaged' – for both painters and biologists, such a flip description seems quite inadequate to the wonder of it. A host of different physiological and anatomical processes have been co-ordinated to evolve colours and shapes that 'fit' the animal into the landscape.

LEFT: Late Rains, Late Morning, Pare Mountains. Calls and Colours of Cuckoos Tint the Caterpillar Season. RIGHT: Ngarunga Gorge and the Badlands – Early Afternoon, Early Dry Season Owls and Butterflies.

Appearances are important. For both prey animals and their predators, appearance or disappearance can mean the difference between survival and extinction. Although external appearances are only a tiny fraction of any living animal's physical being, they account for most of what we can portray or visualise about them. Painting is also about appearances, and most of a successful picture's message lies in the wordless associations, the hidden meanings behind a choice of subject or theme. You have to learn to read a map or a picture but once you can interpret all those lines, symbols and blobs of colour, a more substantial world opens up to you.

Coloured marks and smears are signatures, no less than written autographs, so much so that some collections of paintings can resemble glorified autograph albums.

As for maps, one of their appeals is their lofty viewpoint. They emancipate the viewer from earthbound gravity. With maps, we are no longer hedged in by the boundaries of horizons. We see the structure of organisms differently from

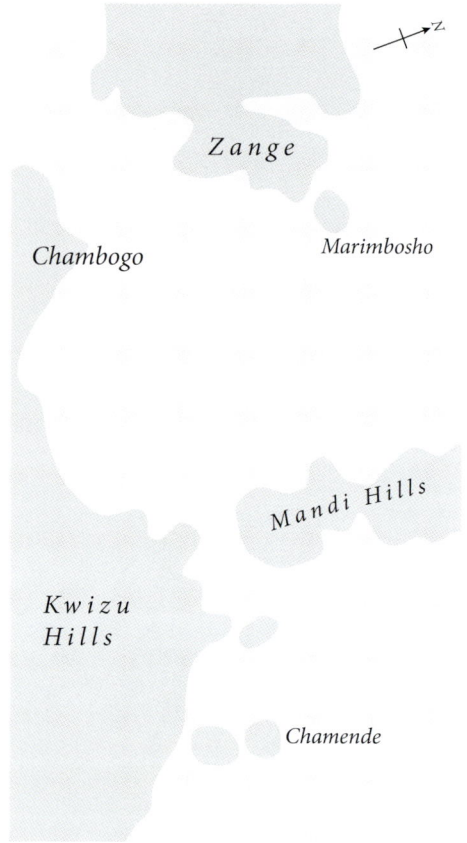

LEFT: Nocturne. *Nightjars and zebras on the burnt-out stubble of Zange sumpland. Little white simba close to Simba Plot. Moonlit baobabs in the foothills.* RIGHT: Nocturne *mapped.*

above. The endless battle of animals and plants against gravity disappears. We no longer see the props, struts, trunks, limbs and towers that rear living bodies (and buildings) up from their ultimate fate – collapse. Linearity and verticality become blobs, and a pattern of blobs resolves into vegetation, termitaries, herds, towns and crowds, all unified into a pattern where flow seems to be the main dynamic. The flow of streams and rivers, cutting into the flows, cracks and bucklings of a restless Earth. The flow of traffic, from elephant and cattle paths to highways and bush-fires. Above it all, the flow of air, sometimes laden with rain or dust, sometimes lazily filling valleys with mist or sculpting mountains with clouds. Migrating flocks of birds and butterflies also join these aerial flows. Fragments of their passage, a broken wing or feather, may float down to earth, losing the complexity of dynamic three-dimensional structures, but their exquisite details are also flat mappings of a life that once fluttered across this broad land, mementos of hidden tempos.

During the dry season large herds of zebras trek back and forth along Mbono valley. After a hasty evening swill from their watering pond, Dindera, at the north end of Mbono, the zebras hurry south to spend the night crowded together on the open Zange sump-lands, where they are safer from Lions. After the fires had swept the valley, moonlight highlights the land with black shadows, all-a-zebrid.

The Landscape of a Lark's Plumage.

BELOW LEFT: Zebra Herds in Bone Valley (Before the Fires). Morning.

BELOW RIGHT: Kisima Airstrip, Kangas and Kudus. Midday, Dry Season.

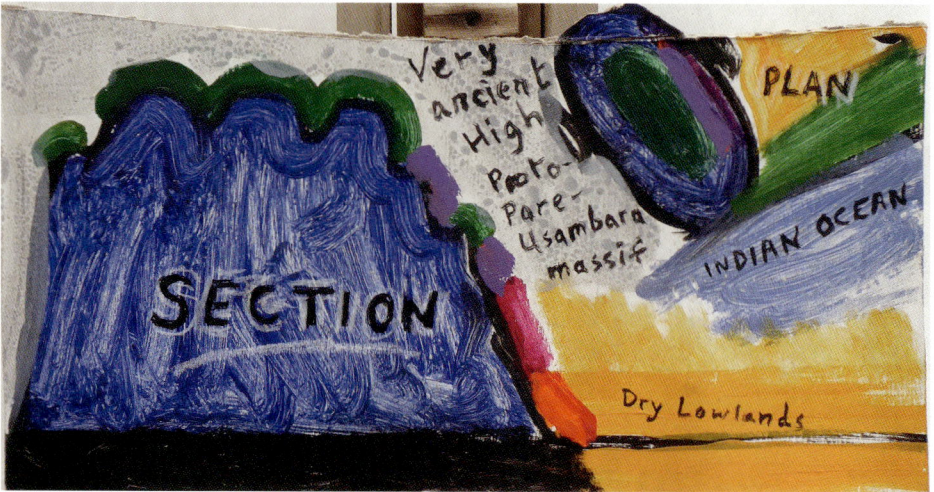

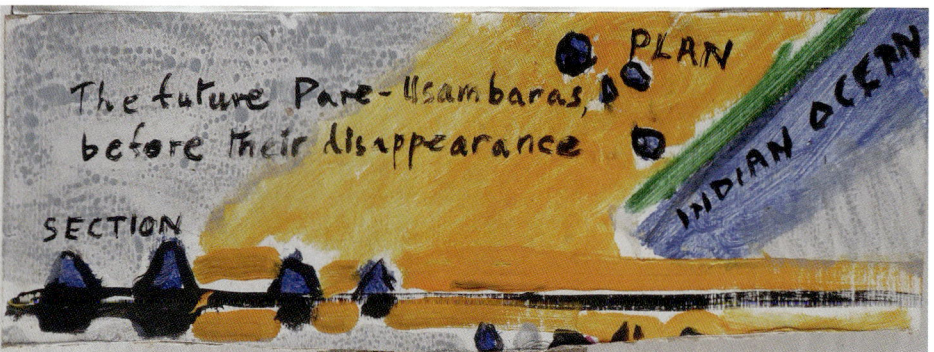

Gondwana's Upended Edges: *(top) section and plan of ancient Eastern Arc Mountains;*
(middle) present eroded remnants; (bottom) erosion will ensure their eventual disappearance.

Tracks between the zebras' two dry-season grazing lands mapped out the cartoon of a giant bone (we started calling Mbono 'Bone Valley' for its elongated shape, wedged between eroding hills, and for its scatterings of bleached skeletons). Zebras, their graceful figures alert in the distance, their stripes crowding out our binocular lenses, became icons that dominated my 'Mkomazi Mind and Memory Maps'.

On a reconnaissance flight out of Kisima, landscapes were seen through a whirling propeller while kanga butterflies spun all awry in the back-draught. Ring-doves clattered away through the grey *Commiphora* bush as we gained height over red rocky hills, spotted with black algae and over-streaked with white fossilised dassie piss.

My memory map of the wet season is enlivened by repetitive volleys of clanking and clinking castanets, which signify dusk for the Vulturine Guineafowl. I borrowed the clean, clear blue of their breast feathers and their elongated zebroid neck feathers (which help shed rain during the wet season). Vertical stripes help emphasise fowl tallness, like City bureaucrats in pin-striped trews. Tinted guineafowl spots colluded in what evolved into a landscape of what we chose to call 'evolutionary hot-spots' (*see* p. 209).

The Mkomazi expedition was privileged to witness an ancient and narrowly stratified complex of ecosystems, in which one square metre of 'scrub' yielded 1,600 individual arthropods.

Each year this immensely old chain of hills (once mountains) continues eroding into the dry lowlands around it. The last remnants are mere outcroppings of hard rock, while the heights have montane rainforest and lush banana gardens. In between lies every permutation of altitude and vegetation.

Desert roses bloom in the rubble of well-drained screes, while cicadas and lichens are nursed by mist and dew on the upper crags. Every nook and cranny of this ancient landscape is inhabited – a living fabric embroiders great vistas of plain and mountain. From elephants and baobabs to earwigs and fungi, each living thing pursues its own unique life history. Much of their lives, like those of strangers in a city, can only be guessed at, but the vitality and variety of natural life in Mkomazi dwarfs the human bustle of any city. Its riches will outlive any city – after all, it has inhabitants whose lineages have been present for tens of millions of years.

Or will they?

All my fellow scientists were impressed by the steady erosion of Mkomazi's fabric under the influence of people – fire, felling and flocks. Every year the fires burned deeper into the woods and thickets. Shade and moisture depart as more trees fall, and the pangas of pole-cutters harvest young growth in the understorey while voracious cattle and goats cut back the remaining herbage. Perhaps our team had the privilege of glimpsing a vanishing world. If so, does this mirror the quality of Cain's civilisation, that his children must burn, kill and consume the biological riches of our planet and plunder our continent's treasures? In this sense, Mkomazi and Africa are the world.

Stump Tombstones for Mkomazi's Forests.

Makerere Art School frontages as drawn by Emily Sempaya.

Elias Jengo and I were once joint participants in one of modern Africa's most distinctive, dynamic and least known art movements. The University of East Africa might seem an unlikely setting for original art, but in its day it was a major cultural influence in Africa. To be a Makererean at the beginning of the 1960s was to belong to a very local indigenous institution and yet be fervent believers in a global Uhuru movement. The optimism of boom-time coincided

A spoof on colonial perceptions, led by Elias Jengo (first porter) and Jonathan (seventh porter) in the lower image, all armed with paint brushes (Makerere, 1961).

with the dismantling of a discredited Europe's colonial empires. The pessimism and austerities of wartime had been replaced with the spirit of Uhuru – Freedom.

For a short while, newborn independence meant respect for all cultures and acceptance of the many faces of cultural and personal expression. It was in this tolerant climate that open minds, civil rights and self-discovery flourished. Elias and I shared that global atmosphere as well as the more local one of love for your own backyard and family folk. In our work, we challenged the assumption that nationality, race or class are the overriding contexts for art. Just as we learnt from scientists and technicians on the Makerere campus, our expedition's biologists and geographers directed our eyes and minds to many wonderful details of the Mkomazi environment, asking endless questions and revealing some truly amazing stories, amazing interdependencies.

As at Makerere, we too sought to show the scientists how meanings, expressed in more personal ways, without words or graphs, could be teased out of our shared studies of landscapes and their inhabitants. If maps are the ultimate scientific artefact, the geographer's pride, we, instead, appropriated much older concepts of charting thoughts about memorable places.

One of our first collaborations (as fledgling teacher and eager student) dated back to what we called 'The Uhuru Show', which celebrated independence in

1961 for Tanganyika, our shared birth-place. Feeling that we should commemorate the place and the project that had brought us together again, we sought permission from Mkomazi's chief warden to leave our mark in Mkomazi. We trekked up to a gully above our camp known as 'The Amphitheatre', where erosion had exposed great rocks with flat surfaces and an overhanging rock roof.

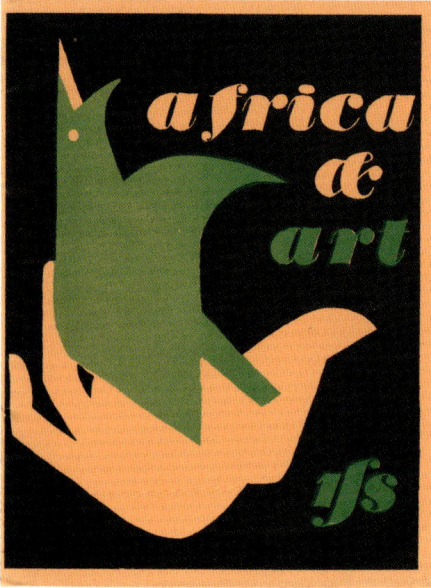

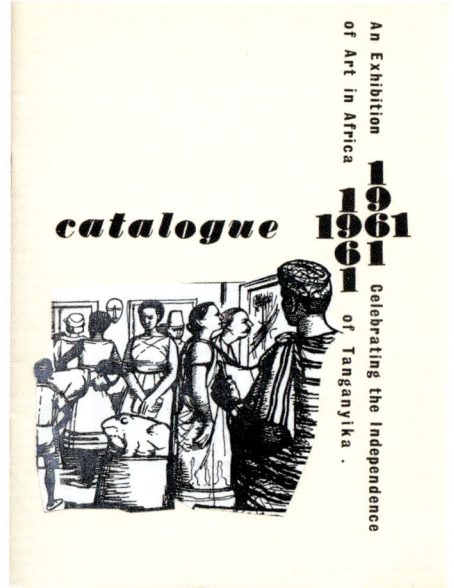

The Bird Flies Free. *Tanzanian Uhuru Show catalogue, cover and title page.*

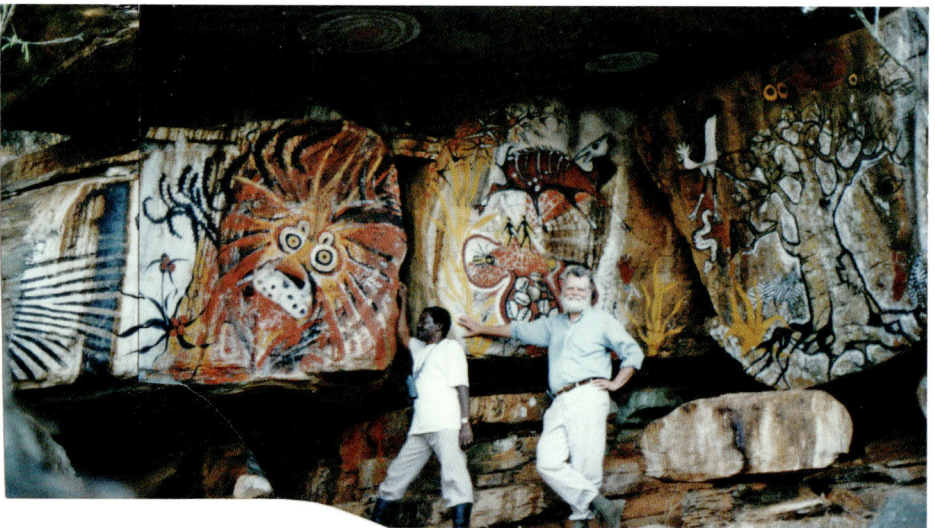

Jengo, Kingdon and the rocks they painted (1995).

We took raw earth oxides: red, yellow, white and carbon black. We emulsified them with eggs, gum and water and started to apply them with local *mswaki* toothbrushes. Our subjects were drawn from our own observations, enthusiasms and experiences of the place, but also from the objectives of our project. Fauna, flora, inter-relationships, natural processes, we wanted to imply them all, past and present. To unify the whole, each of us left our images unfinished. We changed places so that Elias could complete my frontal Lion portrait while I added roots to his baobab. To sceptics we answered that our omelettes were now un-de-scrambleable.

During a future wet season, the rocks we painted will slither into the gully below, yet before that the hyraxes will have overpainted our images with their urine. Even so, as we wrote in the catalogues for our shows in Dar es Salaam and London, 'For a while there will be a record, like fallen feathers from migrant birds, that, in remote 'Bone Valley', two artists joined a team of scientists whose task was to proclaim to a wider world that Mkomazi, one of Tanzania's most precious assets, is a place to be celebrated and must be preserved.'

The African continent is the repository for most of human history. This plain, unvarnished fact needs to become a major underpinning of what we learn and teach in our schools, universities and institutions. If we are to earn the world's respect, we must face down Cain's intellectually impoverished progeny and continue writing ever wiser, ever more truth-seeking Books of the Generations.

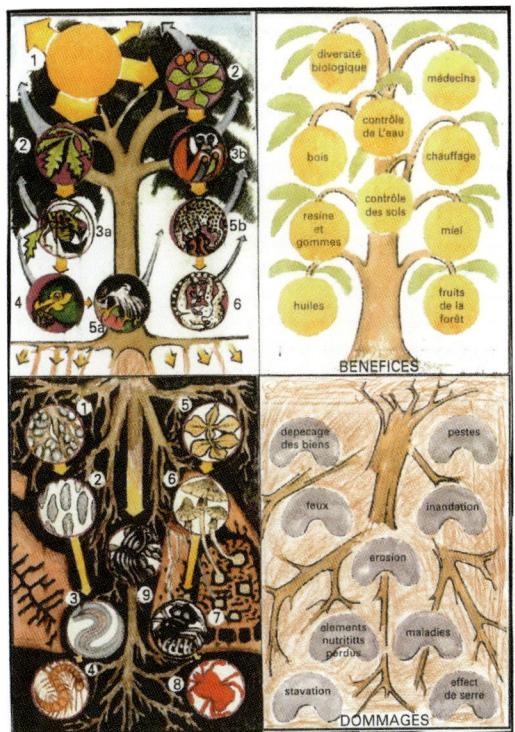

A visual essay trying to comprehend the dynamics of life in an African forest: its benefits versus the ills of deforestation.

PROCESS AS PRINCIPLE

22

*In which the processes that govern all life on our planet have long
inspired the practice of science and natural history. In which a
fabled inheritance is being squandered. Today, the alarums of
scientists and citizens must inform policies and rein in the excesses
of Industry, The Party, so-called 'Development Agencies' and
whatever groupings aspire to assert their own over-riding primacy.
Happy* Habilis *Day.*

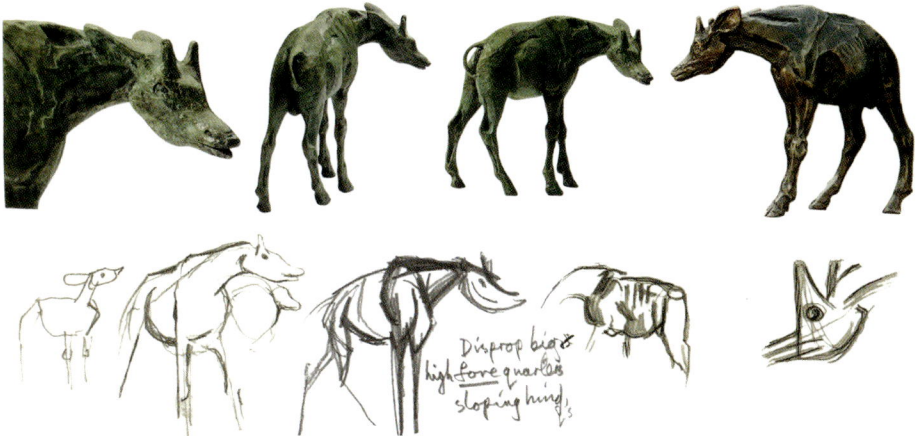

Drawing with a pencil, drawing with clay. An Okapi.

Her carer reported that she was never so tranquil and relaxed as during the
worst of the bombing. She was a penned Okapi, and he was her keeper in
London Zoo during the Blitz of 1943.

A major rainstorm in equatorial Africa, especially over the Okapi's Congo
basin, is accompanied by volleys of thunder and lightning that dwarf any blitz-
krieg. Deafening, drenching storms, sustained month after month, year after
year, are enough to breed indifference to explosions in any Congo Okapi, any
Congo primate, any human in the Congo.

As a four-year-old, I and all my family were hit by a thunderbolt during just
such a tropical storm. Under torrents of rain I was the first to regain consciousness, to see Dorothy and Teddy, sprawled, shiny wet, filthy limbs all akimbo,
in the oddest of postures, grimacing as they came to. With an anxious Teddy
pounding her chest, Grandma Minnie was the last to wake, making funny little
yelps like a puppy. Later she thanked God for saving us – others may well have
wondered if he wasn't trying to kill us.

OPPOSITE: *Kilimanjaro's crater, plus elephants below, as seen through propeller and clouds.*

Nocturne *in stained glass, Rondo Chapel, south Tanzania.*

Teddy, whose father was an electrical engineer, likened thunderclouds to giant wet-air torch batteries. 'Lighter, warmer water, cooling as it wafts upwards, creates a positive pole while negative bits of heavy mist collect along the base of the cloud. Like combing your hair can give out sparks, churning winds rub the two charges against each other to create lightning. Super-heated air around that giant travelling spark makes that big bang – lucky it doesn't thunder every time you comb your hair!'

'Shango's Axe', or thunderbolts, still require more explanation than Teddy could summon up, but by removing divine retribution or reward from the most frightening moment of my small, short life he helped reinforce a vision of life as existence within a dangerous but explicable universe: a universe in which the mind-discipline we call science has become the single greatest bequest from our evolutionary heritage as human animals. Science allows us to examine the world we grow up in as the product of comprehensible processes. I see many benefits and artefacts accruing from early engagement with science.

As a 19-year-old electrical apprentice, Teddy's father served on a cable-laying ship that connected Eurasia with North America. I like to think of his teenage adventure in symbolic terms. I imagine Zachary Senior riding a wave of applied science as it swept out of Eurasia to the furthermost reaches of the world.

My father, his father and their various mentors rode some such waves as part of an ever-wider dissemination of scientific education and research. It now permeates all societies and it will be our only salvation as a global community, but scientists, worldwide, must assert intrinsic as well as instrumental values if we are to survive.

Our planet as a fish: (left) blue and healthy; (right) assaulted by niche-thieves,
close to death (paintings by Afra Kingdon for her book Green Makes White, *2007).*

In the 80 years since Teddy had the audacity to reduce Shango's Axe to an explosive, animated aerial torch battery, our own human numbers (rude appetites,
capricious games, toxic economies and effusions) have come to pose the greatest
challenge to science, medicine, politics and to survival itself. At the time of our
Tarime thunderbolt the planet supported some 2 billion people – now there are
about 8 billion and we are increasing fast. Such numbers of animals, let alone
wastrel humans and their ordure, have an impact on virtually every natural process – global soils, seabeds and climates among them. This also dangles an ever-
enlarging bait over any organisms that can outwit our many defences and colonise, even consume, our combined biomass. In spite of numerous earlier epidemics, some global, some more local, our unpreparedness for Covid-19 highlights the
exaggerated hubris of our times – our vulnerability to the unseen and unknown.

As for our propensity for tribal warfare, in quarrels that employ ever more
destructive weapons, in many respects we know more about Mars, God of War,
than we know about our own planet, our origins, our prehistory and our ability
to govern and educate ourselves. We send eyed buggies to Mars and men to the
moon, marvelling at the images their cameras beam back to us. Our rich imag-
inations, in thrall to science fiction, technophilia and our own technological
prowess, label satellite photographs of our planet 'Spaceship Earth'.

OK, we need metaphors but I prefer my own daughter Afra's vision of Earth,
not as a man-made spaceship but rather a resplendent globular fish swimming
through space. I am moved by her seeing our planet as a living entity, and by
her anguish over that fragile fish-in-the-firmament, being fried in the fumes of
all our fires.

Afra's concern is now global but local. National and provincial affairs are now
so dominated by the actions of actors, often maverick actors, within the dominant
economies of the world that our immediate concern must be to try to understand

Roller theme.

BELOW: *Olokun Janus mask for the Agbo festival, Ogun region, Nigeria.*

the underlying dynamics of both nature and culture as they operate, right now, before our gasping globular fish dies in the heat of our own bungling hubris.

In 1996, I was invited to join the Centre for Environmental Policy and Understanding at Green (now Green Templeton) College, Oxford University, where I presented a paper called 'The Process Principle'. It began with the premise that humans were very recently evolved organisms inhabiting a small planet, which they had only just begun to comprehend via the medium of science.

How could teachers inspire their students with the accessibility of science? As science found expression in ever more diverse institutions and applications, each with diverse histories, I wondered whether others were as confused and intimidated as I was by the Hydra that science presented in 1996.

On what principles did the grand progress of science as a whole cohere? All science seeks to understand how natural processes work, so I took 'process' as the founding principle of all sciences. Disciplines, as they existed in 1996, could be grouped into at least four research divisions – four conceptual categories with the potential of becoming process-oriented data banks. They were:

1) **Physiology**, or the workings of organisms.
2) **Evolution**, or the transformation of organisms through time.
3) **Ecology**, or the interactions of organisms within habitats.
4) **Geophysiology**, or the interactions of Earth and astronomical systems.

If such databases could be open to, contributed to and led by the world's scientists, with their findings open to and flowing out into the education of our citizenry, I felt that science could be more readily seen and taught as humanity's greatest single achievement.

Physiology embraced genetics, biochemistry, microbiology, reproduction and ontogeny.

Among **Evolution**'s many subdisciplines were biology, speciation, biodiversity, biogeography and palaeontology (as well as genetics once again).

Ecology could call upon studies in community dynamics, population biology, succession and sustainability.

Geophysiologists map out, explain and explore tectonics, global climatic cycles, soil sciences and study of the drivers behind biotic, even astronomic, change.

By the last years of the 20th century, databases serving these disciplines were already contributing greatly to the conduct of research. Even so, those pre-existent databanks had the potential to reorganise knowledge in order to make scientific facts more teachable, as well as more relevant to political priorities. Even, perhaps, less vulnerable to sabotage or perversion by the Pharaohs of politics.

Re-ordering our systems of knowledge could change the way in which questions were asked and solutions to problems sought. The format within which new knowledge was planned and presented could be reshaped to increase accessibility and comprehension. Process parameters could help define and defend environmental policies at national, continental and global levels, by clarifying criteria in terms of understanding natural processes.

Actively running programmes and publications such as *Science* and *Nature* stimulate continuous inputs to their own stores, at the same time as helping solve problems and guide initiatives with their outputs. Such journals are both heralds and journeymen for a better-educated future.

In 1996, I tried to examine then current institutions from the perspectives of process, and each from the point of view of sustainability, biodiversity, teachability – each envisaged as a potential division of a process data bank. Take 'Physiology' as title for, say, Division One of a global Process Research Programme. In nature, genes and many other entities do not exist outside their containers – organisms. Genes and symbionts have their own biogeographies, in which very precise preconditions must be met before the replication, growth and maturity of microorganisms (or that of their hosts) can take place. Although the conditions under which physiological processes operate have long been amenable to analysis, their functions need much more comprehensive study in terms of global sustainability.

Earlier on I touched upon the stromatolites that grow in African rift valleys, as well as the Persian Gulf, Yucatan and western Australia. They are significant for understanding Earth's ecological history, and it has even been proposed that we owe to them the early generation of oxygen around our planet. The fossil record shows that stromatolites were once extremely widespread and abundant. Yet it is easy to envision a local enterprise, shrimp farm or fuel spill casually destroying such rare living links with planetary history, even before their physiology and origins had begun to be fully understood.

An important dimension of physiological processes is individual life history. Some trees and marine organisms are known to live 5,000 years while the ecological role of, say, 'Standing Deadwood' may extend up to 8,000 years. Cycads (gymnosperm plants that were a dominant form of life 300 to 400 million years ago) have similarly long lives, and at least one South African species, *Encephalarctos woodi*, was known to have been reduced to a single individual by about 100 years ago.

Increasingly, such last individuals have to be taken into custody, but where the lives of such protected individuals can last several thousand years, conservation must develop very long-term perspectives.

There are much shorter-lived species where self-replicating populations depend on their own behaviour or physiology. Some African birds and certainly some bats only breed and survive in huge, dense flocks or roosts. Understanding such processes is essential for maintaining biodiversity and sustainability. In the context of understanding the matrix from which pandemics or 'pests' emerge, this neglect is a reproach to contemporary practice.

A second Process Research Division would study and concentrate on 'Evolution'. Remember that until very recently the most blatant absence in all international conversation was an almost universal avoidance of any mention of evolution, particularly evolution by natural selection *as a process*. Yet evolution is the process to which we humans and every other living being owe their existence. At any given time in any given place, ecological communities are evolution's communal expression. In 1996 I insisted that there must be deliberate engagement with these 'elephants in the living room' and with all the other processes intrinsic to life itself.

This is a concept that has been triggering negative knee-jerk reactions in way too many people for way too long. Now, many more educated people link their concern for the extinction of animals with charismatic naturalists, led by Sir David Attenborough, who has always framed his passion for natural history in evolutionary terms. Eloquent in his horror of impending extinction for countless animals and plants, he has shown that the crisis is now on a global scale. And he has also identified, in vivid imagery, who are the exterminators.

Over very extensive areas, industrial agriculture's felling, ploughing and poisoning has exterminated entire ecosystems (including vast subterranean communities that had yet to be studied, even in the most cursory way). Attenborough has reminded his audiences that, for an ever-widening circle of species, the prospects of survival decline by the year.

Extinction removes not only another component from a living ecosystem, it always removes an influence that once shaped the evolutionary history and potentially the future of many other species (including humans). Such subtractions progressively favour an ever-smaller number of generalist survivors, which soon outcompete most remnants of the more complex communities they replace. (A similar reductionist dynamic progressively reduces the number of basic foodstuffs, engendering vast, chemically dependent monocultures.)

Nowhere are the consequences of this dynamic more obvious than in tropical rainforests, coral reefs and Africa's Cape, where Earth's biodiversity and biocomplexity peak. Here fast, exponential rates of extinction follow the initial collapse of key species, notably large, long-lived plants, mammals or fish.

Large-scale ecosystems, once widespread, included numerous super-successful giants drawn from a relatively small number of lineages, such as some big tropical trees, elephants, whales, sharks and squid (reduced by vandals to lumber, ivory, oil, soap, soup and chicken feed).

Understanding the many dimensions of any species' evolutionary success has acquired great urgency as contemporary 'development' continues to undermine, degrade and even exterminate the upper ranges of scale in nature.

Taking the evolution of gigantism *as an evolutionary process*, researching the conditions under which giants evolved should be a major consideration for setting aside reserves for large-sized and long-lived organisms. It may be too late for many species but that principle could and should guide designation of many and extensive forest and woodland areas as out of bounds to logging or agriculture, especially so in tropical Africa.

I remember Gregory's gentle reprimand. 'I don't want you to domesticate that terrifying giant. I need real, untamed elephants to shrink me, not them, down to size.'

The survival of large and/or highly specialised species belonging to unique, lonely lineages should be an urgent contemporary priority, as should conserving the localities where such rare endemics survive. Today new threats arise from genetic tinkering, whereby genes are 'borrowed' (more accurately appropriated or just plain stolen) from successful wild biota to enhance the fitness of domestic or cultivated species, a practice known as genetic modification or GM. There are, of course, benefits as well as profits, but also lethal dangers in GM. Dealing with the dilemmas it poses presents unprecedented challenges that are central to a contemporary education and to global culture.

As with all utilitarian, profit-driven practices, the theft, privatisation and transfer of properties acquired through natural selection puts dead end-products before the living processes that allowed the 'product' to exist in the first place – even before the process itself is fully understood.

The enormous profits from patenting organisms have spawned powerful lobbies and generated sci-fi visions of a planet in which all organisms become man-made and privately owned, their sole purpose being to serve human appetites at a profit-generating price.

Outside Africa, just such meddlers have commandeered entire ecosystems – they have arrived and seek to commandeer Africa, but our continent's ecological diversity is of such huge significance for our own self-knowledge and too intricately patterned and structured for us to allow any such sorcerers' apprentices free rein here. For a start, the wholesale felling of our forests by logging companies must be stopped.

A third Process Research Division could operate under the banner of 'Ecology'. For a short while, East Africa, Oman and Costa Rica were global pioneers in assessing and studying their own resources in nature and ecology, but the paucity or absence of follow-up has been deeply disappointing. In Australia, AUSLIG and NATMAP (Australian Surveying and Land Information Group and the National Mapping Division) have published detailed analyses of vegetation types past and present. Yet powerful lobbies still demand access to ecosystems that they have no knowledge of nor interest in, other than profit for themselves.

More positive developments include the World Conservation Monitoring Centre, United Nations Environment Programme (UNEP), the US Conservation Science Programme and the Environmental Systems Research Institute (Washington), all of which have been active in mapping the distribution of species and eco-regions, and in assembling databases. National initiatives such as these need to go global, or at least continental.

It may not be possible to regain much knowledge from the vestiges of already lost habitats, but it is important to try. A rounded knowledge of ecology and evolutionary history relies on models of the full spectrum of natural ecosystems. Natural parameters such as longitude, latitude, altitude, temperature, rainfall, geology, nutrients, soil and so on have long provided criteria by which natural environments are classified and understood. To these should be added the long-term influences of climate changes and past biota, especially the impact, past and present, of prolific animals such as termites, ants, fungi, microbes, herbivores and hominins.

Ecological processes are *not* optional adjuncts to the management of economically productive landscapes. The political and industrial lobbies of arrogant agriculture have, for too long, claimed exclusive priority to land-use and over policy at all levels. Their claim that all lands suitable for the growth of crops or 'animal husbandry' should be reserved for these purposes and come under *their* purview *must* be challenged. Instead, every settlement, from farm and school to city and village, needs its parks and gardens. Wherever possible tracts of vegetation native to every region should be reserved or restored on every landholding and provide the refuges and corridors so many organisms need. All self-described 'owners' of land are mortal. We can be sure that their stewardship will pass on to better educated generations, less entranced by ploughs, poisons and gigantic machines.

Ear-marking all the fertile areas of the world for immediate or future conversion effectively excises all our richer natural ecosystems from having any future. It is no longer acceptable to find demolition notices erected around all Earth's richer habitats. Africa is just one region of the world in need of ecological education at all levels. Already hosting UNEP, Africa has initiated more conservation programs than any other continent. As the region/continent best endowed and most qualified, our scientists will *lead* the world in setting up and developing intelligent ecological policies. Our authority derives from the fact that the human species evolved here and we, the people of Africa, are the contemporary, temporary, yet pioneering guardians as well as expositors of that much-neglected and oft-contested legacy.

A fourth Process Research Division could operate under the title 'Geophysiology', a word that was coined to describe the Earth as a system governed by natural processes. The quest for an integrated approach to understanding Earth Sciences as a whole led to foundation of the Geophysiological Society, which aims to bring together established knowledge as well as fostering and promoting further research on the planetary environment. The interactions of land, seas, atmospheres, geochemistry and climate are obvious and, in as much as these systems are understood so far, that knowledge has, for the most part, emerged from the practicalities of weather prediction, shipping, defence, aviation, geography and so on. Among the many national and international climate studies, CLIMANZ has been a pioneer in exploring past and present climates as a major factor in understanding the biota of Australasia. This initiative was conceived and is run by CSIRO, out of Canberra. Africa needs just such initiatives, operating at a continental level.

'The Process Principle' began by asking what current practices claimed to study or influence natural processes. Among multiple relevant organisations in 1996, few escaped the oversight of growling watchdogs guarding particular interests – national, industrial, factional. Fewer still were led by individuals free from such constraints. In 1996, newly fledged and fledgling non-governmental organisations (NGOs) were anathema to most official bodies, because they existed outside established political and economic control. That hostility itself reinforces and augments the need for just such bodies. Scientists and science must assert their independence, defend the integrity of their science and insist upon their role in politics as guardians and developers of knowledge itself. They must not bow, like serfs, to expedient politicos and dictators.

Were Africa united, I believe that our union could become best placed among all continental powers to assert intrinsic values over 'instrumental' or economic values. The history of slavery, plantation and mining corporations in Africa has taught us not to trust the savage values of those that allowed or imposed such atrocities upon us. In the meantime, we have learned that we are the only custodians of our vast intrinsic wealth. Not only that – we are custodians of humanity's origins. Our soils hold the clues to our very existence as a species. We have the right to set the terms on which any external agency seeks to share our bounties.

I am not advocating greedy gate-keeping. Instead, I am insisting that there are circumstances where the 'instrumental' values that markets, profits and armies impose upon us *can* and *must* give way to intrinsic values (sometimes profitless), wherever exploitation threatens survival itself; the survival of people, of species, but above all, the survival of entire and intact natural communities. Wherever these two opposing value systems come into conflict, our new, imperfect but ever-growing knowledge of natural processes should invest science, scientists and African communities with an authority we have never asserted before.

We have the right to negotiate substantial support for our schools, universities, scientific institutes, museums, parks and conservation estates from those that seek to profit from Africa's largesse.

We have rather oblique precedents in those few, far-sighted (sometimes rich and conscience-stricken) Americans that funded and founded numerous similar great institutions and Parks in their fledgling nation. In Africa we will probably need to be more coercive, because our continent has a long history of exploitation without any trace of conscience by various deeply entrenched and unscrupulous colonial, commercial and political hegemonies.

In 1997 many globally oriented institutions were already major sources and trustees of expertise, data and authority. Their fragmentation and diversity had and still has many practical virtues, their potential as mechanisms for implementation being perhaps the most important. All natural processes are of great complexity and form part of or impinge upon still larger processes. The origins, character and viability of all or any organism, including humans, need to be seen in relation to other organisms.

In 1997 the survival of viable habitats and ecosystems (particularly those within which *Homo sapiens* evolved) seemed one such priority because too few people realised what planetary parvenus, what johnny-cum-latelys, modern humans are. Nor was it then understood how deeply and fundamentally damaging modern industry, agriculture and industrial fisheries had become. Their malignity is now horrifyingly obvious and in urgent need of reform, even prohibition.

Earlier chapters have tried to illustrate, even reconstruct, stage by stage or step by step, our journey to sapienthood. Each stage had its own ecology and geography, each attended by its own community of evolving animals and plants. Today, these are being appropriated, poisoned and obliterated by logging, industrial agriculture and by aggressively promoted monocultures. Any reconstruction of our own evolution will become ever more difficult as Africa's diverse communities die, or struggle to survive.

We are on a protracted quest to find the very localities where some of our Eves and some of our Adams or Abrahams first emerged. What if we wake to find each of a succession of Edens has become sugar, sisal, pulp-wood or oil-palm plantations, a bombing range or a bankrupt beef-lot? Right now, greedy, ignorant gangs, protected by powerful political and industrial lobbies, are demolishing the last nurseries of our race – the human race. Inhabiting those forests are forest people and older relatives from other branches of our family tree. We should have grown out of exterminating 'First Peoples' and cannibalising our closest cousins but, no, that is exactly what proudly 'modern' people, in possession of electronic toys, are doing. Chronically undervalued, under-studied and over-abused, Africa's ecological communities were (and remain) essential for the reconstruction of every stage of our own evolution.

Peering backwards into human origins, all we see (through a dense fog of ignorance) is a pathetically inadequate number of initiatives, all set up by incandescently independent-minded individuals and sustained by the most marvellously stubborn and highly motivated of people in the face of fierce opposition, ignorant hostility or bland indifference.

Every practising student of human evolution and ecology, every environmental idealist, every media-savvy campaigner is a gallant pioneer, a secular songster in our species' Hosannah for dawning self-knowledge and for the beauty of a living, not a dying, planet.

If process principles are to have any traction, they depend upon all those institutions and intellectual traditions that already exist. Early glimmers of global biological thinking were articulated at the International Zoological Congress in Graz in 1910, but it was World War I and the poison of unfettered nationalism that triggered a convulsion of conscience and responsibility among thinking scientists. Albert Einstein, Marie Curie and Alfred Zimmern were among those who set up CICI, the International Committee on Intellectual Co-operation, under the unprecedented 'League of Nations'. That great Australian, Gilbert Murray, became its first chairman, but Hitler and Stalin soon destroyed it. Their only global thinking was nationalistic, military, self-aggrandising, criminal. They have their would-be contemporaries.

After another World War and another convulsion of conscience, the idea of a league was revived, and renamed 'United Nations'. CICI became the Educational, Scientific and Cultural Organisation and its first Director General was Julian Huxley. Under the institutional base of UNESCO, Huxley was one of the creators of the International Union for the Conservation of Nature, IUCN. He was also a driving force behind another UNESCO offshoot, 'Man and Biosphere', launched in 1963. During his tenure Huxley asserted a level of independence for scientists that was soon reined in by industrial, national and political interests. Ever since, their oppressive oversight has tended to eviscerate too many initiatives.

One post-war effort to maintain, protect, study and monitor processes was the International Council of Scientific Unions (ICSU), which set up the International Geophysical Year in 1957. This was somewhat of a role model for the first systematic application of scientific models to the classification of natural environments. In 1962 ICSU set up the International Biological Programme (IBP) to survey, analyse and interpret the entire range of environments under the more utilitarian objective of studying the biological dimensions of productivity and human welfare.

IBP enunciated process principles in its survey of large numbers of readily measurable factors. The organisation explained that 'these could then be grouped and analysed in relation with the utmost practicable freedom of bias for different purposes, at different levels of generalisation or detail and for use with different classes of data'. One priority was to apply new computer modelling to the description, classification, interaction and working of ecotypes. In spite of such promising initiatives, the Brandt Commission warned, as late as 1980, that the Earth's 'natural resources base' would degrade fast if strong conservation measures were not implemented immediately. Today, it degrades ever faster.

Understanding both present and past extremes of climate, especially in Africa, is central to the study of all biota, past and present, in our comprehension of *their* evolution as well as *our* evolution. The destinies of entire human populations now depend upon a deeper, science-grounded understanding of climate dynamics. Already, there are localities where people are drowning, dying of thirst or being cooked by summer temperatures of unprecedented heat. That industries and agriculture have helped precipitate these climatic upheavals can only be denied by the ignorant or wilfully mendacious.

Immediate concerns over global warming led thousands of scientists to form and contribute voluntarily to a panel that became the IPCC (Intergovernmental Panel on Climate Change) in 1987, and this in turn led to the climate convention of 1992. Since then, populist and nationalist politicians, even bureaucrats, have tried to infiltrate and/or sabotage virtually all such initiatives. Too many powerful countries are led by smug liars who cultivate and thrive on cultures of distrust-in, even contempt-for, expertise. Somehow, the oversight of too many rogue-states, too many rogue-leaders and too little independent scientific input must be fearlessly offset by new generations of impassioned scientists.

Climate science and evolutionary biology are reluctant Cassandras of our age. Whether humanity can act on such discomfiting prophesies hinges on who makes political decisions. An actively changing climate screams for global rather than local responses. Guided by independent science, we need more research and more action, unhobbled by entrenched political parties, governments and generally recalcitrant industries. The time for real change, real action, has come and Africa *will* lead on several of many fronts.

'Our Common Future' (The Bruntland Report) of 1987 warned that very harmful long-term effects would follow if industry's present polluting practices continued – a warning that went unheeded by industry and its in-pocket-politicos. These environmental initiatives were followed by 'The State of the Environment' (OECD 1991), which, in turn, was followed by the Earth Summit in Rio de Janeiro in 1992. Here, a programme of sustainable development was posed and a biodiversity convention set up to catalogue 'natural resources'. Leading nations pledged 0.7 per cent of their gross national product as aid towards the promotion of sustainable development in the world's poorer countries. Few of them actually delivered, and, for some, 'Aid' offered a stalking horse to further the interests of already powerful nations, industries and their carpetbaggers.

One grandchild of post-war heart-searchings was the launch, in 1970, of the United Nations Environment Programme (UNEP). Given that Africa has the most to gain from an enlightened approach to natural processes (and the most to lose in their disregard), it was heartening that Nairobi, in equatorial Kenya, was chosen for its headquarters. UNEP made an early start on monitoring natural environments, reinforced by the World Conservation Monitoring Service, based in Cambridge. Ten years later the World Conservation Strategy (WoCoSt) was founded with three stated objectives for global action. The first was to maintain essential ecological processes and life-support systems. The second was to preserve the world's genetic diversity. The third was to ensure that species and ecosystems exploited by humans were sustainable. At the time, these objectives and the WoCoSt plan of action received token endorsement by governments, scientists and an educated public, but any such gains were soon undermined by the manoeuvres of assorted autocrats – today's Pharaohs and Pharisees.

In conservation's search for common cause with expanding commercial enterprise (actually more like tenants begging concessions from self-appointed landlords) important distinctions have become lost. The sustainable production of particular commodities, even within multi-species suites, might be manageable. The contrived maintenance of an intact natural ecosystem, say within tropical rainforests while centuries-old trees *are being actively felled wholesale*, is NOT. These are the perspectives in which we should remember that rainforests hosted significant periods of human prehistory.

We urgently need to know much more about the dynamics of rainforest communities, worldwide, *as processes*: processes that once embraced whole stretches of our own evolution yet are now permitted to become the province of primitive log barons, land-hungry industrial agriculture, mostly corrupt banks, palm-oil oligarchs and purveyors of third-rate furniture.

So what can still be done?

Significant proportions of a less dishonest pricing of tropical timbers should be directed towards fundamental research. Understanding the operation of evolutionary adaptations should become a defining parameter of process-informed policies. Power must be prised away from Big Farmers' (and Big Pharma's) baneful claws.

Turning from one land-grab to another, we have already seen how damaging the industrialisation of livestock has been. The burdens of maintaining exotic livestock in a hostile environment must give way, albeit by stages, to more intensive husbandry and an informed embrace and research into indigenous animals, their biology, ecology and their rational exploitation. This requires massive research, precisely the research that the mega-scale livestock industry has so assiduously obstructed or eviscerated up to now.

Most tropical ecosystems embrace long life-histories and slow, complex, drawn-out interactions. This means that many conservation programmes have little future without the operation of larger, longer-term processes being respected, researched, understood and allowed for, by public bodies, authorities and youth that are becoming environmentally better-educated by the day.

Africa, guardian of the greatest overall diversity of forest biota and home to the habitats within which our species evolved, has the most to lose and is losing it right now.

Utilitarian approaches to sustainability mask ignorance and allow displays of simple managerial skills and fake-solidarity to masquerade as authority. It is insufficient for such technicians to map spatial and temporal patterns in disintegrating landscapes, without concerted efforts to understand why patterns and biodiversity exist in the first place.

In 1993 the Brookings Institution, the World Resources Institute and Santa Fe Institute launched the '2050 Project', seeking key variables in the modelling of how systems change or adapt. The 2050 project was travelling in the right direction but 50 odd years is a very short period of time in terms of ecological or evolutionary processes.

A very general inability to see nature in any other time-frame than that of our pathetically short lives makes systematic studies of complex natural processes an urgent priority. Most tropical ecosystems embrace long life-histories and slow, complex, drawn-out interactions. In a land-hungry world, the longer the delay and the later the start, the fewer viable systems will be left to study or preserve and the later the start of what will have to become multi-generational programmes.

In 1997 I could list NERC, ORSTOM, CNRS, CSIRO, WCMC, WRI, ICSU and WCS as a few promising consultants or instruments in formulating and setting up process databanks, while both UNEP and IBP had gathered much relevant material and information. Even then, there was a growing awareness of the public's helplessness in the face of such self-inflicted crises as climate change, mass extinctions, desertification, soil erosion, unregulated GM, AIDS, 1918 Pesadilla and now Ebola and COVID-19 viruses.

In 2007, under the banner 'The Planet is not at risk – WE are', a world leader, Vaclav Havel, wrote, 'We must return, again and again, to the roots of human existence and consider our prospects in centuries to come ... We will either achieve an awareness of our place in the living and life-giving organism of our planet, or we will face the threat that our evolutionary journey may be set back thousands or even millions of years.'

This was no empty rhetoric and, by 2015, havoc from skies and waters was cooking or flooding enough people for all but a few maverick nations to sign up for the Paris Agreement on climate change. This life-raft programme, initiated by scientists, must become central to global and continental politics and culture.

In the larger context of understanding human origins, African scientists are best placed to research and teach awareness of our niche in nature. Revealing that evolutionary journey is in our African hands and we cannot allow those who have pillaged other continents to continue to plough and bulldoze away all living traces of human history as their appetites and those of their proxies grow, degrading and transforming African landscapes along the way.

We have allies among fellow scientists from all over the world, and with them we are building new research institutions, museums, universities, natural history societies and nature parks all across Africa. Most important of all, we must build schools that inform our youth of Africa's central role in the origins of humanity.

When I first elaborated a scoring system for some unusual rainforest biota in the 1960s, I naïvely believed that the evolutionary, ecological and biogeographic processes that had generated 'Centres of Endemism' or 'Hot-spots' of biodiversity could be assumed to have a generally accepted *intrinsic* value. An early death for that forlorn hope convinced me that, until the long-term factors that cause local endemism have been correctly identified and are respected in their own right, too many environment-oriented initiatives will rest on too-weak foundations.

As a student and young man I was among many fervent Pan-Africanists, all of us convinced that the aggressive, racist, exaggerated nationalism that had culminated in two World Wars and one Cold one should have been enough to invite unity among thinking people and especially among young Africans. Then the brutal political murder of our most charismatic Pan-Africanist, Patrice Lumumba, in 1961 so enraged all Africans that by 1963 Kwame Nkrumah and Haile Selassie had huge support and a done deal in founding the Organisation of African Unity, OAU, in Addis Ababa.

We longed for the dissolution of arbitrary colonial borders and expected the creation of institutions quite other than those bequeathed by colonialism. We quarter-expected the rise of one or more great powers, even a United States of Africa that might embrace great regions of our continent 'from sea to shining sea'.

In 2002 the OAU became the African Union, AU, just as HIV/Aids brought African scientists together like nothing before. Under AU auspices, scientists drew up a 'roadmap' to co-ordinate practical responses to Malaria, TB, Leprosy and AIDS. Few noticed one crucial detail – scientific authority *should,* no, *must* be able to hold governments accountable! Accountable for concealing facts and data, for obstructing health measures or for cavalier rejection of their responsibilities.

Since then, the leaders of some 55 countries, caged within the colonial borders they have inherited (and too many beholden to external patrons) have had mixed success in fostering science. Much more promising than the squabbles of nationalist politicians has been the quiet background work of African scientists.

During the 1960s my then colleague and friend, David Wasawo, a dedicated teacher, biologist and conservationist, helped shape OAU scientific activities. David, a quiet foe of imperialism but also aware of the myopia of most farmers, embellished our Makerere motto 'We build for the Future' with his own addendum – 'to open the eyes of our people to what is going on around them'.

David saw that national or political leaders like his close and brilliant friend, Tom Mboya, on their own could never foresee or forestall the damage inflicted by advances in desertification, mass extinctions, collapse of coastal fisheries and the irruption of uncontrollable diseases. Only science and conscientious scientists offer any hope of mitigating consequences from the extinctions, epidemics and climate crises that are now upon us and can be predicted to become ever more frequent, ever more damaging.

Thanks to their pioneer efforts, experienced African scientists like David Wasawo have initiated pan-African measures to try to contain some of the horrors of pandemics, starting with HIV/Aids and Ebola and now Covid-19.

I have already indicated how deeply embedded diseases are in the web of biological existence. In 2020 some early outlines emerged of how our future welfare might fare in the hands of conscientious scientists and medics with co-operative instincts and pan-African perspectives. In Addis Ababa, AU chair Cyril Ramaphosa and Dr John Nkengasong, Director of AU's leading Health Centre, set up and co-ordinated vigorous pan-African responses to Covid 19 that have saved many lives, most notably in dense but well-ordered societies such as Rwanda (but we must remember that Covid-19 has not yet run its cruel course, nor will it be the last pandemic).

For some, there has been a reluctance to admit that some natural phenomena are so slow, so complex, so old or on such a large scale that they may be beyond total comprehension and are certainly beyond 'management'. The deeper time perspectives attempted in this book have emphasised the age and complexity of Africa and forced confession that our collective ignorance and impotence are more significant than the boastful knowledge and disproportionate power of that global parvenu we call *Homo sapiens.*

If we are to assign a genuine priority to processes over products, effectively valuing living over dead-thing-products (which most accept in theory), then the practical programmes and institutions devoted to both learning about and maintaining processes need to be restructured towards that end.

The time to implement such policies has arrived – it is now.

The bad news is that, up until now, all efforts to institute process-oriented approaches have met fierce opposition. The profits from exploiting large-scale 'free' resources (the bodies of trees and other 'wild' plants, of oceanic fish and other 'wild' animals), browse, graze and LAND have all remained disgracefully and unacceptably high.

For too many generations, profits have gone into entrenching the political power of dynasties and legitimising land-grabs as the land 'holdings' of individuals, companies, empires and countries. Meanwhile, the prices paid for products from tropical timber, land-sterilising monocultures and livestock industries, not to mention piratical industrial fisheries, have all remained scandalously low.

Entrepreneurs and providers own ever-more-gigantic machinery – boats, trawlers, harvesters, freezers, processors, sawmills, bulldozers and mass-producing factories. To feed such hungry machines, the owners now have the audacity to claim priority access to Africa's resources. Such brigandry must be challenged, discredited, and denied.

It is precisely these pirates of such momentarily 'free' products that should be subject to laws and levies sufficient to fund the creation of the sort of databases being advocated here. All could and should also support substantial research and conservation levies, but the standard response has been to pervert, infiltrate or suppress research and marshal political support, on both right and left, for the status quo under all manner of contrived banners and slogans.

The good news is that sophisticated computing has made environmental modelling a lot more feasible, and invites experiments in more holistic approaches to natural processes. The digital age has opened up new channels of communication, inviting illiterate or otherwise disadvantaged people into conversations about the future of our continent and of our planet.

The present era of multiple crises could be particularly appropriate for both experiment and for revolution. As global information networks are being developed, there are unique opportunities to institute process-oriented methodologies integrated into process-oriented databases, dedicated to the propagation of process-oriented knowledge.

I bring up all these initiatives in the context of Africa because our continent deserves to be the natural and logical centre for action at all levels. Furthermore, we have been an open field for the 'Locusts of the Anthropocene' for way too long, and such piracy, currently at its highest level ever, has become intolerable.

Scientists in the African Union and the United Nations have articulated all these concerns, the latter in a Report authored by Ivar Baste and Robert Watson and issued by the Secretary General in 2021. In this Report the extermination of species, the climate crisis and pervasive, life-threatening pollution are three areas where action is now imperative. Africa has the greatest incentive to embrace this report, to initiate wide-ranging changes and to try to present a united front to ecologically less sagacious but still disproportionately influential nationalities.

Artificial subsidies for industrial interests, including livestock and mechanised agriculture, must be redirected towards greener, smaller-scale modes of food production (some closer, perhaps, to the well-tested subsistence farming of pre-colonial times).

Now is the time to break decisively with our still all-too-colonial past. We must institute enforceable laws, regulations and financial incentives that force pillagers of our resources, both home-grown and foreign, to desist and/or pay the real costs of their actions.

Process-based research ought to provide more objective foundations for the many political and strategic judgements that are having to be made as human numbers reach Malthusian proportions. If environmental policies are to have any traction over uninhibited, trumpeted *laissez-faire*, rational use or conservation of lands and landscapes will need to be backed up by process-oriented rationales.

To reinforce memory of our own evolutionary journey, why not institute calendars in which *Ardipithecus* Day ('Ardi-Day'?) or 'Lucy Day' are celebrated with fireworks and parades in Addis Ababa? Let '*Homo habilis* Day' excite street parties in Arusha and Dar, while Kampala lays on the *waragi* for *Morotopithecus* or plain 'Moroto-Day'. Just such benchmarks could be calibrated against the dates on which our scientists made their momentous discoveries.

Often, indeed usually, these discoveries have been made in close collaboration with fellow-scientists from other countries and continents. Thus every holiday would combine remembrance of our evolutionary journey with the essential unity of our species and pride in the more beneficent and illuminating aspects of contemporary science and the natural history of humanity's origins.

My dream has been for every school, from primary upwards, to possess its own 'school park' – even if only a few nestboxes hung on street trees, or a roadside pond, but, ideally, variably sized patches of vegetated, watered ground where children and their teachers can follow the fortunes of beehives, chicks, spiders, froglets, perhaps monkeys, identify wild flowers and record the passage of seasons and the free comings and goings of plants and animals. Thus, together with field guides in youthful hands, the seeds of a deeper respect for nature will be sown, and with it a respect for our own origins in nature and the emergence of a new pride in being African, in being human, that is loaded with so many, as yet unrealised, potentials.

My dream has also been to see natural history emerge as a respected and universal subject on the curriculum, from primary school onwards. Schools and pupils all across Africa should be competing for annual prizes that reward excellence in natural history as both academic subject and as exhilarating recreation.

These are not idle dreams – they are shared and actioned by dedicated and imaginative naturalists such as Wangari Muta and Wanjira Maathai, Christine Kabuye, Paula Kahumbu, Noah Mpunga, Sophy Machaga, Sylvanos Kimiti, Tiassa Mutunkei, Kaddu Sebunya, Alfred Lokasola, Eddie Ayensu – I could go on and on, adding local initiatives that provide models of the mechanisms that will have to be developed. Wildlife clubs in Kenya, the Southern Africa Nature Foundation, the Botswana Bird Club and Tanzania's experience in pioneering ecologically representative National Parks and reserves offer models and mechanisms for similar initiatives in other regions or nations.

Wildlife clubs now flourish across several countries, and proposals that natural history become a recognised subject at all scholastic levels are no longer a pipe dream. To date, Tanzania has inaugurated 13 National Parks, prising some of them out of the bumbledom of colonial-era 'Livestock & Fisheries'. There are more parks to come, many more schools to be built. Meanwhile, every year, our scientists induct students, local and foreign, into the thrills of digging the pregnant soils of Afar, of Olduvai Gorge, the Songwe valley, Gorongosa and more.

Education is itself a process with no finality to learning, teaching, revising or knowing. In this, several African countries can boast exceptional progress as they bring radically new African perspectives to a world trying to shrug off way too many mendacious tyrannies.

Canticle to the Sun.

In these pages I have tried to share something of the inspiration that nature in Africa offers to open minds and observant eyes. I have tried to show how the saga of human origins in Africa is an inescapable, concrete reality for all humans with curious minds, unshackled from the strictures of both imported and indigenous ideologies. We are slaves to no one.

The time to implement new, experimental policies has arrived – it is now.

I envisage a world in which we consciously acknowledge our dependence on the female principle as our passport to the future – our origins and the better angels of our nature can only flourish in healthy wombs, among healthy, creative families, not in the factories, armies or propaganda schools of tyrants. Like any other mammal, there is no pathway into the future without mothers and their feminine insights.

Humanity should not be made subordinate to commercial, economic or military ends. These are intolerable impositions that must give way to more humane and ecologically educated modes of self-governance.

Those that deny our animal roots in Africa cultivate a wilful ignorance that dooms them to ever-increasing contradiction (and their culture's eventual extinction) because the appetite for knowledge and a taste for truth is endemic in children. I therefore take heart in the knowledge that adults with closed minds are no match for the independent intellect of children.

Covid-19's most emphatic reprimand for humanity is to take the study of natural processes more seriously. One African leader, modelled on an American president in his contempt for science, medicine, education and conservation, died of Covid as a direct consequence of his bulldozing hubris.

We must take children's passion for truth and knowledge seriously. Every level of education needs to present the natural history and prehistory of becoming human as an expression of intrinsic, truthful values embedded in an always-growing, never-ending, but ultimately manageable store of knowledge.

The fount of our intellect, our creativity, our curiosity, our capacity for inspiration, even our sense of humour and rhythm, arose in Africa, probably long before we became sapient. All need exploration as never before, all need cultivation. Integral to that end, we must acknowledge our mother-continent's nurture of our species and make our continent what it has always been – nursery, school and university of human origins.

In our children's hands, armed with an ever-expanding knowledge that can be tested for its truthfulness, Africa becomes the true primary source of our natural history, our origins – the ultimate setting for future universities of human creative thinking, creative making, creative living.

Demography.

FURTHER READING

The COVID-19 pandemic left me stranded in southern Mexico for many months. Because most of this book was written there in the absence of references of any sort, I have written largely from memory, but with some recourse to several of my previous books, most of which were copiously referenced.

I hope the ideas explored in *Origin Africa* will stimulate others to pursue the literature, tramp the trails and unearth the soils of Africa in search of our great ancestral saga. I, therefore, invite readers in search of follow-up to consult my previous works. Substantial parts of *Origin Africa* draw from or summarise information and ideas that were explored in previous publications. The earlier and more extensive of these began with my seven-volume *East African Mammals. An Atlas of Evolution in Africa*, published by Academic Press between 1971 and 1982.

Island Africa: The Evolution of Africa's Rare Animals and Plants, published in 1990 by Collins in the UK and by Princeton University Press in the USA, remains a pioneer in African biogeography. My field and pocket guides to the mammals of Africa, together with their maps, are in some instances primary references for anyone interested in the geography and evolution of mammals. *Self-made Man*, published by Simon and Schuster in the UK and by John Wiley in the USA in 1993, has some 400 references. Each chapter of *Lowly Origin. Where, When and Why our Ancestors First Stood Up*, published by Princeton University Press in 2003, concludes with extensive references relevant to the chapter topic. *Mammals of Africa*, a six-volume work published by Bloomsbury Press in 2013, includes thousands of references.

The single most comprehensive and analytical survey of contemporary knowledge of mammalian and human social behaviour that I know of is Tim Clutton-Brock's *Mammal Societies*, published by Wiley Blackwell in 2016. I can but echo and paraphrase Tim's concluding question. Given ever-increasing threats to the environment, to the natural world and to human well-being, even our very existence, all generated by human demographic success, are we capable of recognising, let alone implementing, all the ethical, political and economic actions that are needed?

The most recent, most detailed and well-referenced account of Africa's unique ecology as the setting for human evolution is Norman Owen-Smith's *Only in Africa. The Ecology of Human Evolution*, published in 2021 by Cambridge University Press.

Two scientific journals, *Nature* and *Science*, have been perennial and very extensive sources of data, scientific information, discovery and inspiration. Several popular science magazines keep up running commentaries on scientific discovery, with writers trawling the world's literature for new revelations and emergent ideas about the nature of Nature. *National Geographic* has an admirable history of richly illustrated essays on human evolution as well as much inspirational geography and natural history. Finally, the internet, most notably Wikipedia, are rich sources of information and extensive referencing.

Independent thinking about human origins, about the ecological impact of Nature's premier 'niche-thief' (Tim Flannery's *Future Eaters*?) and speculation about that future range way beyond the academies. The inexorable march of events, the pace of scientific discovery and the dynamics of all natural processes dwarf the follies of our era mocking the sanctities of books, sacred or otherwise, and most of our pretentions to knowledge and foresight.

On a more positive note, the quest for knowledge, especially our origins and our place in Nature, has created a fellowship unlike all others. Amen.

INDEX